CHRISTOPHER JOSEPH is the Founder and Editor-in-chief of *Chicane*—the world's first carbon-neutral motorsport magazine—and has covered Formula One since 2010.

Chicane is centred on the three key areas of Innovation, Technology and Sustainability and how they relate to Formula One in a wider context. It seeks to illustrate that Formula One is far more than a sport and much more than a business.

Christopher writes about "the business of the sport and the sport of the business" and can be followed via @chicanef1. His passions within F1 are the finding of the unconventional, the pursuit of technological excellence and exploring the "iceberg" nature of the spectacle.

It is the lost souls of Formula One such as Ascari, von Trips, Rindt and Cevert that form the basis of his next project.

The Fastest Show on Earth

The Mammoth Book of Formula 1®

CHICANE

RUNNING PRESS
PHILADELPHIA · LONDON

ROBINSON

First published in Great Britain in 2015 by Robinson

Copyright © *Chicane*, 2015
The braking diagrams and data are courtesy of Brembo
1 3 5 7 9 8 6 4 2

The moral right of the author has been asserted.

A CIP catalogue record for this book
is available from the British Library.

ISBN 978-1-47211-047-3 (paperback)
ISBN 978-1-47211-052-7 (ebook)

Typeset in Great Britain by Hewer Text UK Ltd, Edinburgh
Printed and bound in Great Britain by CPI Group (UK) Ltd, Croydon CRO 4YY
Papers used by Robinson are from well-managed forests and other responsible sources

MIX
Paper from
responsible sources
FSC
www.fsc.org FSC® C104740

Robinson
is an imprint of
Little, Brown Book Group
Carmelite House
50 Victoria Embankment
London EC4Y 0DZ

An Hachette UK Company
www.hachette.co.uk

www.littlebrown.co.uk

First published in the United States in 2015 by Running Press Book Publishers,
A Member of the Perseus Books Group

Books published by Running Press are available at special discounts for bulk
purchases in the United States by corporations, institutions and other organizations.
For more information, please contact the Special Markets Department at the
Perseus Books Group, 2300 Chestnut Street, Suite 200, Philadelphia, PA 19103,
or call (800) 810-4145, ext. 5000, or email special.markets@perseusbooks.com.

US ISBN: 978-0-7624-5622-2
US Library of Congress Control Number: 2015931298

9 8 7 6 5 4 3 2 1
Digit on the right indicates the number of this printing

Running Press Book Publishers
2300 Chestnut Street
Philadelphia, PA 19103-4371

Visit us on the web!
www.runningpress.com

CONTENTS

ACKNOWLEDGEMENTS AND SOURCES

The editor has made every effort to locate all persons having any rights in the selections and materials appearing in this edition and to secure permission from the holders of such rights. The editor apologizes in advance for any errors or omissions inadvertently made. Queries regarding the use of material should be addressed to the editor c/o the publishers.

This publication is unofficial and is not associated in any way with the Formula One group of companies. F1, FORMULA ONE, FORMULA 1, FIA FORMULA ONE WORLD CHAMPIONSHIP, GRAND PRIX and related marks are trade marks of Formula One Licensing B.V.

Formula One in the United States
Key Contributor: Will Saunders for The History of Grands Prix in The United States and America's First Champion—Phil Hill.

Amy Hollowbush and Patty Reid at the Office of Mario Andretti for Mario Andretti and Andretti Race Record.

Teams and Drivers in Their Own Words
These chapters were written, abridged and compiled based upon information supplied and used with permission by the following copyright holders:

Williams
© Williams Grand Prix Engineering Limited

Manor
© 2015 Manor Grand Prix Racing Ltd. All rights reserved.

Mercedes
© 2015 Mercedes-Benz Grand Prix Limited. All rights reserved.

Lotus
© Lotus F1 Team Limited

McLaren
© 2015 McLaren Honda

Ferrari
© 2015 Ferrari S.p.A—all rights reserved.

Red Bull
© 2014–2015 Red Bull

Toro Rosso
© 2014–2015 Scuderia Toro Rosso S.p.A con socio unico

Sauber
© 2015 Sauber Motorsport AG, all rights reserved.

Force India
© 2014 Force India Formula One Team Ltd

Carlos Sainz Jr
© 2014 Carlos Sainz Jr—all rights reserved.

Max Verstappen
© 1996–2015 Verstappen Info Page

Kimi Räikkönen
© 2015 Kimi Räikkönen, all rights reserved.

Marcus Ericsson
© 2014 Marcus Ericsson

Glossary
A-Z Glossary

Allianz

Circuit Diagrams and Fast Facts
FIA

Brake Diagrams
Brembo

The following chapters utilized significant technical information supplied by and used by permission from the following copyright holders:

Aerodynamics
The text of this article is courtesy of Willem Toet, Head of Aerodynamics Sauber F1 Team Sauber Motorsport AG.

Brakes
Brembo

Fuel & Lubricants
Total

Powertrain
Mercedes High Performance Powertrains

Safety
Allianz

PHG Vision International
© 2015 PHG Vision International Ltd – PHG U.K. Ltd, all rights reserved.

Tyres

Pirelli
© 2015 Pirelli & C. S.p.A.—Pirelli Tyre S.p.A.

Anthology

Bernie Goes West and Meets his Match © Terry Lovell. Chapter 9 from *Bernie's Game*, Metro Publishing, 2003. Printed by permission of the author.

1961 © Michael Cannell. Chapter 10 from *The Limit: Life and Death in Formula One's Most Dangerous Era*, Atlantic Books, 2011. Printed by permission of the publisher.

Speed of Flight © Clyde Brolin. Chapter 1 from *Overdrive: Formula 1 in the Zone*, Vatersay Books, 2010. Printed by permission of the author.

Late Blossom At Lotus August to December, 1969 © David Tremayne. Chapter 15 from *Jochen Rindt Uncrowned King: The Superfast Life of F1's only Posthumous World Champion*, Haynes Publishing 2010. Printed by permission of the author.

Innovating: The drive for continual change © Mark Jenkins, Ken Pasternak and Richard West. Chapter 9 from *Performance at the Limit*, Cambridge University Press, 2009. Reprinted by permission of the publisher.

1993 © Steve Matchett. Chapter 6 from *The Mechanic's Tale: Life in the Pit-Lanes of Formula One*, Orion, 1999. Printed by permission of the publisher.

Welcome To The Piranha Club © Tim Collings. Chapter 1 from *The Piranha Club*, Virgin Books, 2001. Printed by permission of the publisher.

The all-nighter: par for the course © Michael Oliver. Chapter 3 from *Tales from the Toolbox*, Veloce, 2009. Printed by permission of the publisher.

Insights on Leadership © Mark Gallagher. Chapter 2 from *The Business of Winning*, Kogan Page, 2014. Printed by permission of the publisher.

All other content © 2015 Chicane Ltd./Christopher Joseph.

Additional writing from and special thanks to Nathan Hughes, Trisha Telep, Philip Conneller, James McKeown, Craig Young, Phil Yardley, Gameboy and The Snowman, Paul Hamilton, Rogan Jeans, Kate Walker, Will Saunders, Chris Mullins, Brian the Ring, Steve the Shirt and all at The Tipperary, especially Dr Vanda Collins and Mrs Doyle aka Sex on Legs.

Inspirational thanks to Graham Mitchell, Christian Vogt, Nick Downes & Interstate, Keith Sutton, George Baker Woods, Katja Heim & Tom Cooney, Sarah Webster, Tom Webb, Mark Alexander, Adam Parr, Matteo Bonciani, Pat Behar, Bradley Lord, James Francis, Andy Stobbart, Tracy Novak, Anthony at Mediatica, Phillipe et Carl Gurdjian, Alejandro Agag, Mark Gallagher, Alex Wurz, Simon Fitchett, Jonathan Neale, Nigel Geach, James Allen, Mark Abbs, Liam Clogger, Andrew Eaton, Isabel Juarez, Arnaud Boetsch, Bob Epstein, Rad Weaver, Mario Andretti, Zak Brown, Andreas Sigl, Jack Murray, Stewart Hosford, Alex Thomson, Mike Gascoyne, Charles Pic, Graeme Lowdon, Matt Strachan, Paul Hembery, Richard Goddard, Paul Steentjes, Julia Wurz, Monisha Kaltenborn, Cyril Abiteboul, Carlo Boutagy, John Bailey, Kirsty Andrew, Geoff McGrath, Jean-Claude Biver, Simon Collingwood, Graham Hackland, Laurence James, Michael Carrick, John Rhodes, Charles Pic, Kamarudin Meranun, Natalie Cerny, Andrew Westacott, Brendan McClements, Mokhzani Mahathir, Salvador Servia, Jonathan Hallett, Fiona Smith, Alexandra Schieren, James Moy, The Vancouver Millionaires and 23 Rawle Street.

Commercial thanks to Oliver Gadney, Mehul Kapadia, Sven Jarby, Bob & Pat Bondurant, Jochen Braunwarth, Yan LeFort, Joel Leonoff & Angharad Couch, Tim Bampton, David Barzilay, Alex Bishop, the UKTI and HMS President 1918.

Overriding gratitude for exemplary patience to Duncan Proudfoot, Emily Byron and Clive Hebard at Little, Brown—as well as top marks to David Lloyd for a copy-edit extraordinaire.

Origin of the Species award to Trisha Telep at Pretext Agency.

INTRODUCTION

"THIRTY MINUTES TO LIGHTS OUT"
An Introduction to Formula One

The thirty-minute period before the start of a Formula One Grand Prix Race is undoubtedly one of the quintessentially unique periods in world sport. The short half-hour lead-up to the race is both tense and frantic but most importantly it is indicative of the complicated nature of the sport and all that lies beneath the surface of a complex spectacle.

00:30 At thirty minutes before the start of the Formation Lap (warm-up lap), the pit lane opens and remains so for fifteen minutes. The cars are allowed to leave the pit lane and complete a reconnaissance lap while team personnel are allowed to get out onto the grid with the necessary equipment needed to start the race.

00:25 At the end of the reconnaissance lap the cars must stop at the rear of the grid, cut their engines and be pushed through the assembled race personnel, media and VIPs to their starting positions on the grid. Here they must remain with their engines turned off.

00:24 The cars are plugged into data monitoring and cooling systems. The race engineer gives the "one minute to fire up" instruction.

00:23 The ignition is turned on while the rear wheels are removed but kept in their heating blankets and the engine is started. Temperatures are measured and monitored before the engine is turned off again.

00:17 A warning signal is given that the end of the pit lane is to be closed in two minutes.

00:15 A second warning signal is given and the pit lane is closed. At this point no car will be permitted to enter the grid and must start from the end of pit lane. The front wing flap settings are cross-checked both on the grid and on the spares in the garage. The mechanics perform a wheel gun check to test the hydraulics.

00:11 Another ignition test and the drivers are summoned to their vehicles.

00:10 The ten-minute signal is given. Everybody except drivers, officials and team technical staff must leave the grid. The drivers suit up, fit their helmets and HANS (head-and-neck safety) devices, and are strapped into the cockpit by one of the mechanics.

00:05 Four red lights are illuminated on the gantry and an audible warning is given five minutes before the start of the formation lap.

00:03 Three lights are illuminated on the gantry and the three minute signal is given. All cars must have their race tyres fitted by this point.

00:01 Two lights are illuminated on the gantry when the one-minute signal is shown and by this time engines must be started and all team personnel must leave the grid, taking all equipment with them before the fifteen-second signal is given.

00:00:15 There is one light left illuminated on the gantry and the launch map is initiated by the driver.

00:00:00 The red lights go out and the green lights are illuminated to begin the formation lap led by the pole sitter.

The cars return to the grid after the formation lap and stop within their respective grid positions with their engines running. The Safety Car takes up its position at the rear of the grid.

A standing start will be signalled by the permanent starter (usually Mr Charlie Whiting of the FIA) by activating the gantry lights.

Once all the cars have come to a halt, the five-second light will appear followed by the four-, three-, two- and one-second lights until five red lights are illuminated on the gantry. At any time after the one-second light appears, the race will be started by extinguishing all five red lights.

The race to the chequered flag begins . . .

THE INSIDERS' GUIDE TO THE FASTEST SHOW ON EARTH

In the quest for ultimate speed Formula One combines the human, the technological and the financial in an exciting and compelling global circus watched by half-a-billion avid fans around the world.

As the above race countdown shows, even within the confines of the three-day race weekend, there are many intricate details, subtleties and minutiae to be explored in Formula One, not to mention the wider season, the industry at large and sixty-five years of heritage.

With this in mind I have divided this edition into two very different yet complementary sections.

In the first section, I have drawn on expert analysis from within the industry to examine the effects that the revolutionary regulation changes of 2014 have had on the engines, tyres, brakes, aerodynamics, fuel, safety and data in what is to be a new era of racing.

This extensive guide also includes the official FIA Circuit Diagrams, Team Histories, Driver Profiles, Circuit Fast Facts, Technical Braking Profiles, a special section on the History of Formula One in the US and the most comprehensive Glossary of Formula One terms ever printed. At every opportunity I have allowed those closest to the action to express their thoughts and ideas from their own special perspectives.

Despite the best-laid plans of mice, men and myself it has not been physically possible to include every aspect of Formula One in this edition nor has it been my desire to be encyclopedic in approach. The extensive glossary that is included in Chapter 11 provides a comprehensive and easy-to-access guide to all of the crucial aspects of the sport from A to Z.

However, for an expert and excellent source of further information readers are referred to the "Understanding F1 Racing" section of www.formula1.com, the official website. Here you will find in-depth information on topics such as logistics, suspension, testing, the steering wheel, the gearbox, racing control, pit stops, flags, driver fitness, overtaking and the DRS, cornering and race strategy.

For those of you who wish to access the annually changing rules and regulations that we have referred to in the technical chapters, these can be found in their entirety at www.fia.com. In the "Regulations" section you will be able to download the complete Sporting Regulations and Technical Regulations. Be warned, these documents run to fifty-seven and eighty-nine pages respectively. You will also be able to view the first set of proposals from the F1 Commission that advises on changes to both the Sporting and Technical Formula One Regulations for the 2016 season.

In the second part of this edition I have compiled what I believe to be a great collection of insightful pieces from some of the most experienced writers who uncover the inner workings of the wider Formula One world both on-and-off the track.

From the wealth of intrigue that exists in the cutthroat world of *The Piranha Club* to the unsung heroes of pit lane in *The Mechanics Tale* and *Tales from the Toolbox*, the glamour, glory, success and failure of Formula One is expertly captured.

The spectre of tragedy that hung over the sport in bygone eras is skilfully evoked in *The Limit* and *Jochen Rindt: Uncrowned King*, while the never-ending quest by drivers to find the mysterious "zone" is explored in *Overdrive*.

Much more than just a sport and far more than a business, it is a world of fascinating personalities and business-savvy individuals who demonstrate their leadership skills in *The Business of Winning* and their ability to innovate in *Performance at the Limit*. There is also a window into the private world of the ultimate ring-master in *Bernie's Game*.

I hope my approach will provide something for the casual observer being introduced to the sport, the race-going spectator, the avid fan wanting to know more as well as the aficionado hungry to delve deeper into all that Formula One has to offer.

Christopher Joseph
London, 2015

2015 F1 CALENDAR AND REGULATION CHANGES

FIA FORMULA ONE WORLD CHAMPIONSHIP

The 2015 FIA Formula One World Championship calendar is confirmed as follows:

15 March	AUS	Grand Prix of Australia
29 March	MYS	Grand Prix of Malaysia
12 April	CHN	Grand Prix of China
19 April	BHR	Grand Prix of Bahrain
10 May	ESP	Grand Prix of Spain
24 May	MCO	Grand Prix of Monaco
7 June	CAN	Grand Prix of Canada
21 June	AUT	Grand Prix of Austria
5 July	GBR	Grand Prix of Great Britain
26 July	HUN	Grand Prix of Hungary
23 August	BEL	Grand Prix of Belgium
6 September	ITA	Grand Prix of Italy
20 September	SGP	Grand Prix of Singapore
27 September	JPN	Grand Prix of Japan
11 October	RUS	Grand Prix of Russia (Sochi)
25 October	USA	Grand Prix of USA (Austin)
1 November	MEX	Grand Prix of Mexico
15 November	BRA	Grand Prix of Brazil
29 November	ADE	Grand Prix of Abu Dhabi

2015 F1 SPORTING REGULATIONS

The following regulations have been amended from the 2014 version:

Points
Points for both titles will no longer be doubled for the final Event of the Championship.

Standing Restarts

After consultation with the Teams who raised a number of safety concerns, Articles 42.7 and 42.8 on standing restarts have been rescinded.

Virtual Safety Car (VSC)

Following tests of the VSC system at the final Events of 2014, the introduction of the system has been approved for 2015. The VSC procedure may be initiated to neutralise a race upon the order of the clerk of the course. It will normally be used when double waved yellow flags are needed on any section of track and competitors or officials may be in danger, but the circumstances are not such as to warrant use of the Safety Car itself. The full text of the article is available in Annex I.

Suspending a Race

When a race is suspended, the pit exit will be closed and all cars must now proceed slowly into the pit lane, not the starting grid. The first car to arrive in the pit lane should proceed directly to the pit exit staying in the fast lane, all the other cars should form up in a line behind the first car.

Team Personnel or Equipment on Grid

If any team personnel or team equipment remain on the grid after the fifteen-second signal has been shown the driver of the car concerned must start the race from the pit lane. A ten-second stop-and-go penalty will be imposed on any driver who fails to do this.

Power Unit Penalties

The replacement of a complete power unit will no longer result in a penalty; instead, as specified in the current regulations, penalties will be applied cumulatively for individual components of the power unit. If a grid place penalty is imposed, and the driver's grid position is such that the full penalty cannot be applied, the remainder of the penalty will be applied in the form of a time penalty during the race (not at the next race as was previously the case) according to the following scale:

- 1 to 5 grid places untaken: A penalty under Article 16.3(a) will be applied.
- 6 to 10 grid places untaken: A penalty under Article 16.3(b) will be applied.
- 11 to 20 grid places untaken: A penalty under Article 16.3(c) will be applied.
- More than 20 grid places untaken: A penalty under Article 16.3(d) will be applied.

Time Penalties

In addition to the existing five-second penalty (Article 16.3a), a new ten-second penalty (Article 16.3b) will also be introduced, to be applied in the same manner.

Unsafe Release

If a car is deemed to have been released in an unsafe condition during a race a ten-second stop-and-go penalty will be imposed on the driver concerned. An additional penalty will be imposed on any driver who, in the opinion of the stewards, continues to drive a car knowing it to have been released in an unsafe condition.

Qualifying Procedure

The qualifying procedure was clarified: for cases when twenty-four cars are eligible, seven will be excluded after Q1 and Q2; if twenty-two cars are eligible, six cars will be excluded after Q1 and Q2; and so on if fewer cars are eligible.

Safety Car: Lapped Cars

Once the last lapped car has passed the leader the Safety Car will return to the pits at the end of the following lap, the race director will no longer have to wait for all the lapped cars to reach the back of the pack behind the safety car.

2015 F1 TECHNICAL REGULATIONS

- The weight of the car, without fuel, must not be less than 702kg at all times during the Event (up from 701kg).

- Changes have been made to the rules governing Wind Tunnel Testing and with regard to the aerodynamic reporting periods for 2015 and 2016.
- Any suspension system fitted to the front wheels must be so arranged that its response results only from changes in load applied to the front wheels.
- Any suspension system fitted to the rear wheels must be so arranged that its response results only from changes in load applied to the rear wheels.
- The Zylon anti-intrusion panels on both sides of the survival cell have been extended upwards to the rim of the cockpit and alongside the pilot's head.

CRITERIA FOR THE ISSUING OF SUPER LICENCES
A proposal on the conditions of attribution of the Super Licence was approved for 2016, on the basis of the following criteria:

1 Safety criteria
The following changes have been made compared to the current regulations:

- There is a minimum age requirement (18 years of age).
- There is a verification of knowledge of the F1 Sporting Regulations/ISC rules.

2 Experience criteria
With the following changes compared to the current regulations:

- There is the 300 km in F1 TCC or TPC as a minimum requirement.
- There is a two years minimum running in minor Formulas.

3 Performance criteria
With the following changes compared to the current regulations:

- There is a point system requirement, based on the driver results in previous Formulas.

FORMULA ONE IN THE US

THE HISTORY OF GRANDS PRIX IN THE UNITED STATES

The United States Grand Prix may have recently laid the foundations of a permanent home in Austin, but the long history of Formula One races held in the US is a peripatetic patchwork tapestry; a legacy of tentative footholds tainted by circumstance (Watkins Glen, Indianapolis) and half-baked follies around concrete jungles (Detroit, Dallas, Phoenix).

Despite its status as an "emerging market" for Formula One, the United States has in fact hosted sixty-five championship Grands Prix at eleven different circuits during the sixty-five-year history of the Formula One World Championship. This places the US fourth on the list of countries who have hosted the most Grands Prix, behind only Italy (ninety-two), Germany (seventy-five) and Great Britain (sixty-eight). Of these, thirty-six races have been run under the "United States Grand Prix" moniker, with further Grands Prix in destinations including Dallas, Detroit and Las Vegas titled according to the host city.

Indeed, the US has nominally played host to Formula One Grands Prix since the inception of the championship in 1950. From 1950 to 1960 the Indianapolis 500 counted towards the final championship standings, but the race was largely ignored by the F1 teams and drivers at that time and contested almost exclusively between American racers—making something of a folly of its championship inclusion.

It wasn't until 1959 that the first "proper" United States Grand

Prix was held, when the Formula One circus pitched up in Sebring, Florida, for the concluding and deciding round of that year's world championship.

SEBRING AND RIVERSIDE—TENTATIVE FIRST STEPS

The Sebring International Raceway in Florida was a converted Air Force training base that had hosted motorsport since 1950. The first running of its most famous event, the 12 Hours of Sebring, took place two years later. By 1959, Formula One was searching for a genuine "American" race to supplement the Indy 500, and Sebring was chosen as the setting for the ninth and final round of the championship season.

The race, organized by Russian-born promoter Alec Ullman, saw an expanded roster of American entrants, with six of the nineteen-strong field racing under the stars and stripes. The American entry list included F1 regulars Phil Hill, Harry Schell, Dan Gurney and Masten Gregory—with reigning Indianapolis 500 winner Rodger Ward invited to add a genuine "American" flavour, only to be hopelessly outclassed at the wheel of a Kurtis Kraft-Offy Midget.

The real story of the weekend though was the three-man title shootout between Jack Brabham (Australia, Cooper), Stirling Moss (GBR, Cooper) and Tony Brooks (GBR, Ferrari). Cooper had revolutionized the sport since pioneering the rear-engined approach two years previously, and by 1959 rear-engined cars held a clear performance advantage in Formula One. Brabham held a substantial lead in the standings, with Moss requiring at least a second place and Brooks nothing less than a win to overhaul him.

In the event, the title battle was something of a damp squib. Moss led the early running from pole position, but retired with a gearbox failure on lap five. Brooks was hit by Ferrari teammate Wolfgang von Trips at the first corner and took a long pit stop to repair the damage. Brabham ultimately finished fourth after running out of fuel on the last lap and pushing his car over the line, securing his first World Driver's Championship—and the first for an Australian— as well as the Constructors' Championship for Cooper.

Despite the exciting climax to the season, the race was a

disaster for Ullman. The attendance was barely half the size of that at the 12 Hours of Sebring, and after paying out the $15,000 purse (which included a plot of land for the race-winner, New Zealander Bruce McLaren), Ullman barely broke even. Sensing a need to enhance the commercial viability of the United States Grand Prix, Ullman switched coasts for the 1960 event.

The 1960 United States Grand Prix took place at the Riverside International Raceway in California. Again, the venue was an established setting for American sportscar racing, but was a completely new frontier for Formula One. Unlike the previous year, the United States GP would not play host to a championship showdown, with Jack Brabham having sewn up a second consecutive championship in Portugal two races previously.

British driver Stirling Moss romped to victory for Lotus, heading Lotus teammate Innes Ireland (GBR) and Bruce McLaren (NZ, Cooper) over the line. Of wider concern though was the fact that the race had once again failed to ignite public interest—and promoter Ullman was again left out of pocket once prize and appearance fees had been paid. Formula One wouldn't return to California for sixteen years, but in the meantime the first permanent home for the race established the United States Grand Prix as a regular fixture on the Formula One calendar.

WATKINS GLEN—TRIUMPH AND TRAGEDY
In 1956, Cameron Argetsinger, a sports car enthusiast and head of the recently formed Watkins Glen Grand Prix corporation, purchased 550 acres of land outside the tiny village of Watkins Glen in picturesque upstate New York—with the intention of building an international-standard permanent racing facility to replace the temporary street circuit which had hosted motorsport of every class since 1948.

Within a couple of years of its construction, Watkins Glen was hosting Formula Libra road races and these attracted such luminary names as Stirling Moss, Jo Bonnier, Phil Hill and Dan Gurney. Argetsinger though had designs on the ultimate American road-racing event, and after the commercial disappointment of the first two United States Grands Prix was able to take the race to Watkins Glen for 1961.

The inaugural race would mark the start of twenty years of unbroken Formula One racing at The Glen. The circuit proved one of the most popular in Formula One history, playing host to an annual autumn jamboree among the fall foliage and bearing witness to championship coronations, classic races and shocking tragedies during its tenure as host of the United States Grand Prix.

The first configuration of the Watkins Glen Grand Prix Race Course was a short, 2.35-mile blast around undulating terrain, with just eight corners encompassing one of the quickest and highest-speed laps on the Formula One calendar.

By the time of the maiden race in 1961, the US had already achieved an important milestone: Phil Hill had been crowned the first American World Champion at the wheel of the dominant "shark-nose" Ferrari 156. Unfortunately, Hill's Ferrari team declined to attend the United States Grand Prix after their driver Wolfgang von Trips and fifteen spectators were killed in an accident at the previous race in Italy, so the only chance the crowds would have to see their hero on track was during the pre-race parade as an Honorary Race Steward.

A total paying crowd of 60,000 fans across the weekend made the first race a guaranteed success from a commercial perspective, and the event brought about a popular conclusion on-track too, with the Briton Innes Ireland claiming his first and only Grand Prix success—which was also the maiden victory for the Lotus team.

The race quickly became tremendously popular with both the F1 fraternity and the paying public. The large starting and prize money on offer frequently exceeded the money available for all of the other races put together. The race was also well run and administered, winning the Grand Prix Drivers' Association award for the best-staged GP of the season in 1965, 1970 and 1972, and by the mid-60s the United States Grand Prix was an entrenched staple of the Formula One season.

American successes through the early years were few and far between though. A podium finish for Dan Gurney's Eagle in 1967 and Mario Andretti's sensational debut pole position in 1968 aside, the United States Grand Prix remained very much the preserve of European racers during the 1960s.

Despite victories for such racing greats as Jim Clark, Graham Hill, Jackie Stewart and Jochen Rindt, Watkins Glen wouldn't play host to a championship decider until 1970. The circumstances were laced with tragedy though, as Ferrari's Jacky Ickx attempted to chase down the points advantage of Lotus driver Jochen Rindt—who had been killed during the previous race in Italy. By finishing fourth, Ickx failed to take the victory required to keep the title fight alive, and Rindt was crowned Formula One's first, and only, posthumous World Champion.

The late 1960s and early 1970s was a treacherously perilous era for Formula One racing. Increasing speeds as cars became more powerful and aerodynamically sophisticated were not matched by improvements to the safety of the circuits—many of which were simply open roads bordered by trees, verges or other natural obstacles, with run-off areas and guardrails in their infancy. Between 1966 and 1970, eight drivers were killed at the wheel of Formula One cars. In an effort to improve both the safety and the spectacle of the Watkins Glen circuit, significant upgrades were made to the facilities ahead of the 1971 race.

The entire lower section of the track was torn up and redrawn, and a new "boot" section was added—lengthening the lap distance by over a mile to 3.4 miles. The improvements cost $2.5 million, and transformed the circuit into one of the most demanding challenges in racing: a gruelling up-and-down course where almost every corner was banked and long. Entries and prize-money hit record levels, with thirty drivers competing for a record race-winning pot of $267,000, and the first race on the new configuration was won by dashingly charismatic Frenchman François Cevert at the wheel of a Tyrrell as the popularity of the United States Grand Prix reached its zenith.

Unfortunately, tragedy would mar The Glen over the coming years. The 1973 event was dominated again by a narrative surrounding Cevert, but this time there was no happy ending. With his third World Championship sewn up and retirement due after the weekend of his 100th Grand Prix, Scotland's Jackie Stewart was ready to pass the baton of Tyrrell team leader to Cevert by letting the Frenchman take victory at the US GP in return for his "obedient" support during the season. Neither

Tyrrell would take the start though; Cevert was killed during a violent high-speed crash at The Esses during Saturday qualifying, and Stewart withdrew from what would have been his career swansong. Cut down on the cusp of his prime, Cevert's death shocked the Formula One fraternity—and further catastrophe was to follow in 1974.

Against the backdrop of a three-way championship showdown between Jody Sheckter (South Africa, Tyrrell), Emerson Fittipaldi (Brazil, McLaren) and Clay Regazzoni (Switzerland, Ferrari), Austrian Helmuth Koenigg, in just his third Grand Prix, crashed at the fast, long Outer Loop—with the young Austrian killed instantly as his Surtees car split the Armco barrier in half. Fittipaldi took the title, but a second fatality in consecutive years cast a long shadow over The Glen.

Off-track, issues were beginning to mount too. The circuit and facilities were deteriorating from a lack of substantial investment, and increasingly rowdy spectators—who cared little for motorsport—had taken to drunkenly setting fire to cars in the infamous "Bog" infield section. With Long Beach now hosting a rival "American Grand Prix" at a cosmopolitan street circuit in Los Angeles, The Glen's reputation waned rapidly through the late 1970s.

In 1978, an ultimatum from the FISA European motorsport governing body forced belated safety improvements, but the momentum was irreversible. In 1980, against a backdrop of mounting debts, Watkins Glen played host to its final Grand Prix. The race was a thriller, with Australian World Champion Alan Jones charging through the field to win after falling as low as seventeenth at the start, but with the promoters unable to either repay debts or secure loans to put on a race in 1981 it would prove to be Formula One's curtain call at The Glen.

LONG BEACH—THE WEST COAST MONACO

Set against The Glen's declining popularity was the rise of the United States Grand Prix West at Long Beach. A small city within Greater Los Angeles, Long Beach had a vastly different appeal to Watkins Glen—that of a cosmopolitan, urban, sun-kissed jaunt alongside the Pacific waterfront in Southern California.

After a trial Formula 5000 race in 1975 proved a successful demonstration run, Formula One made its debut at Long Beach in 1976. As with many street circuits, Long Beach was narrow, gruelling and predominantly slow, with its 2.04-mile lap encompassing thirteen corners—of which twelve were tighter than ninety degrees.

Set in early spring to provide a counterpoint to the autumnal race at Watkins Glen, the United States Grand Prix West proved a success from the off. The Swiss Clay Regazzoni won the first event for Ferrari in 1976, although the most popular victory was claimed by Mario Andretti in 1977—becoming the first and only American to win a Formula One championship Grand Prix on home soil (excluding winners of the Indianapolis 500 in the 1950s).

Changes to the circuit throughout its lifespan resulted in the curious quirk of the race being run to seventy-nine-and-a-half laps due to the pits and start-line unusually being at opposite ends of the track, but for the most part the layout remained constant until 1982. Sweeping changes ahead of the 1982 edition shortened the lap, and further amends ahead of the 1983 race saw the circuit truncated almost beyond recognition.

However, 1983 did see one of Formula One's most remarkable races, with Belfast-born John Watson recovering from qualifying twenty-second to charge through the field and take a scarcely believable victory—a drive that set a record for winning from a lowly starting position that stands to this day.

Watson's win was to prove Long Beach's swansong, however. Race organizer Chris Pook felt F1 was too expensive, and after an approach from CART (Championship Auto Racing Teams— known as Champ Car after 2004) decided to replace the Grand Prix with an Indycar race from 1984. The domestic series contained leading "home" drivers including Mario Andretti, and to this day the Long Beach Grand Prix remains one of the most popular motor races on the American calendar.

LAS VEGAS—A SHORT STAY ON THE STRIP
Perhaps Formula One's most bizarre foray in the US came in the early eighties, with a Grand Prix in Las Vegas run around a

makeshift circuit in the car park of the Caesars Palace casino. The loss of Watkins Glen was an opportunity for Formula One to expand its offering in the States to new frontiers, and Vegas, which had been a non-starter in 1980, was confirmed as a second west coast race to close the 1981 season.

Constrained by the tight confines of the Caesars Palace perimiter, the circuit was a bizarre-looking sequence of three repeating "fingers", a short, flat, 2.2-mile anti-clockwise sequence of fourteen turns winding around a crooked "M" shape.

Although the track surface was relatively smooth, the circuit's anti-clockwise configuration caused significant problems for the drivers—leading to a common complaint of "Las Vegas neck". Combined with the dry desert heat, the Caesars Palace circuit became an ultimate endurance test for a generation of drivers whose notions of diet and fitness were far removed from the standards taken for granted today.

John Watson memorably described the circuit as "a racetrack made up of canyons of concrete", and the high average speed of over 100 mph failed to alleviate the boredom for drivers as the course repeatedly doubled back on itself to make the best use of the available space. The circuit was as uninspiring for spectators and television viewers as it was for the drivers, with the barren backdrop of dust, sand and highways failing to convey any of the associated glamour of Vegas.

Another unique feature of the race was its nomenclature. It was the first bespoke Formula One Grand Prix not to be named after the host country, with the race instead officially known as the Caesars Palace Grand Prix.

One benefit for race organizers of assuming Watkins Glen's slot on the calendar was that Caesars Palace played host to the season finale, and both editions of the race saw the title settled at the very last.

In 1981, Nelson Piquet (Brazil, Brabham) pipped Carlos Reutemann (Argentina, Williams) and Jacques Laffite (France, Ligier) to the title in a three-way showdown, overcoming exhaustion and severe neck strain—throwing up in his helmet during the race—to bring the car home in fifth place and beat Reutemann by one point.

For 1982, the equation was equally knife-edged, with Keke Rosberg (Finland, Williams) and John Watson (GBR, McLaren) battling for the Drivers' Title. A fifth place behind maiden winner Michele Alboreto (Italy, Tyrrell) was enough for Rosberg, but a paltry crowd of 30,000 illustrated the extent to which the race had failed to capture the imagination.

Rosberg's coronation was to prove the final act at Caesars Palace. The 1982 season had seen three street races in the US, at Long Beach, Detroit and Vegas, and many within the sport felt that this was a saturation point—especially given the fact that none of these races had proven as well attended or as popular as Watkins Glen. After hosting two title deciders that were enthralling in spite of the setting rather than because of it, the race was pulled ahead of the 1983 season.

DETROIT, DALLAS AND PHOENIX—CONCRETE JUNGLES

Perhaps the least remembered of the "glasscrete" urban circuits that adorned the US in the 1980s, Detroit was a notoriously tricky car-breaker set in the heart of Motor City.

A 2.5-mile, seventeen-corner "point and squirt" style circuit, renowned for being exceptionally bumpy, slower than Monaco and containing the only level crossing in F1 history, Detroit was a fixture of the calendar from 1982 to 1988, displaying remarkable staying power compared to Dallas, Phoenix and Las Vegas.

Organizational problems dogged the event from the first running in 1982, with Thursday practice and Friday qualifying cancelled before John Watson provided a stunning show to win the race from a lowly seventeenth on the grid.

The race though did at least prove popular, with over 70,000 fans enjoying Michele Alboreto's victory in the 1983 event. A demolition derby in 1984 saw only eight finishers, but typically Detroit was unusually low on attrition for a street race during its lifespan. The race gained a reputation as a haven for street circuit specialists during the mid-eighties, with Nelson Piquet, Keke Rosberg and Ayrton Senna all tasting victory.

By 1988, the race was still attracting crowds of 60,000-plus, but the slippery track surface of the temporary circuit was proving troublesome—and the race organizers came in for fierce criticism from the drivers. Negotiations to upgrade the facilities and save the event failed, and the race was moved cross-country to Phoenix for 1989.

While Detroit may have had its problems as a destination for Formula One, the United States Grand Prix's most inglorious folly was a one-off stop in Dallas in 1984.

The layout of the temporary circuit in downtown Dallas' Fair Park was itself well-received, despite its tight and twisty nature, but the issues for which the Dallas Grand Prix are remembered were more deep-rooted.

A lack of run-off areas were and remain commonplace at street circuits, and the 100-plus-degree heat was an occupational hazard of scheduling a mid-summer race in Texas. The track surface, however, was a unique variable: a bubbling volcanic potion of metamorphic tarmac that flared and fractured throughout the weekend—much to the detriment of the low-ride height Formula One cars expected to traverse sixty-seven laps on raceday.

With rumours of cancellation abounding throughout the weekend, and a drivers' boycott proposed on Sunday morning, the entire weekend operated under a cloud—a mood heightened by Martin Brundle's leg-breaking accident during Friday practice.

Waved off by Larry Hagman and under the watchful eye of former President Jimmy Carter, when it eventually started the race itself was an attritional farce, with the unpredictable surface and abrasive conditions accounting for eighteen of the twenty-five-strong starting field.

Keke Rosberg won for Williams, attributing his success to a specially water-cooled helmet, allowing him to literally keep a cool head while those around were losing theirs—famously including pole-sitter Nigel Mansell, who unnecessarily tried to push his Lotus home after a late breakdown, ultimately collapsing onto the track with exhaustion and dehydration.

Citing the oppressive heat and farcical conditions, Formula One beat a hasty retreat from Texas after only one outing, replacing the Dallas Grand Prix with the infinitely more popular Australian race in Adelaide from 1985.

The last of Formula One's concrete jungle tour of uninspiring urban landscapes across America saw the circus set up camp in Phoenix for three seasons from 1989–91. Bizarrely, considering local ostrich races were rumoured to attract more fans than the Grand Prix, Phoenix was chosen to ring in the new season with a whimper in both 1990 and 1991.

Despite the unpromising portents, the 1990 United States Grand Prix was an absolute classic, thanks to the career-making performance of a hard-charging French-Sicilian by the name of Jean Alesi. Driving for Tyrrell, Alesi remarkably challenged the supreme McLaren-Honda of Ayrton Senna—impetuously passing the Brazilian for the lead before ultimately falling back into Senna's clutches and finishing second.

The 1991 race was a dismal procession, with Senna winning easily. Poor attendances and limited local interest (not a single American driver had started a race in 1990, no American team had entered a race since 1986, and no American had won a Grand Prix for more than a decade since Mario Andretti triumphed at the 1978 Dutch GP) saw Formula One beat a retreat from the United States—and the sport wouldn't return for almost ten years.

INDIANAPOLIS—TO THE BRICKYARD

The 1990s saw Formula One's global popularity and commercial power accelerate at an incredible rate. Although the decade was marred by the tragic death of Ayrton Senna, F1's biggest star, at the 1994 San Marino Grand Prix, the sport as a whole grew to reach new audiences on new frontiers—with a generation of new drivers, headed by Michael Schumacher, becoming global stars in their own right.

The United States though remained a closed door. Various attempts at engineered cross-pollination, including Nigel Mansell's intercontinental Indycar adventures in 1993–4, CART legend Michael Andretti's doomed stint racing for McLaren in 1993 and the efforts of Formula One supremo Bernie Ecclestone to parachute in leading "American" talent such as Canadian 1996 CART champion Jacques Villeneuve, failed to set pulses racing in the US.

In 1998 however, a deal was brokered between Ecclestone and Tony George, President of the Indianapolis Motor

Speedway, to bring Formula One to the spiritual home of motorsport in North America. Two years of renovation and construction followed to adapt the venue for Formula One, during which time a 2.6-mile course was built on the infield of the famous oval—although F1 cars would still run on the iconic banking for half of the lap. A further concession to Formula One was the direction of the circuit, with the Grand Prix becoming the first-ever race at Indianapolis to run in a clockwise direction (as per the general practice in F1). The scheduling of the event also played into F1's hands, with the Grand Prix organized for early autumn to avoid any clashes with the major dates on the American racing calendar.

To the relief of all concerned, the event proved spectacularly popular. With American open-wheel racing disastrously split into the IRL and Champ Car factions, fans flocked to Indianapolis to see what had become unquestionably the premier open-wheel motor racing series in the world. An estimated 225,000 fans saw Michael Schumacher's Ferrari take victory at the first United States Grand Prix in nine years, setting a new record for a crowd at a Formula One race in the process.

Although incredibly successful from an attendance perspective, the history of the race was unfortunately tarnished by external circumstances and internal controversies. In 2001, the race was the first international sporting event to be held in the US since the September 11th attacks, and a sombre event, marked by many drivers bearing messages of support on their helmets and flanks, saw double-World Champion Mika Häkkinen take his twentieth and final Grand Prix victory. In 2002, the dominant Ferraris of Michael Schumacher and Rubens Barrichello attempted a staged "dead heat" finish—with Barrichello taking a last-gasp win but the team accused of disrespecting fans by manipulating the race result.

The most contentious event of all though, and indeed one of the most controversial races in Formula One history, came in 2005. After several tyre failures in practice, Michelin, who provided tyres to seven of the ten teams, declared that their race tyre could only safely last for ten laps unless measures were taken to slow the cars at the banked Turn Thirteen. The sport's governing body, the FIA, stated that any measures to slow the cars would present a competitive disadvantage to the three teams supplied by rival manufacturer Goodyear—who after all had arrived at the race with fully functional tyres.

The Michelin teams, unable to find a compromise with the FIA, were left with no choice but to withdraw from the race. The fourteen Michelin-shod cars peeled into the pitlane after the formation lap to howls of derision from the packed grandstands —who had not been kept informed of the developing situation. Soundtracked by a chorus of boos, the six Goodyear cars, led by the Ferraris of Schumacher and Barrichello, took the most bizarre start in Grand Prix history.

The race, won by Schumacher, was a sideshow to the PR disaster—with many questioning whether F1 could ever return to Indianaoplis after such a farce. Subsequent races in 2006 and 2007 passed off without incident, but the damage to Formula One's reputation was irrevocable. After the 2007 race, Tony George stated publicly that he was unable to match Ecclestone's financial demands, and the United States Grand Prix once again drifted off the F1 calendar—leaving behind it a familiar whiff of missed opportunity, greed and poor treatment of the most belittled group in sport: the fans.

AUSTIN—A FRESH START

The disgrace of 2005 may have lingered in the minds of fans, but to those with a commercial interest in Formula One the American market remained too important to leave dormant for long. In due course, 2009 and 2010 were soundtracked by a plethora of proposals, denials, plans and negotiations to restore the United States Grand Prix. Despite various statements from potential race venues including New Jersey and Weehawken and Monticello in New York, it was announced in May 2010 that Austin, Texas, had been awarded the Grand Prix on a ten-year contract, starting in 2012.

The purpose-built Circuit of the Americas (COTA), constructed on 800 acres of land to the east of Austin, opened its doors to Formula One in November. The late season Texan sunshine and Austin's cool, youthful vibe and buzzing nightlife rendered the race an instant success with teams and fans alike.

Of the three races at COTA to date, the spoils have been shared by Lewis Hamilton (two wins) and Sebastian Vettel (one win). The circuit has been universally praised, offering iconography such as the dauntingly steep Turn One, an "esses" sequence of bends that recalls the famous Maggots and Becketts complex at Silverstone and a triple-apex fast 180-degree corner that apes the infamous Turn Eight at Istanbul Park.

Respectful touches such as the "tailgating" areas alongside the circuit and the cowboy hats sported on the podium have shown an awareness of what a United States Grand Prix has to do to genuinely connect with American audiences. For the first time in the history of the United States Grand Prix, the race has a purpose-built home and a commitment from all parties to making the event a long-lasting and commercially viable success on and off the track. The relationship between Formula One and the United States has long been a difficult one, but Austin's fledgling success makes it seem as though the love might, finally, be reciprocally consummated.

AMERICA'S FIRST CHAMPION—PHIL HILL

Of the thirty-two drivers to have won the Formula One World Championship, only two have raced bearing the Stars and Stripes on their flanks.

While 1978 World Champion Mario Andretti is one of the most renowned figures in motorsport history, the charismatic head of a dynastic family who enjoyed popular success on both sides of the Atlantic, America's "other" champion remains one of Formula One's lesser-known legends.

Phil Hill may not be a name familiar to many Formula One fans, but he was a true titan of F1 and sportscar racing in the late 1950s and early 1960s—and a man fully deserving of his place on the roll call of Grand Prix racing history.

A LIFELONG PASSION

Phil Hill was a child of the Roaring Twenties, born into a prominent family in Miami in 1927 and subsequently raised in Santa Monica, California. The son of a postmaster general, from an early age Hill was distant from his parents and displayed tendencies towards social introversion and self-doubt—but his passion for automobiles saved him from an unhappy childhood.

At the age of twelve, Hill's favourite aunt bought him a Ford Model T—which her chauffeur taught him to drive and Hill dissected repeatedly in order to understand how it worked. After graduating from high school Hill enrolled at the University of Southern California to study business administration, but he quickly tired of studying and dropped out to become an assistant mechanic in a Los Angeles garage.

The proprietor of the garage was an amateur racer, and Hill's first exposure to auto racing came helping other drivers tune and set up their performance vehicles. In 1947, Hill bought an MG-TC two-seater, modifying it himself so he could start racing on the amateur circuit. He won his first professional race, a three-lap event at the Carrell Speedway in July 1949, and in the same year Hill's skills as a mechanic took him to England as a trainee with the Jaguar racing outfit—but buoyed by on-track success his aspirations were increasingly turning to a career behind the wheel.

After both Hill's parents died in 1951, he used his inheritance to buy a 2.6-litre Ferrari—beginning a lifelong love affair with the scarlet racers from Maranello. Although Hill was a regular winner, his emotional intelligence and preternatural introversion made him constantly worry about and question the dangers of racing— even leading him to quit driving for ten months after developing severe stomach ulcers.

Upon returning to full health, Hill forged a reputation as the best sportscar racer in America—winning a succession of races in Ferraris run and entered by their wealthy owners. Hill's successes brought him to the attention of Enzo Ferrari, and he was invited to join the Prancing Horse's roster for the 1955 24 Hours of Le Mans.

FIRST SPORTSCARS TO FORMULA ONE
Unfortunately for Hill, his promotion to the Ferrari race team could not have come at a worse time. The 1955 24 Hours of Le Mans saw the worst disaster in motorsport history, with eighty-three spectators killed when the car of Mercedes driver Pierre Levegh somersaulted into packed grandstands. For the sensitive Hill, the disaster caused fresh personal turmoil—but the American continued to race for Ferrari.

By now Hill had turned his aspirations to Formula One, but promotion to the Ferrari's Grand Prix outfit was a long time coming as Enzo Ferrari felt Hill lacked the appropriate tempera-ment for single-seater racing. As it transpired, Hill's Grand Prix debut would come at the wheel of a Maserati, finishing a credita-ble seventh at Reims in the 1958 French Grand Prix.

Hill was subsequently elevated to the Ferrari Formula One team after the deaths of Luigi Musso and Peter Collins, and almost instantly proved his class with late-season podium finishes at the Italian and Moroccan Grands Prix.

By 1959 Hill was racing full-time for Ferrari, driving alongside a stable of thoroughbred teammates including Frenchman Jean Behra, Briton Tony Brooks and Hill's countryman, Dan Gurney. Hill finished the season fourth in the standings thanks to a strong return of three podiums from five races finished, helping to establish Hill as one of the leading drivers on the grid.

In 1960 the Ferraris endured a frustrating year, racing an obsolete front-engined Dino 246 against the rear-engined Coopers. Hill could only finish fifth in the championship, but he took his first Grand Prix win at the Italian Grand Prix in Monza. Starting from his maiden pole position, Hill also claimed fastest lap as he led home compatriot Richie Ginther and Belgian Willy Mairesse for a rare Ferrari one-two-three. Hill's victory was the first Grand Prix triumph for an American driver for over forty years, and the first win by an American in the "modern" era since the codification of the Formula One championship in 1950.

Hill would subsequently enter his home Grand Prix for Cooper after Ferrari declined to travel to the United States Grand Prix in Riverside, California. The American finished sixth on home turf, but much better was to follow in 1961.

1961: HILL REACHES THE SUMMIT

Hill's 1961 season would see him claim the Formula One World Championship, and it was a victory as much indebted to Hill's prodigious skill behind the wheel as his car: the iconic "sharknose" Ferrari 156.

The 156 was Ferrari's first rear-engined effort, and the Maranello outfit's response to the changing regulations for the 1961 season which reduced engine displacement from 2.5 to 1.5 litres. The "sharknose" epithet came from the dramatic air intake "nostrils" at the front of the car, and the 156 quickly became one of the most revered machines in Formula One history.

The three full-time Ferrari drivers for the 1961 season were Hill, fellow American Richie Ginther, and Wolfgang von Trips, a charming and flamboyant German of noble aristocratic heritage. While Hill travelled to races with a bulky tape recorder and would spend his time between sessions alone in his room listening to cassette recordings of old operas, von Trips would take debonair swings through the local nightclubs. Only Hill had tasted Grand Prix victory before, and started the season as de facto team leader.

The Ferraris were the class of the field from the outset. Only an inspired Stirling Moss denied the scarlet 156s victory at the opening race in Monaco, but a Ginther-Hill-von Trips two-three-four demonstrated the speed and reliability of the scarlet cars. The following race, the Dutch Grand Prix, saw Hill start a sequence of five consecutive pole positions, although von Trips took the chequered flag by less than a second for his first Formula One victory.

The tables were turned at round three in Belgium, with Hill heading home von Trips and Ginther for a Ferrari one-two-three—and the battle at the top of the standings already looked to be a straight fight between Hill and von Trips. Mechanical issues suspended the championship battle in France, but the race was still a Ferrari success with Giancarlo Baghetti taking victory in a privately entered 156.

The fifth race of the season at Silverstone saw Ferrari clinch their first ever Constructors' Championship, with von Trips heading home Hill and Ginther for another crushing one-two-three. The wet German Grand Prix at the fearsome Nürburgring saw Stirling Moss take his second win of the season in treacherous conditions—but second place for von Trips ahead of Hill in third put the German four points ahead in the standings with just two races left.

Heading to Monza, von Trips was on the cusp of destiny. A victory would have given the German the World Championship, but the race was to be marred by one of the most tragic accidents in Formula One history. On lap two, von Trips was duelling with Jim Clark's Lotus on the approach to the Parabolica corner when the two cars touched. Von Trips' Ferrari lurched to the left, launching up a verge and into a packed spectator area before becoming airborne as the car flew back across the track. Von Trips was flung from his car and killed instantly, along with fourteen spectators who had been struck by his careering Ferrari. Surprisingly the race continued, supposedly to help the emergency services' recovery effort, and Hill went on to take a comfortable victory.

With von Trips' death, the victorious Hill could not be overhauled in the standings, and the win sealed a maiden World Championship for the American. It was a tarnished title, won amid tragic circumstances, and Hill's celebrations were understandably muted on one of Formula One's darkest days. Hill went on to be a pallbearer at von Trips' funeral, remarking, "I never in my life experienced anything so profoundly mournful."

A VERSATILE TALENT

Throughout his Formula One career Hill retained a vested interest in other forms of racing, dovetailing his burgeoning skill in single-seaters with continuing success in top-class sportscar races.

As was common among leading drivers of the day, Hill's racing was not exclusively constrained by the schedule of the Formula One calendar. He entered Formula One as the reigning 24 Hours of Le Mans champion, having become the first American-born driver to win the race driving for Ferrari alongside Belgian teammate Olivier Gendebien. Hill, who displayed a lifelong affinity for wet-weather racing, drove most of the night in torrential conditions to help secure a famous victory.

Hill had also taken a maiden victory at the 12 Hours of Sebring alongside Peter Collins in 1958, and for a time was the pre-eminent sportscar driver in the world alongside his fledgling Formula One career. Further prestige victories were secured in the 1000km Buenos Aires (1958, 1960), the 12 Hours of Sebring (1959, 1961) and the 24 Hours of Le Mans (1961, 1962). Hill also had the distinction of bookending his career with victory; having won his first professional race in 1949, he also won his final event, the Brands Hatch 1000km, in 1967 at the wheel of a Chaparral Chevrolet.

The skillset that enabled Hill to race with such distinction in sportscars was born of his love for and appreciation of a car's internal mechanics. As much as Hill was a chronically nervous and frenetic presence before a race, pacing endlessly and feverishly chain-smoking, he was a notably calm and careful driver—mechanically sympathetic and famously easy on his cars. In an era of high mechanical fallibility, Hill's style allowed him to frequently bring his cars home to the finish—and his natural speed meant this was invariably near the front of the field.

ELDER STATESMAN

Hill's triumphant success in Formula One saw him finally slay his inferiority complex, but his introspective turmoil about the nature of racing and his personal demons endured. Hill once claimed that, "racing brings out the worst in me. Without it, I don't know what kind of person I might have become. But I'm not sure I like the person I am now." He was also outspoken on the dangers of racing, openly stating after claiming the World Championship that, "I no longer have as much need to race, to win. I don't have as much hunger anymore. I am no longer willing to risk killing myself."

Hill came back with Ferrari in 1962 to defend his title, but after starting the season with three consecutive podiums, results tailed off drastically—and Hill left the outclassed Scuderia before the end of the season. The end of his relationship with Ferrari marked the conclusion of Hill's time as a frontline Formula One driver, and ever-diminishing results during 1963 with ATS and 1964 with Cooper yielded only a single points-scoring finish at the 1964 British Grand Prix.

Although he officially retired from single-seater racing in 1964, Hill's final Grand Prix entry came at the 1966 Italian Grand Prix with a one-off entry at the wheel of an Eagle. Hill though failed to qualify, and his Formula One career concluded with something of a whimper.

Hill continued intermittently to race sportscars through the 1960s, but by this time his lifelong interest in restoring classic cars had developed into a lucrative business, Hill and Vaughn, which Hill ran alongside business partner Ken Vaughn. Selling the business made Hill an independently wealthy man, allowing him and wife Alma to raise their three children in comfortable surroundings in Southern California.

Hill remained in the public eye through his media work, acting as a television commentator for ABC's *Wide World of Sports* and enjoying a decades-long and distinguished association with *Road & Track* magazine. In 1989, Hill was inducted as the sole sportscar racer in the inaugural class of the Motorsports Hall of Fame of America, and he was also inducted into the International Motorsports Hall of Fame two years later.

Always an eloquent speaker who was self-effacingly open about his inner demons, Hill seemed to reach a semblance of personal peace in his later years. "In retrospect it was worth it," he said. "I had a very exciting life and learned a lot about myself and others that I might never have learned. Racing sort of forced a confrontation with reality. Lots of people spend their lives in a state that is never really destined to go anywhere."

After being diagnosed with Parkinson's disease in later life Hill withdrew from public life, and he died from complications related to the disease after a short illness in August 2008, aged eighty-one.

MARIO ANDRETTI

Not many have driven a race car better than Mario Andretti. He could make a bad car competitive and a competitive car victorious. He won the Indianapolis 500, the Daytona 500, the Formula One World Championship and the Pikes Peak Hillclimb. He won the Indy Car National Championship four times and was a three-time winner at Sebring. He won races in sports cars, sprint cars and stock cars—on ovals, road courses, drag strips, on dirt and on pavement. He won at virtually every level of motorsports since he arrived in America from his native Italy at age fifteen. He is a racing icon, considered by many to be the greatest race car driver in the history of the sport.

Assessing his legacy is easy: he drove the careers of three men; he drove with a passion and joy that few have equalled—and he won. Mario Andretti took the checkered flag 111 times during his career—a career that stretched over five decades. And he was competitive all of those years: He was named Driver of the Year in three different decades (the sixties, seventies, eighties), Driver of the Quarter Century (in the nineties) and the Associated Press named him Driver of the Century in January, 2000.

The admiration for Andretti has been for his achievements on the racetrack. He is in the very elite, top-superstar category of his game. Yet, if you look at his whole life, he has seen a world that most people will never see. And the journey he has made is what made him a very rich man.

THE 1940S

Mario was born in Montona, Italy (now Croatia), about thirty-five miles from the northeastern city of Trieste. World War II broke out around the time he was born, at the beginning of 1940. When the war ended, the peninsula of Istria, which is where Montona was located, became part of Yugoslavia. So the Andrettis were inside a Communist country. The family stuck it out for three years, hoping that the only world they had ever known would right itself. In 1948, they decided to leave. Their first stop was a central dispersal camp in Udine. About a week later they were transferred to Lucca, in Tuscany.

For seven years, from 1948 to 1955, the Andrettis lived in a refugee camp in Lucca. They lived in one room with several other families—with blankets separating each one's quarters.

Unfortunately, it was not just "pick up and go". Formal requests for visas had to be submitted to the American Consulate. Only so many people got visas. And when they reached a quota, there were no more granted. And many people were trying to leave. Some were going to Argentina, some to Australia, some to Canada. Everyone followed their imagination. And some people stayed. It was a case of whatever you thought was your best bet.

THE 1950s
The Andrettis waited several years for US visas. When they were finally granted, the family of five left all their belongings behind and began their new life in America. On the morning of 16 June 1955, the Italian ocean liner *Conte Biancamano* arrived into New York Harbor. Settling in Nazareth, Pennsylvania, the family of five had $125 and didn't speak English. Mario and his twin brother, Aldo, were fifteen.

In 1954, a year before arriving in the United States, the twins had gone to Monza to watch the Italian Grand Prix. They were only fourteen years old, but motor racing was becoming more and more of a dream for both boys. In those days, it was more popular than any other sport in Italy. That was especially true in the 1950s, when Ferrari, Maserati and Alfa Romeo were the top players in Formula One. And the world champion at that time was Alberto Ascari—Mario's idol.

So imagine the thrill when, a few days after arriving in America, the Andretti boys discovered a racetrack—right near their home in Pennsylvania. It was a half-mile oval track, which was different to what they had seen in Europe. And the cars were modified stock cars, not sophisticated grand prix cars. But there was a lot of speed. And it looked very, very do-able to them. The twins were now on a mission.

To become a racer, you need your own car. And there are only three ways to obtain a car: Steal it (which was against their

principles). Buy it (except they didn't have any money). Or build
it yourself (their only real option). They set out listening to
anybody who had even the slightest knowledge of a race car. They
found out about engines, shocks, spring rates, suspension, chassis
setup, and on and on.

With some local friends they built a car—a 1948 Hudson
Hornet Sportsman Stock Car—and raced it for the first time in
March of 1959. Four years after arriving in America, Mario and
Aldo were racing. While Aldo didn't have the same good fortune,
Mario's career flourished as he won twenty races in the sports-
man stock car class in his first two seasons.

THE 1960s
Mario's "first victory of consequence" came on 3 March 1962, a
100-lap feature TQ Midget race at Teaneck, New Jersey. On
Labor Day in 1963, he won three midget features on the same
day—one at Flemington, New Jersey and two at Hatfield,
Pennsylvania.

After joining the United States Auto Club in 1964, Mario
finished third in the sprint car point standings, capped by a
dramatic victory in a 100-lap race at Salem, Indiana. He also
drove in his first Indy Car event at Trenton, New Jersey on 19
April 1964, starting sixteenth and finishing eleventh in the
100-mile race, and earned $526.90 on his professional debut.

Mario won his first Indy Car race in 1965, the Hoosier Grand
Prix, and finished third in the Indianapolis 500, earning him
Rookie-of-the-Year honors. He went on to win his first Indy Car
Championship that year—with twelve top-four finishes—and
became the youngest driver (at age twenty-five) to win that title.
In 1966, he won eight Indy Car races, his first pole at the Indy
500 and a second straight national championship.

In 1967, Mario's passion for racing saw him compete and win
in the Daytona 500 stock car race, take his second pole at the
Indy 500, claim his first of three career victories in the 12 Hours
of Sebring endurance race, finish as runner-up in the Indy Car
national championship and be named Driver of the Year for the
first time.

He even tried drag racing in 1968—driving a Ford

Mustang—and earned eight more Indy Car victories en route to second place in the Indy Car point standings. Realizing a lifelong dream, Mario qualified on the pole in his very first Formula One race at the 1968 US Grand Prix at Watkins Glen, but was forced out of the race with a clutch problem. But Grand Prix racing was in his blood and the decade of the Seventies would see his dream come true.

Mario's celebrated win in the Indianapolis 500 came in 1969. He led a total of 116 laps and established fifteen of twenty new records set during that event. Mario scored a total of nine wins and five pole positions that season and went on to win his third national Indy Car title. He ended the decade with a total of thirty victories and twenty-nine poles out of 111 Indy Car starts.

THE 1970s

The 1970s proved to be a decade of successful versatility for Andretti, beginning with his second victory in the 12 Hours of Sebring in 1970, followed by his first Formula One triumph in South Africa, driving for Ferrari in 1971. His mastery of endurance racing was at its zenith in 1972, when he co-drove a Ferrari 312P to victory with Jacky Ickx at the 6 Hours of Daytona, 12 Hours of Sebring, BOAC 1000 km at Brands Hatch and Watkins Glen 6 Hours.

He continued his attack on the open-wheel series, winning a total of seven Formula 5000 events in 1974 and 1975, while finishing second in points in both seasons. He also took the USAC National Dirt Track Championship title in 1974, with three wins.

Andretti returned full-time to the Grand Prix circuit in the mid-seventies. His quest for the world title began in earnest in 1976, racing for the legendary Colin Chapman at Team Lotus. Their first taste of success came in the year's final Grand Prix in Japan, a race Andretti won in a monumental downpour. Conditions, in fact, were so atrocious that Niki Lauda pulled into the pits and forfeited his chance to retain the championship.

In 1977, Mario was third in the world standings with seven poles and four wins, including Grand Prix victories in his native Italy (Monza) and again in the United States (Long Beach).

The culmination of his international career came in 1978, when he won the World Championship driving for Lotus, making him the first driver in motor racing history to win the Formula One and Indy Car titles. Mario dominated the scene with nine poles and six wins in the revolutionary "ground effect" Lotus, which he had worked so hard to develop, and joined Phil Hill (1961) as the only American ever to capture the world title. He was again honored by being selected Driver of the Year, in recognition of his accomplishments.

Andretti topped the sport's best again in 1979, taking the International Race of Champions (IROC) series. As the decade came to an end, his full-time return to Indy Cars was imminent.

THE 1980s

In 1980, Mario competed for one last season with Team Lotus but was plagued by mechanical problems. He switched to the Alfa Romeo team in 1981, in what was to be his last full-time stint as a Formula One driver.

When the call came from his old friends at Ferrari to replace the injured Didier Pironi in 1982, Andretti put the car on pole at Monza and finished third, much to the delight of his Italian fans. The final Grand Prix start of his career was in the last race of the season at Las Vegas; however, a mechanical failure caused his day to end early. All told, Mario earned twelve victories and won eighteen poles in a total of 128 Grand Prix starts.

As he returned to the States in the early 1980s to concentrate on Indy Car competition, Andretti teamed up with his son Michael and Philippe Alliot to compete at the 24 Hours of Le Mans in 1983, where they qualified and finished third, the highest finish for a non-factory team. It was also Mario's first season with the newly formed Newman/Haas Racing team.

The 1984 Indy Car season proved to be a memorable one for Andretti who, at forty-four, won his fourth national championship by winning six events, eight pole positions and setting ten track records. The season was capped with his third Driver of the Year selection, bestowed for the first time by unanimous vote, making Mario the only man to ever win the trophy in three different decades (1969, 1978, 1984).

As the eighties progressed, Andretti continued to make racing history with some personal milestones. With his son, Michael, they established the first-ever, father-son front row in qualifying for the 1986 Phoenix Indy Car event a feat they accomplished a total of ten times before the close of the decade.

In 1987, in the debut of the Chevrolet-powered engine, Mario sat on the pole eight times, including his third pole at the Indy 500. He went on to lead 170 of the first 177 laps before engine failure cut his day disappointingly short.

Mario won his fiftieth Indy Car race at Phoenix and his fifty-first at Cleveland in 1988. As the decade came to a close in 1989, Andretti took on what would be the ultimate teammate—his son, Michael. It marked the first father-son team in Indy Car history.

THE 1990s
With his two sons (Michael and Jeff) and his nephew (John Andretti), Mario attained another "first" as the four family members competed in the same Indy Car race at Milwaukee, 3 June 1990.

The following year, the four Andrettis raced against one another for the first time in the Indianapolis 500. Jeff was voted Rookie of the Year, joining Mario (1965) and Michael (1984) as the only three members of the same family to win the award.

In 1992, Mario achieved two new milestones. He became the oldest Indy Car pole winner when he earned his record-setting sixty-sixth pole at the Michigan 500 and, at Cleveland, he set an all-time record for most Indy Car race starts with 370 (Mario finished his career with 407 starts). He was also named Driver of the Quarter Century by a vote of all former Driver of the Year winners and a panel of twelve journalists.

As he began his thity-fifth year of professional racing in 1993, Mario continued to make headlines with his fifty-second Indy Car victory at the Phoenix 200, making him the first driver to win Indy Car races in four decades and the first driver to win races in five decades. This race also marked his 111th major career victory. Records continued to be made in 1993, when Mario set a world closed-course speed record (234.275 mph) in qualifying for the Michigan 500, as he earned his sixty-seventh pole. This record stood intact until 1996, a year after the track was repaved.

Mario decided that 1994 would be his final year of competition as an Indy Car driver. The season-long farewell campaign, entitled Arrivederci, Mario featured special tributes, salutes and honours at every race venue. As he sped around an Indy Car track for the last time on 9 October 1994 at Laguna Seca Raceway, the legend of Mario Andretti assumed its place in the record books and in the hearts of his many fans.

Although officially retired, for the next few years Mario continued to seek the one major trophy missing from his mantle, the 24 Hours of Le Mans. He competed an additional four times in the world's most prestigious endurance race, winning the WSC class and finishing second overall in 1995.

MARIO TODAY

Today Mario is a spokesman, associate and friend to top executives around the world. He works with Bridgestone Firestone, MagnaFlow, Mattel, Phillips Van Heusen, Honda and GoDaddy.

Healthy and fit, he looks as though he could slip right back into the cockpit of a race car and often does, in the two-seater which allows for a passenger to sit behind the driver and truly experience the speed and pressure that comes with open-wheel racing. Mario remains vibrant, pursuing other passions, still working at a number of personal business ventures including a winery and petroleum business. He plays tennis, enjoys waterskiing and flying his ultralight. He was in the first *Cars* movie. He's on Twitter and Facebook, carries a tablet computer and stays current in the digital era. He isn't just any seventy-five-year-old, any more than he was just any racing car driver. He remains one of the most popular interviewers in racing and the most respected voice in motorsports. He is a much sought-after source by journalists from all over the world.

ANDRETTI—CAREER HIGHLIGHTS

- Four-time IndyCar National Champion (1965, 1966, 1969, 1984)
- Formula One World Champion (1978)
- Daytona 500 winner (1967)
- Indianapolis 500 winner (1969)
- Three-time Indianapolis 500 pole winner (1966, 1967, 1987)
- Pikes Peak Hill Climb winner (1969)
- Three-time 12 Hours of Sebring winner (1967, 1970, 1972)
- USAC National Dirt Track champion (1974)
- IROC (International Race of Champions) Champion (1979)
- Only driver to be named Driver of the Year in three different decades (1967, 1978, 1984) Named Driver of the Quarter Century (1992) by vote of past Drivers of the Year and a panel of twelve journalists
- Named Driver of the Century by The Associated Press (10 December 1999)
- Named Driver of the Century by *RACER* magazine (January, 2000)
- Named Greatest American Driver Ever by *RACER* magazine (May, 2002)
- All-time leader in IndyCar pole positions won (67)
- All-time IndyCar lap leader (7,595)
- All-time leader in IndyCar race starts (407)
- All-time leader in wire-to-wire IndyCar victories (14)
- Second all-time in IndyCar victories (52)
- Only driver ever to win IndyCar races in four decades
- Only driver ever to win races in five decades
- Oldest race winner in IndyCar history, with 1993 victory at Phoenix at age fifty-three
- Only driver to win the Indy 500, Daytona 500 and the Formula One World Championship
- From 1961 to 2000, competed in 879 races, had 111 wins and 109 poles (includes all forms of motorsports)

Mario's 111 career wins

52 IndyCar victories (USAC and CART)
12 Formula One victories (FIA)
9 Sprint car victories (USAC)
9 Midget victories (ARDC, NASCAR and USAC)
7 Formula 5000 victories (SCCA/USAC)
7 World Sports Car victories (FIA)
5 Dirt track victories (USAC)
4 Three-Quarter Midget victories (ATQMRA)
3 IROC victories
2 Stock car victories (NASCAR and USAC)
1 Non-championship race

Formula one victories (12)

1971 South Africa (Kyalami); Questor Grand Prix (non-championship race at Ontario Motor Speedway)
1976 Japan (Mount Fuji)
1977 United States (Long Beach); Spain (Jarama); France (Dijon); Italy (Monza)
1978 Argentina (Buenos Aires); Belgium (Zolder); Spain (Jarama); France (Le Castellet); Germany (Hockenheim); Holland (Zandvoort)

Formula one pole positions (18)

1968 United States (Watkins Glen)
1976 Japan (Mount Fuji)
1977 Spain (Jarama); Belgium (Zolder); Sweden (Anderstorp); France (Dijon); Holland (Zandvoort); Canada (Mosport); Japan (Mount Fuji)
1978 Argentina (Buenos Aires); Belgium (Zolder); Spain (Jarama); Sweden (Anderstorp);
Germany (Hockenheim); Holland (Zandvoort); Italy (Monza);
United States (Watkins Glen)
1982 Italy (Monza)

Indycar victories (52)
1965 Hoosier Grand Prix
1966 Milwaukee 100; Langhorne 100; Atlanta 300; Hoosier Grand Prix; Milwaukee 200, Hoosier 150; Trenton 200; Phoenix 200
1967 Trenton 150; Indianapolis Raceway Park 150; Langhorne 150; St. Jovite 100; St Jovite 100; Milwaukee 200; Hoosier 100; Phoenix 200
1968 St Jovite 100; St. Jovite 100; DuQuoin 100; Trenton 200
1969 Indy 500; Hanford 200; Pike's Peak Hill Climb; Nazareth 100; Trenton 200; Springfield 100; Trenton 300; Kent 100; Riverside 300
1970 Castle Rock 150
1973 Trenton 150
1978 Trenton 150
1980 Michigan 150
1983 Elkhart Lake 200; Caesars Palace 200
1984 Long Beach Grand Prix; Meadowlands Grand Prix; Michigan 500; Elkhart Lake 200; Mid-Ohio 200; Michigan 200
1985 Long Beach Grand Prix; Milwaukee 200; Portland 200
1986 Portland 200; Pocono 500
1987 Long Beach Grand Prix; Elkhart Lake 200
1988 Phoenix 200; Cleveland Grand Prix
1993 Phoenix 200

Indycar pole positions (67)
1965 Langhorne; Indianapolis Raceway Park; Phoenix
1966 Phoenix; Trenton; Indy 500; Milwaukee; Langhorne; Atlanta; Langhorne; Milwaukee; Trenton; Phoenix
1967 Trenton; Indy 500; St Jovite; Trenton
1968 Trenton; Indianapolis Raceway Park; St Jovite; St Jovite; Indiana Fairgrounds; Sacramento; Michigan 250; Phoenix

1969 Hanford; Milwaukee; Langhorne; Trenton; Kent
1970 Phoenix; Milwaukee; Indianapolis Raceway Park; Sacramento
1972 Milwaukee
1973 Texas
1974 Trenton
1980 Michigan 150; Phoenix
1981 Watkins Glen
1982 Michigan 500
1983 Cleveland; Elkhart Lake
1984 Long Beach; Portland; Meadowlands; Cleveland; Michigan 500; Elkhart Lake; Mid-Ohio; Laguna Seca
1985 Long Beach; Milwaukee; Meadowlands
1986 Phoenix; Mid-Ohio; Laguna Seca
1987 Long Beach; Phoenix; Indy 500; Meadowlands; Pocono 500; Elkhart Lake; Laguna Seca; Miami
1992 Michigan 500
1993 Michigan 500

Endurance race record

Date	Race	Car	Co-Driver	Start/ Finish	
2/6/66	Daytona 24 Hours	Ferrari	Pedro Rodriguez		3
3/66	Sebring 12 Hours	Ferrari 365P2	Pedro Rodriguez	DNF	
6/19/66	24 Hours of Le Mans	Ford GT Mk II	Lucien Bianchi	DNF	
2/5/67	Daytona 24 Hours	Ford GT Mk II	Lucien Bianchi	DNF	
4/1/67	Sebring 12 Hours	Ford GT Mk IV	Bruce McLaren	1	1
6/11/67	24 Hours of Le Mans	Ford GT Mk IV	Lucien Bianchi	DNF	
2/68	Daytona 24 Hours	Alfa Romeo T33	Lucien Bianchi		5
3/22/69	Sebring 12 Hours	Ferrari 312P	Chris Amon		2
4/25/69	Monza 1000 km	Ferrari 312P	Chris Amon	DNF	
2/1/70	Daytona 24 Hours	Ferrari 512S	Arturo Merzario and Jacky Ickx	1	3
3/21/70	Sebring 12 Hours	Ferrari 512S	Nino Vaccarella and Ignazio Giunti		1

7/11/70	Watkins Glen 6 Hours	Ferrari 512S	Ignazio Giunti		3
3/20/71	Sebring 12 Hours	Ferrari 312P	Jacky Ickx		2 DNF
7/24/71	Watkins Glen 6 Hours	Ferrari 312P	Jacky Ickx		DNF
11/9/71	9 Hours of Kyalami	Ferrari 312PB	Jacky Ickx	1	2
1/9/72	Buenos Aires 1000 km	Ferrari 312PB	Jacky Ickx		10
2/6/72	Daytona 6 Hours	Ferrari 312PB	Jacky Ickx	1	1
3/25/72	Sebring 12 Hours	Ferrari 312PB	Jacky Ickx	1	1
4/16/72	BOAC 1000 km at Brands Hatch	Ferrari 312PB	Jacky Ickx	1	1
7/22/72	Watkins Glen 6 Hrs	Ferrari 312PB	Jacky Ickx	1	1
4/25/74	Monza 1000 km	Alfa Romeo 33TT	Arturo Merzario	1	1
7/74	Watkins Glen 6 Hours	Alfa Romeo 33TT	Arturo Merzario		DNF
6/83	24 Hours of Le Mans	Porsche	Michael Andretti and Philippe Alliot	3	3
2/84	Daytona 24 Hours	Porsche	Michael Andretti	1	DNF
6/88	24 Hours of Le Mans	Porsche 962C	Michael Andretti and John Andretti	3	6
2/89	Daytona 24 Hours	Porsche 962	Michael Andretti	6	DNF
2/91	Daytona 24 Hours	Porsche 962	Michael Andretti and Jeff Andretti	6	5 (DNF)
6/95	24 Hours of Le Mans	Courage Porsche C34	Bob Wollek and Eric Hélary	3	2
6/96	24 Hours of Le Mans	Courage Porsche C36	Derek Warwick and Jan Lammers	9	13
6/97	24 Hours of Le Mans	Courage Porsche C36	Michael Andretti and Olivier Grouillard	31	27 (DNF)
6/00	24 Hours of Le Mans	Panoz LMP-1 Roadster S	David Brabham & Jan Magnussen	4	16

THE BEST OF THE REST—*OTHER NOTABLE US F1 DRIVERS*

DAN GURNEY

Dan Gurney is highly regarded for his Formula One contributions, both on and off the track. He is the only driver to score the first F1 victory for three different constructors: Porsche (1962), Brabham (1964), and Eagle (1967). He himself had built the Eagle chassis. As a result, Gurney's name is still associated with race car aerodynamics. He was the first person to use what is now known as a "Gurney flap" on the wing of his car. He is also credited as being the first driver to spray champagne on the podium. Gurney made his F1 racing debut with Ferrari in 1959, finishing in the top three in two of the four races he entered. His second season racing for BRM was much less successful, with his car failing to finish most races. His only race finish with BRM saw him end up in tenth place. Between 1961 and 1965, Gurney drove for three different teams and was classified in the top six in the Drivers' Championship each year, but he would never finish better than fourth in the title race. Officially, he left the sport in 1968 but made a brief return with McLaren after the death of founder Bruce McLaren. Gurney won just four races of the eighty-six he started.

PETER REVSON

Revson started four races for Lotus in 1964, but returned to the United States to drive Indy Cars and closed-wheel sports cars. He made a guest appearance for Tyrrell at the 1971 Formula One United States Grand Prix. While he failed to finish the race, his driving was sufficiently impressive that he was signed by McLaren to drive for the 1972 season. He stood on the podium at four of the nine races he attended, and stayed with the team for an even more successful year in 1973. Revson won the 1973 British Grand Prix and the 1973 Canadian Grand Prix, both in wet conditions. However, he made the fateful decision to join the Shadow team when McLaren offered him a third car for the next season. While testing their car in preparation for the 1974 South

African Grand Prix, Revson suffered a fatal accident when the front suspension of his vehicle failed.

EDDIE CHEEVER
Cheever entered a few F1 races in 1978 before starting full time with Osella in 1980. However, the new team was unable to provide him with a competitive car and he only finished one of the races he entered that year. He joined Tyrrell in 1981, and had improved results, but still failed to secure any podium finishes. With Ligier in 1982 he finished in the top three at three races before changing teams once again. Cheever enjoyed his most successful season in 1983 with Renault, mounting the podium four times. However, he never tasted victory throughout his F1 racing career, which ended in 1989. He finished no higher than third in the 132 races in which he competed.

RICHIE GINTHER
Ginther started fifty-two races during the 1960s, finishing on the podium fourteen different times. He won the 1965 Mexican Grand Prix driving for Honda and finished third overall in the 1963 season, racing for the BRM team.

BILL VUKOVICH
Vukovich competed in five Indy 500 races when they were part of the Formula One World Championship. In 1951 he retired after just twenty-nine laps and finished a disappointing seventeenth the following year. However, he returned for the 1953 event, started from pole and went on to win the race. He won again in 1954 and was leading in 1955 when he crashed into a marker. The collision pitched the car into, then over, a concrete wall, fracturing Vukovich's skull. He died at the scene. Statistically Vukovich won 40 per cent of the Formula 1 races in which he competed, but drivers who competed only at the Indy 500 events are often omitted from the history of the sport.

THE REST
The following drivers started at least ten Formula 1 races during their driving careers: Michael Andretti, Tony Bettenhausen,

Ronnie Bucknum, Mark Donohue, George Follmer, Masten Gregory, Jim Hall, Brett Lunger, Jim Rathmann, Harry Schell, Scott Speed, Danny Sullivan and Rodger Ward. Two other drivers, Bob Bondurant and Skip Barber, competed in F1 racing during the 1960s and later formed two of the most respected and successful driving schools in the USA.

"THE AMERICAN"—BOB BONDURANT

Bob Bondurant recovered from a horrific near-death crash in 1967 to become a driving instructor for Hollywood actors such as Paul Newman and Tom Cruise. Today, his high-performance driving school has graduated over half a million people.

"I went as high as the treetops," recalls Bob Bondurant incredulously. "I remember looking down at the embankment from the air and thinking, 'Uh-oh, this is going to be a bad one!' And that's the last thing I remember."

It was 27 June 1967, the United States Road Racing Championship, and Bondurant's McLaren MK 1C had just careered around the 150-mph corner when disaster struck. As he neared 200 mph, the front steering arm broke and he slammed into the embankment. It was one of the most shocking crashes in the history of motor racing. Somehow, in a split-second, Bondurant was able to switch the engine off, take a deep breath, relax the muscles in his neck, shoulders, hands and wrists and pray for the best. He should have been killed that day. The car flipped eight times. All that was left of it was the rear suspension.

Bondurant woke up in hospital in casts. "How soon am I getting out?" he asked anxiously.

"Well, young man," said the doctor severely, "I saw your accident and it's the worst I've ever seen. Do you want the good news or the bad news?"

"The good news," suggested Bondurant.

"You have a mild concussion and three broken ribs and two broken legs below the knees, but they will heal. The bad news is I cannot allow you to sit up because the lower vertebrae in your back is damaged and if you sit up there's a risk you could become paralyzed. So don't sit up."

"Okay," he agreed.

"And you broke nearly every bone in your feet and ankles and you'll never walk again."

"Are you *sure*?" asked Bob Bondurant.

* * *

The Bob Bondurant School of High Performance Driving sprawls across 450 acres of scorched earth in Arizona's Sonoran Desert, eleven miles from Phoenix, on land leased from the Gila River Indian community. The fifteen-turn road course covers 1.6 miles and was designed by Bondurant himself, serving wannabe racers, the odd Hollywood film star, the occasional Black Ops team needing a crash course in covert, high-speed manoeuvres, as well as everyday drivers looking to upgrade their road safety skills. There's also a four-day Grand Prix course ($4,999 if you're interested).

Bondurant is a sprightly eighty-one-year old, although the hearing in one ear has been damaged from racing cars in the 1950s, in an era before health and safety, when the engine was mounted directly under the driver's ear. On most days, alongside his crack team of driving instructors—which includes Andy Lee, World Challenge Rookie of the Year in the 2012 GTS Championship—America's uncrowned world driving champion takes to the track at 180 mph, often with gleeful passengers who pay him for the pleasure. It's how he stays in shape, he says.

"When I put my helmet on," says Bondurant, "I'm like a different man."

Bondurant has fond memories of his time in Europe, racing Formula One, when his colleagues were the likes of Jochen Rindt and Jackie Stewart, and he remains a very proud member of the hallowed Club International des Anciens Pilotes de Grand Prix F1 (aka CIAPGPF1 Grand Prix Drivers Club). For a kid with the heart of a racer, the arrival of Formula One on the world motorsport scene when he was seventeen made a huge impression on the young Bondurant. At the time, he was cutting his teeth on an Indian Scout motorcycle near his family home in Los Angeles. At twenty-three, he was racing sports cars and winning races as well as accolades. But it was the Cobras that defined Bob Bondurant's career. In 1963, he joined Caroll Shelby's Cobra team and began chalking up an impressive number of wins, taking Europe by storm and securing a permanent place on the team. In 1964, he won the GT class at the 24 Hours of Le Mans. In 1965, while racing in Reims, France, where he helped to win the World Manufacturers' Championship for the US in his now trademark Cobra, the impressive American had his first brush with Formula One.

Enzo Ferrari summoned the man the Italians were calling "Bondurante Sir Cobra" to Italy for a meeting. John Surtees, reigning F1 World Champion, did the translating.

"Ferrari said, 'Would you like to live in Italy?'" recalls Bondurant. "And I said, 'Si, if I'm driving in Formula Uno.'"

But Ferrari was not to be rushed.

"When will you let me know?" asked Bondurant.

"When I decide," said Ferrari.

"One week? Two weeks?" Bondurant persisted.

"He looked at me very sternly and he repeated, 'When I decide.' So I shut my mouth."

Bondurant returned to digs in England, thinking he'd blown it, that he'd pressed the great man too hard. Three days later, however, the phone rang. It was Luigi Chinetti at Ferrari, requesting Signore Bondurante's presence in Italy. He was to report to the factory to be fitted for the Ferrari 158V8 that he'd be driving that October in the US Grand Prix in Watkins Glen, New York.

It was his Formula One debut. He began fourteenth on the grid, got a good start and soon made it up into sixth place. Then the track was beset by wind and rain. Bondurant says, "You race in Europe, you get used to racing in the rain, so you just do it." But halfway through the race, the elastic on his goggles came unstuck. The rain had stretched the elastic band and they kept blowing down over his eyes. His first time in a Formula One car, racing for Ferrari, he recalls ruefully, and this happens! What are the odds?

Bondurant soldiered on, pulling the goggles back up over his eyes and holding them there, driving one-handed for the rest of the race, sometimes at 170 miles per hour. To shift gears, he was forced to put his knee on the steering wheel to free up his right hand. He finished ninth, but it wasn't enough to make the grade with Ferrari. While throughout 1966 he continued to race for Ferrari in prototypes, at Sebring, Daytona and Le Mans, at the following Mexican Grand Prix, he drove a Lotus 33 for Reg Parnell, and after that he competed in five Grands Prix for Team Chamaco Collect, finishing fourth at Monaco.

As a rainstorm plagued the Belgian GP of 1966, Bondurant worked frantically beside Graham Hill to prise an immobilized Jackie Stewart out of his demolished—and dangerously leaking—car. Stewart was soaked in fuel and trapped in the car for over thirty minutes, his steering column pinned to his leg. The pair were eventually able to free the Scot using spanners from a spectator's toolkit. This was the incident that provoked Stewart's one-man crusade to improve driver safety in the sport.

About ten years ago, Bondurant received a phone call from a stranger who claimed he'd been at that ill-fated race in 1967. He was seventeen at the time, he explained, and had witnessed the crash and captured some of it on camera. Would Bondurant like to see the negatives, he asked.

"A lot of drivers never got to see the accidents they were in back then," says Bondurant. "Blowing up those negatives and looking at those pictures gave me so much closure. To see how badly the car was damaged—and that I survived it."

The photographs are now hanging in the museum at the Bondurant School of High Performance Driving, among Bondurant's lovingly archived collection of motor racing memorabilia.

The crash finished his racing career, but just eight months later, through sheer force of will, he was walking again, albeit painfully. He'd also just opened his first driving school at Orange County International Raceway.

"I just never gave up," he says. "I *never* give up."

Back in 1966, Bondurant had acted as technical director on the cult film *Grand Prix*, teaching its star, James Garner, how to race. He'd enjoyed teaching and realized a new vocation. "It's like the man upstairs said, 'Before you get maimed or killed I'm going to take you out of racing and help you teach other people.'" says Bondurant. It was the same bloody-minded determination that had enabled him to achieve so much as a racing driver that got him back on his feet and succeeding in business. While still in a cast and wheelchair, he was tirelessly canvassing sponsors for the school—despite, as he admits, not being the greatest of advertisements for a new driving school.

Bob opened his school with three Datsuns, a Lola T70 Can-Am

car and a Formula V. His first class consisted of just three students. The second week there were two students, Robert Wagner and Paul Newman, who were training for the film *Winning*. Bob was a technical advisor, camera-car driver, and actor-instructor in the film. Newman told Bob that he had turned down two better-paid films to do *Winning*, just because he wanted to see if he had the guts to be a racing driver. Bob was impressed by his daredevil attitude and a mutual respect and enduring friendship developed between the two men. Legend has it that there were only three people that Newman would ever call on the telephone, one of whom was Bondurant.

A move to Ontario Motor Speedway in California followed— and the addition of a few cars from Porsche—then to Sears Point in Sonoma. The school finally settled in Arizona in 1989 with a Ford partnership, switching to GM and Goodyear in 2005.

Along the way, Bondurant had started racing again, and was inducted into the Motorsports Hall of Fame of America. In April 2012, he expanded with a satellite school at Pikes Peak International Raceway (PPIR) in Colorado. Arizona's summers get too hot to run much of a corporate programme, so the partnership at PPIR takes up the slack. They've trained up to 500 people a day in Colorado.

"We've now graduated nearly 500,000 people," he says. A further expansion of the Bondurant School out to the Circuit of the Americas Formula One track in Austin is on the cards, too. Brand Bondurant represents an American connection to the tumultuous early days of Formula One—a living archive as, just two states to the east, Formula One finds a home in Austin.

MARIO UNCUT

MARIO ANDRETTI'S ON THE PHONE! INTERVIEW
—UNCUT

Mario Andretti is in his office in Pennsylvania when the call is put
through.

"Hello this is Mario Andretti."

Christopher Joseph (CJ): "Hi Mario. I saw that you had some fun
and games on the weekend down in Austin?"

Mario Andretti (MA): "Yeah. Finally we were able to just have a
taste of what it will all be, obviously, the racing part of it, the
surface was quite ready. A few touches here and there. No
breaking points yet. But it was all great to be able to go down
for the first time and sample what it's going to be."

CJ: "Everything is ready?"

MA: "Yes. It's amazing the amount of work that they've done
since the last time I was there which was in June. If you
would've seen it then and then now, it's almost a miracle
happened. But there's work to do yet . . . You could tell the
infrastructure's quite ready, just finishing touches here and
there, as you can imagine. But then again I'm told that 1,100
bodies [are] on site everyday, forty different contractors show
up, there's a lot of action going on. Quite honest, while we were
out there playing on the track, you could look around and see
work going on. They didn't even waste that day by being idle.
But it's definitely ready to go . . ."

CJ: "How did you find it behind the wheel again and on an F1
track and a new track at that? What about up the hill, Turn
One, your Turn One, is it going to be called the Andretti Turn
by the way? How was that?"

MA: "Well, it's . . . [laughs] Who knows? I don't think they've
named any corners yet but I call that a marquee corner when
the track wasn't even paved. Something impressed me about
it, it's going to draw comments from other drivers, I suspect.
Again, you're always anxious to get on and get the first taste
of it. That's what we're all about. Those are the days we always
look forward to. That's our playground, if you will. I had

different choices of cars to go out. The only disappointment of the day was not getting into the R30. It went out for five laps to shake it down and the engine blew. So, I didn't get to drive that."

CJ: "That's a shame but that's racing, I guess."

MA: "That's a shame but they promised me another day so that's good. And they had the 79 there, which was great, so I was able to tool about with that a little bit. And I took out some other cars, Ferraris and things."

CJ: "Did you take Bobby Epstein around?"

MA: "Yeah, Bobby, I took him out in the three-seater formula one car. I don't know if you ever seen those?"

CJ: "Yeah. Is that the Pirelli one you had?"

MA: "Yeah, that's the one. I did quite a bit of driving with that. They had a lot of passengers waiting in line. Then the engine overheated on that too."

CJ: "When you first decided to become the ambassador for the Circuit of the Americas, I was told you took quite a time to decide. You analyzed the role and what you could do? Is that the case?"

MA: "First of all, it's like any new project, you have many questions. And there were potential issues here and there, which you would expect from a project of this magnitude. When you go to the private sector on something like this, and you're trying to compete with venues that have been financed by governments, you wonder whether it's ever going to come off. That was my concern, of course. I was flattered that I was asked to be part of it, but I wanted to confirm many aspects of it. I started with Bernie. He gave me certain reassurances that the game is on. I said, well, that's good enough for me, and then, of course, I had to see for myself. I went down there in June. It was just before the Montreal GP. I was satisfied with what I saw as far as the plans, project and people involved. Then I just reported back to Bernie in Montreal and I said, I think things should be good there. That was it. And I feel strongly that the group behind it is very, very committed."

CJ: "So it's not just the case of Bobby and Red opening up a cheque-book so they can acquire you as an ambassador . . . It actually was a question of you doing your due diligence and going down there to see if these guys were serious and could bring it in on time?"

MA: "Exactly. And the thing that impressed me the most was Bobby. He was there with his work boots on and dirty jeans and construction helmet. And, I mean, he was hands-on, and very passionate about what was going on. Then I got the same feeling again when I met him the second time, the third . . . When you see someone that is investing but also wants to be part of it, keep their hand in it, I think that speaks volumes for me. You can see that they have a lot riding there, you can imagine."

CJ: "When I spoke to him he said it was probably the biggest thing he'll ever do in his life."

MA: "Yeah. I saw so much satisfaction in their faces when they were finally able to present it to the press and the world. That, to me, was very rewarding, in many senses. Of course, you just wish them well because of that. The overall plans, what they'd secured already, MotoGP and so forth, or some of the events they had planned for, and others, outside of motor racing, to keep the facility active year-round, I think, maybe those things make a lot of sense and because of that I hope their business model really works for all the reasons."

CJ: "You've pretty much raced on every US F1 track. Why do you think it will work this time in Austin at COTA as opposed to Indy, Phoenix, Detroit? I guess the last time we had a stable US race was back in Watkins Glen . . ."

MA: "Well if you look back at . . . okay, let's start at Watkins Glen. Why does Formula One leave Watkins Glen? Because there was no real commitment, reinvestment in bringing the facility up to the standards that Formula One required, standards that the rest of the world was adhering to. From there, it went somewhere else, then it went to, I think, Long Beach. So that seemed to be an interesting venue but one thing leads to another, it was still a temporary course, and there was some other issues. And from there it moved from Dallas to Phoenix to . . . You know, all temporary venues. Detroit. And the tempo-rary aspect of it was exactly that. So, then, all of a sudden, I felt very good about Indianapolis, but here's my take on it. There's never been a successful dual venue for road racing. If it's an oval with a road course, the road racing side suffers. Why? Because there's no ambience. I mean, we've seen that even with some circuits like Germany and so forth, where they have a dual purpose . . . They never survive. There's something about it. Take The Daytona oval where they have the road course. Why isn't that 24 hours Daytona successful from the standpoint of attendance?"

CJ: "That's a good question."

MA: "It lacks the ambience of road racing. Indianapolis suffered that tremendously. You know, you sit in a grandstand and what do you see? Cars go by on the straightaway and then you don't see the rest of it, you have to look at a television screen. So, again, that's why it didn't survive. Now, finally, I think, for the first time since, probably, the seventies, it's going to have a proper home, with a circuit built up to the standard that is required today to be competitive with the rest of the world. And the United States does not have another venue like this, not for road racing. Here they have many classics, and a really nice road course, but they're years behind the times as far as what I'm talking about. Again, that's why, for Formula One fans, for road racing fans it's time to rejoice. And quite honestly, I had the first question to cross my mind too is—is Austin going to work? Basically, the location. Well, to be honest with you, I think you have to look at accessibility from South America too. Just incredibly fertile ground for motor racing. You have the proximity of Mexico, you have Brazil and so on and so forth. And you could see from the response for tickets, it was amazing what they got from South America, for instance. It's got a lot of benefits, and it's easy to travel to, to Austin. It's not a big issue from that standpoint."

CJ: "What do you think then that Formula One in Austin and in the States has to do culturally to make this work? Because, I think, the first one is always easy and then you contract for another nine years. There's a cultural difference I think in motor sports between F1 and the rest of motor sports in the States. So what do the guys down in Austin have to do to make this work?"

MA: "I think the one thing that's going to work is the fact that it's a solid home, it's something you can definitely look forward to for the years to come. That in itself should be very strong for the continuity aspects, should be very strong to be able to bring the fans back, and I think the Formula One fan base—I keep saying this always—is underestimated here in the United States. I can't see why the event shouldn't be something that's so welcome by the fans and be attended for the years to come. I think it can only get stronger. That's the way I see it, because, again, you're going to have MotoGP coming back in and, obviously, you're going to have some GrandAm, some sports car racing, and all that. The place will continue to be showcased and I'm sure that the exposure it will receive is also going to be another tool that will be in the minds of the fans, the ones that will not be here this time but want to experience it. I feel strongly in that respect."

CJ: "What do you think F1 has to do itself to make it successful? Because it's a big return after five years, and as you know, it's a strange circus that comes to town when F1 visits."

MA: "Just be there. I don't think they need to do anything in particular, anything they don't do anywhere else. There's only one thing I would like to interject on that—I would love to see Formula One open up to the opportunity to have some guest drives. By that I mean, I know how I started, you can say times are different, this and that. But you know what? Times are different but a lot of things are the same. I got the opportunity by being the third driver on a top team when I broke into Formula One, and I won races as a third driver. Can you imagine we go to some of these new markets and so forth, to have a local hero be invited by McLaren or Ferrari or something to be the third man on the team. You know, after some proper testing or whatever."

CJ: "So, you mean not just driving practice Friday but actually driving in the race?"

MA: "Driving in the race, yes. I mean, it's happened before and I think this could be a big, big promotional item to bring the local fans. Formula One, in so many ways, being so international, it's very partisan too and there's a lot of national pride going into it, and a lot of fans who would love to see, maybe, one of their own, see how they measure up, but not with a weaker team, it would have to be with a top team. And if the talent could reveal itself, it could be here, another launch. The financial aspect needs to be considered but I think again Formula One should be thinking in those terms—as elite as it is, there are some drivers in different countries that I think would belong and make the most of the opportunity. It would do wonders, I think, for the event in that country."

CJ: "Do you think we need an American driver, and if not that, an American team, for F1 to really take hold in the States?"

MA: "I don't think so. I think it's still on the cards though, but I think it would be a lot better, it would be a lot more interesting, a lot more coverage, a lot more interest from the fans, the press standpoint, and so forth, because naturally you will be rooting for your own. When you have such an international event, to be able to fly your own flag has that particular pride to it. So all of these elements, I think, would be a big plus in promoting Formula One in the States. I think by having a venue such as this, those opportunities will open up."

CJ: "Being a businessman, what will we see business opportunity-wise for F1 in the US? This year, for instance, was the first time Unilever came on board, so we had some fast moving consumer goods. What else do you think we'll see?"

MA: "Well, here again I think when you look at the sponsorship in F1, obviously, you have a lot of global companies involved for the obvious reasons, and I'm sure those global companies do a fair amount of business in the United States. They would want to benefit from the exposure here. So then again, and this might also encourage some other companies here that would want to expand in international markets to be able to be in Formula One to have that opportunity to have the visibility. All of this plays in a positive direction. I think this could be the beginning of a lot of good things. There are many areas where you can see there could be more growth, more interest in what we have now because of the venue that's in place."

CJ: "What was your take on the whole Texas versus New Jersey race? Do we need two? Is it beneficial to have two American races?"

MA: "It's beneficial for America for sure. I think of any country in the world, America could easily host two Formula One races. New York, I think it would be a great market, it's a market that Bernie has been eyeballing for years. I hope that happens as well. It's been pushed back another year now. I've been in touch with the principals there, people involved, and they're working very hard . . . they're very professional, trying to make that happen as well. It can only help the fan base by having two events to look forward to. One can also feed off the other."

CJ: "I know that for Austin they've arranged for one-day flights from New York, Miami and Los Angeles direct into the race, which is a really cool idea."

MA: "I think Austin is going to be a very nice host city for this event. I think it will be a pleasant surprise for a lot of people that are not familiar with that particular area."

CJ: "When we wake up on November nineteenth, what do you think the opinion will be of what's happened on that day?"

MA: "I think it's going to be a great weekend. We might even crown the world champion right there."

CJ: "That would be nice."

M: "It would be nice for Austin, yes."

CJ: "How do you think the mind-set of the modern driver differs from your mind set during your period?"

MA: "I don't think the mind-set should differ, mainly because the mind-set is having this positive attitude of coming in with the idea of having a result that particular day. My mind-set was always to enter every event with the confidence of winning regardless of how realistically things looked at that moment. And that's the mind-set of every champion, every proper racing driver, to go out there and feel, you know what, things might not look too good but I still think I have a chance of winning today. The preparation for the champion drivers has always been the same—the ultimate in physical conditioning and the total focus of the driver on his job. And that's it. There's no difference. The only difference is in the metal, that is the car that you're sitting in. Today you have more tools to work with, and that's why you go faster."

CJ: "Do you think it's more car-focused today or just as car-focused as in your period?"

MA: "I feel especially in Formula One, the credit goes fifty-fifty. Obviously that could shift by 10 per cent either way at any given time, but the thing is the driver's still in command, and it's the driver's job to bring whatever equipment he's sitting in to the limit. In the past the cars were slower because they didn't have the cornering capability, but the driver worked just as hard as he does today. I've gone through the decades, and I know how this works because I've had the experience. I felt that I fit very well back in the sixties and I would fit very well right in 2012 if I was a little younger. I wouldn't have to change a damn thing as far as my approach to the sport as a driver."

CJ: "There's a great quote by Dale Earnhardt—'It's not the fastest guy, it's the one who refuses to lose.'"

MA: "I'll go along with that."

CJ: "If I may ask a question sent in by the fans?"

MA: "Sure, go ahead . . ."

CJ: "This one from Dr Vanda Collins. What was your late and great friend Ronnie Peterson like to have as a teammate?"

MA: "From the human side and the professional side, he was the perfect teammate. We had a great relationship on a human side as friends with our families, and we definitely had a respect for one another. We enjoyed the racing together and shared the team. We never had a cross word for each other at any time. When I lost him, it was like losing a brother or a son. No question about it."

CJ: "I think that's a great way to finish."

THE LONG AND WINDING ROAD TO
TURN ONE—*USGP 17:11:2012 1400.01.03*

"I didn't get into this as a Formula One or racing fan. My initial reaction was 'Why Austin? Why Texas?' I didn't get into it with the expectation that I'd be running the whole show one day. I think we're going to do well with Formula One racing. I think we will be here for a long time."
Bobby Epstein

"Red McCombs changed my life and so has this project and I am sure it has been the biggest thing that Bobby has done in his life. Perhaps it is the biggest thing we will ever do."
Rad Weaver

Bobby Epstein crackles over the phoneline in that silky southern Texas drawl that he has and immediately apologizes for being late as he was taking his son to *"soccer"*. Yes, even in Texas they play soccer.

Across the board from management, teams, drivers, sponsors and fans the inaugural United States Grand Prix was universally acclaimed yet not a lot is known about the people who made it happen on 17 November.

Turn One at the Circuit of the Americas (COTA) may yet become an iconic part of Formula One folklore but the first turn for the men who brought the race to Texas was a downturn in the economy, which brought an unexpected result.

"I bought the land. I bought it at the worst possible time and I used to joke that we'd put roller coasters and waterparks there one day because no one was going to build houses as the housing market had died," said Bobby Epstein.

"So when I was told that some marketing executives were looking at various properties in that part of town to bring a Formula One race here I listened. I listened very carefully."

I am listening very carefully now as I wonder what exactly this former bond trader thought was his biggest challenge and his biggest achievement.

"Besides ensuring the financial health of the project and navigating some of the very difficult waters I think I focused mainly on what happens outside the racing lines. I focused on the fan experience, the spectator point of view. I think that's where I knew I had something to contribute."

He continued: "From the start I asked for a great fan experience: Give us a large number of overtaking opportunities was my first instruction to the designers. We gave them more or less a blank canvas. We let them choose the land that they thought would fit from a driver's standpoint as well as give a unique fan perspective.

"Tilke were able to draw upon the best characteristics of the most respected features of different tracks from around the world and incorporate that into something that functions year round but can still be the premier track in the world. I think we ended up with something special."

THE SAN ANTONIO CONNECTION

Texas may be "wide open for business" as the slogan goes, so it is not surprising that it also took a great partnership to get the job done. Former baseball player Rad Weaver, who is a director on the board of McCombs Partners, the investment capital firm founded by the extraordinary Red McCombs, joins into the conversation to explain his role.

"Red and I were lead investors on the project. Bobby and I, in many ways have kind of held each other's hand through this project and worked together to help guide this thing through to race day."

McCombs Partners are canny investors who believe in people first but also in the strength of brands. So where will business both from and outside Texas make the most of the USGP?

"They will all learn how they can integrate their brands and activate their brands around F1 and how to make the most of the race weekend as a corporate event."

Looking ahead to the future, Rad said: "In my opinion a natural maturation process will take place through which the teams, COTA and Formula One as well as the brands will find the best way to maximize the exposure that Formula One provides. I think there's no question you will see the US get even more excited next year and more importantly US brands will get very excited."

While the USGP and COTA were definitely an all-American affair they did have some European assistance. Throughout the process they were guided by KHP Consulting in the most important aspects of cultural, social and economic engagement that are necessary to successfully stage a Grand Prix for the first time.

Katja Heim, their CEO, who has guided Bahrain, Shanghai and Abu Dhabi through the same challenging process, is also optimistic for the future. "Year one is normally when you struggle right up to the last minute to deliver the race on time and you think to yourself, 'My God, is everything ready? Is the paint dry? Is everything in the right place?'

"Year two is when people start to show a lot more interest and look to position their brands especially if you've done a good job as COTA have. You have to manage people's expectations and deliver even better than the first time.

"COTA's impact on Texas should exceed $500 million in 2013 while the Grand Prix itself is estimated to bring in more than $250 million to the local economy.

"Year three for the new Grand Prix is the most difficult year, like the difficult third album in music. You have to retain the fans from the first two years but still gather new fans while improving the quality of the product. Whatever you did right the first two times you have to do again, but as if this is the first race all over again. You might be older and more experienced but you have to be a lot wiser."

So Bobby, what was the most difficult part of bringing the race to Austin?

"Other than a disgruntled former partner, not having a contract for a race while simultaneously having a partially built track, lawsuits, impossible construction schedules, politics, the weather and needing a couple of hundred million bucks, the project was a piece of cake."

TEAMS: IN THEIR OWN WORDS

MERCEDES

When it comes to the history of Mercedes-Benz's involvement in Formula One motor racing it can be divided into two distinct eras: the Daimler-Benz years of 1954 and 1955 and the current team that returned to the Formula One World Championship in 2010.

Of course it is important to remember that prior to the establishment of the Formula One World Championship in 1950 Mercedes-Benz competed in Grand Prix motor racing in the 1930s when the legendary "Silver Arrows" with their exposed aluminium bodywork dominated pre-war Grands Prix. In fact, Mercedes have been active since 1901 when the first-ever Mercedes was raced to victory in the Nice-Salon-Nice race and in 1908 the first victory in a Grand Prix motor race for a Mercedes was secured in the French Grand Prix in Dieppe.

It is also important to point out that Mercedes, initially in association with Ilmor Engineering, has operated as an engine supplier since 1994, achieving eighty-six race wins. What is now known as Mercedes AMG High Performance Powertrains first supplied engines to Sauber in 1994 then to McLaren from 1995 to 2014, Force India since 2009, Brawn in 2009, the factory team since 2010, Williams since 2014 and Lotus from 2015.

1954
Mercedes-Benz at the behest of the Chairman of the Board of Management returned to Formula One in spectacular style with

a triumphant one-two victory by Juan Manuel Fangio and Karl Kling in the French Grand Prix in Reims. The Mercedes W196 R was upgraded after the British Grand Prix and Fangio, who had transferred from Maserati earlier in the season, went on to win the German, Swiss and Italian Grand Prix races. His forty-two Drivers' Championship points clinched the World Championship for Fangio from Jose Froilan Gonzalez and Mike Hawthorn.

1955
In 1955 Fangio, Kling and Hans Hermann were joined by Stirling Moss, Andre Simon and Piero Taruffi. The success of 1954 continued into the new season as the W196 R dominated the rest of the field. Fangio won his home race, the Grand Prix of Argentina, followed by victory in the Belgian Grand Prix and a one-two victory with Moss in the Dutch Grand Prix. Moss then went on to sensationally win the British Grand Prix at Aintree from Fangio. In what turned out to be the last appearance for the 1955 Silver Arrows, Fangio won the Italian Grand Prix unchallenged; it turned out to be his last victory for Mercedes-Benz. Fangio won the Drivers' Championship with a total of forty points to the twenty-three secured by Stirling Moss in second place. All of this was overshadowed, however, by the tragic disaster at the 24 Hours of Le Mans which killed Pierre Levegh and over eighty spectators. The French, German, Spanish and Swiss Grands Prix were cancelled as a result. At the end of the season the team withdrew from racing to concentrate on production car engineering, a decision that had been made prior to the tragedy in Le Mans.

2010
Amid much fanfare Mercedes-Benz returned to Formula One as a constructor for the 2010 season with Michael Schumacher and Nico Rosberg as drivers and Ross Brawn as Team Principal. Daimler AG acquired a 45 per cent stake in the Brawn GP team and, along with Aabar Investments (30 per cent), secured a long-term sponsorship deal with Petronas to enter the Championship as Mercedes GP Petronas Formula One Team. The modern era of the Silver Arrows began with a fifth and sixth place at the Bahrain Grand Prix, followed by fifth and tenth place at the

Australian Grand Prix. By season end the team had achieved three podiums and a total of 214 Championship points to finish fourth in the Constructors' Championship, with Rosberg finishing seventh and Schumacher ninth in the driver standings.

2011

With the purchase of the remaining 24.9 per cent stake owned by the team management, Daimler and Aabar completed the takeover of the team as Mercedes entered its second season in the modern era of the Silver Arrows. The MGP W02 was the product of intensive design and development work but results on the track did not live up to expectations. The team once again finished fourth in the Constructors' Championship with 165 points as Rosberg finished in seventh place with eighty-nine points and Schumacher one spot behind with seventy-six points in the Drivers' World Championship. There were no wins, no podiums and no poles, but the team built for the future with the appointments in the engineering department of Bob Bell as Technical Director, Geoffrey Willis as Technology Director and Aldo Costa as Engineering Director.

2012

For the 2012 season the team strengthened the ties between AMG and Mercedes-Benz, to become known as Mercedes AMG Petronas F1 Team, and at the Chinese Grand Prix Rosberg achieved the first victory since Fangio at the 1955 Italian Grand Prix. This was Rosberg's first victory in his 111th Grand prix and the first by a German driver in a Silver Arrow since Hermann Lang in the 1939 Swiss Grand Prix. However, in what was to be Schumacher's last season in Formula One, the team finished a disappointing fifth in the Constructors' standings on 142 points with Rosberg (ninety-three points) and Schumacher (forty-nine points) placed ninth and thirteenth respectively in the Drivers' Championship.

2013

Lewis Hamilton joined the team in 2013, replacing Michael Schumacher (308 GPs), in a three-year deal to race alongside

Nico Rosberg and immediately made an impact by finishing third in the second race of the season in Malaysia. After a series of four pole positions for the team Rosberg secured victory at the Monaco Grand Prix. More success followed with third in Canada for Hamilton and victory for Rosberg at Silverstone. After four consecutive pole positions Hamilton claimed his first victory for the team in the Hungarian Grand Prix. Three further podium finishes and points in every race in the second half of the season enabled the team to pip Ferrari for second place in the Constructor's Championship by six points in the season finale in Brazil. Ross Brawn retired as Team Principal at the end of the season to be replaced by the pairing of Toto Wolff (Executive Director Business) and Paddy Lowe (Executive Director Technical).

2014

A revolution in engine regulations in 2014 introduced a 1.6-litre turbocharged V6 engine with a built-in energy recovery system. Mercedes made the best use of these new rules to produce a dominant car due to the innovative design of the car's engine. The F1 W05 utilized the PU106A Hybrid which packaged the turbo and compressor at opposite ends of the internal combustion engine, giving the team an aerodynamic, battery and packaging advantage. The F1 W05 Hybrid became the most dominant car in Formula One history, with the team finishing the season 296 points ahead of their nearest rival Red Bull Racing to secure the Constructors' Championship with a massive 701 points. Lewis Hamilton and Nico Rosberg achieved a record breaking sixteen wins (split 11-5), eighteen poles (7-11), twelve fastest laps (7-5), eleven front row lockouts and eleven one-two finishes. Hamilton was crowned World Champion in Abu Dhabi finishing on 384 points, sixty-seven ahead of second-placed Rosberg on 317 points. The team won a reported $102 million prize for winning the Constructors' Title.

Mercedes Formula One World Championship Results

Debut	1954 French Grand Prix
Latest race	2014 Abu Dhabi Grand Prix
Races competed	108
Constructors' Championships	1 (2014)
Drivers' Championships	3 (1954, 1955, 2014)
Podiums	63
Race victories	29
Pole positions	35
Fastest laps	25
2014 position	First (701 points)

THE ARCHITECTS OF VICTORY
More often than not very little is known about the key personnel off the track. Along with a team of more than seven hundred these are the people who are responsible for the 2014 success enjoyed by Mercedes. Of course, the contribution of Non-Executive Chairman Niki Lauda cannot be underestimated and the influence of former team principal Ross Brawn should also be recognized.

Toto Wolff—Head of Mercedes-Benz Motorsport
Nationality: Austrian
Date of Birth: 12 January 1972
Place of Birth: Vienna, Austria
Lives: Lake Constance, Switzerland

Career History:
2013 Appointed Head of Mercedes-Benz Motorsport
2012 Appointed Executive Director of Williams Grand Prix
   Holdings PLC
2009 Invested in Williams F1
2006 Invested in HWA AG; winner of Dubai 24 Hours
2004 Founded Marchsixteen Investments
2002 FIA NGT World Championship, sixth overall with one win
1998 Founded Marchfifteen Investments
1994 Class winner, Nürburgring 24 Hours

1992 Began motorsport career in Austrian Formula Ford
Championship

Career Profile:
After a short career as an amateur driver, including class and
overall wins in endurance racing, Christian "Toto"Wolff founded
the investment company Marchfifteen in 1998, followed by
Marchsixteen in 2004, focusing on strategic investments in medi-
um-size industrial companies and listed companies. These
included the IPO of HWA AG, the company responsible for
developing and racing Mercedes-Benz cars for the DTM
(German Touring Car Championship) as well as Mercedes-Benz
Formula 3 engines. Toto has been a member of the HWA AG
board of directors since 2007.

Further motor racing interests included co-ownership with
Mika Häkkinen of a driver management company and, from
November 2009, investment in Williams F1. Toto became an
Executive Director of Williams in July 2012 before accepting the
role of Head of Mercedes-Benz Motorsport in January 2013;
including taking a 30 per cent shareholding in Mercedes-Benz
Grand Prix Ltd.

In his current role, Toto has full responsibility for the success
of the entire Mercedes-Benz motorsport programme, including
the DTM and Formula 3 engine programmes. He is based at the
Formula One team's headquarters in Brackley, UK.

Paddy Lowe—Executive Director (Technical)
Nationality: British
Date of Birth: 8 April 1962
Place of Birth: Nairobi, Kenya
Lives: Oxford, England

Career History:
2013 Executive Director (Technical), MERCEDES AMG
 PETRONAS
2011 Technical Director, McLaren
2005 Engineering Director, McLaren
2001 Chief Engineer, Systems Development, McLaren
1993 Head of Research & Development, McLaren
1987 Joint-Head of Electronics, Williams

Career Profile:
Paddy Lowe has been Executive Director (Technical) at
MERCEDES AMG PETRONAS since June 2013. After gradu-
ating with an engineering degree from Sidney Sussex College,
University of Cambridge in 1984, he has worked in various
capacities in Formula One since 1987.

Paddy's first experience in the premier class of motor racing
came as Joint-Head of Electronics during a six-year stint with the
Williams team, overseeing the Active Suspension project which
helped Nigel Mansell to the 1992 World Drivers' Championship.

In 1993, Paddy moved to McLaren where he steadily worked
his way up through the hierarchy during the ensuing two decades.
Heading up the research and development department of the
Woking-based outfit over an eight-year period, Paddy oversaw
various innovative control system projects in addition to laying
the groundwork for the simulator facility.

Having been appointed Chief Engineer, Systems Development
in 2001, Paddy then held the position of Engineering Director
from 2005 to 2011, playing a key role in Lewis Hamilton's 2008
World Championship success before his promotion to Technical
Director in January 2011.

Andy Cowell—Managing Director, Mercedes AMG High Performance Powertrains

Nationality: British
Date of Birth: 12 February 1969
Place of Birth: Blackpool, England
Lives: Market Harborough, England

Career History:

2013 Managing Director, Mercedes AMG High Performance Powertrains

2008 Engineering and Programme Director, Mercedes-Benz HPE

2006 Chief Engineer, Engines, Mercedes-Benz HPE

2005 Chief Engineer, V8 Engine Project, Mercedes-Benz High Performance Engines

2004 Principal Engineer, V10 Engine, Mercedes-Ilmor

2000 Principal Engineer, F1 Design and Development, Cosworth Racing

1999 Head of Pre-Development, Formula One, BMW Motorsport

1997 Principal Engineer, F1 Design and Development, Cosworth Racing

1994 Senior Engineer, F1 Design and Development, Cosworth Racing

1992 Project Engineer, F1 Design and Development, Cosworth Racing

1991 Graduate Engineer, Cosworth Racing

Career Profile:

Following a year out from his Mechanical Engineering degree, spent on the Reynard Scholarship Scheme, Andy Cowell's career path was set. He joined Cosworth Racing Ltd straight from university on their graduate scheme, rotating through the company's technical departments before specializing in the design and development of Formula One engines. By 1999, Andy was leading the engineering project group responsible for the top end of the innovative CK engine; as raced by Stewart-Ford in 1999.

After a year spent with BMW Motorsport, where he managed the engineering group responsible for the concept and detail of the 2001 engine used by Williams BMW, Andy returned to Cosworth as Principal Engineer for F1 design and development; managing the new engine projects in 2001 and 2004.

Andy joined the company then known as Mercedes-Ilmor in 2005 as Principal Engineer for the FQ V10 engine project, which delivered the most race wins of any manufacturer that season. He was then Chief Engineer on the V8 engine project before taking on responsibility for the technical and programme leadership of all engine projects, including the KERS Hybrid which made its race debut in 2009. Andy was Engineering and Programme Director for Mercedes-Benz High Performance Engines from July 2008 to January 2013, responsible for technical and programme leadership of all engine and powertrain projects, plus the strategy and organisation of the engineering group.

Since 1 January 2013, Andy has been Managing Director of Mercedes AMG High Performance Powertrains.

Further key personnel Engineering Director: Aldo Costa; Technology Director: Geoffrey Willis; Sporting Director: Ron Meadows; Chief Race Engineer: Andrew Shovlin; Chief Track Engineer: Simon Cole; Race Engineer (Nico Rosberg): Tony Ross; Race Engineer (Lewis Hamilton): Peter Bonnington; Chief Mechanic: Matthew Deane.

RED BULL RACING

LOOKING BACK ON OUR FIRST TEN SEASONS

The 2014 campaign was our tenth in Formula One. It was an up-and-down affair for Infiniti Red Bull Racing. We recorded twelve podium finishes, the highlights of which were three victories for Daniel Ricciardo. For any team in a sport as competitive as F1, that would be a pretty good return—but set against it is the hard truth that we were never in the hunt for either World Championship title. It's a measure of how far the team has travelled in the last ten years that "just" winning races isn't enough.

All season we've been celebrating our tenth anniversary in the sport, culminating with all ten cars being on display at our end-of-year party. It was a good chance to reflect and an opportunity to look back at the last decade in which we've contested 184 Grands Prix, won fifty of them, took another sixty-six podium finishes, fifty-seven pole positions and taken four double world championships.

It all started in 2005 or, more accurately, at the end of 2004 when Red Bull took over Jaguar Racing, its diminutive factory on the outskirts of Milton Keynes and designs for what became the RB1. Christian Horner was recruited as team principal, while the experienced David Coulthard was hired to drive. The second car was shared by two Red Bull Junior programme alumni: Christian Klien and Vitantonio Liuzzi. The team fought hard in the midfield and went on to finish seventh in the Constructors' Championship. We were up and running.

Activities on-track received a boost for 2006 with the arrival of Adrian Newey. Under his guidance the technical departments grew in strength, the RB2 improved as the season went on—but the first hint of what was to come appeared when, in Monaco, "DC" secured the team's first podium.

In 2007 Adrian's first design, the RB3, moved the team up to fifth position in the championship. It was the first Red Bull powered by Renault and the first with Mark Webber behind the wheel. In time, both French horsepower and Aussie Grit would help the team make a giant leap forward, but for 2007 the on-track highlight was Webber's third-place at the Nürburgring.

A lot was going on behind the scenes. Facilities were expanding, talented personnel were recruited or promoted and an infrastructure was being built that could challenge for victories. The highlight for the RB4 came at the Canadian Grand Prix where David took the sixty-second and final podium of a glittering career. Shortly after he announced his intention to retire.

Sebastian Vettel slotted into DC's seat for 2009. The comprehensive set of rule changes that year gave the team the chance to build a radically different car. It proved to be a front-runner, and the drivers wasted no time turning it into a winner. Sebastian gave the team a maiden victory, winning third time out in China. Five more wins followed, including Webber's own maiden victory, and the team sealed second in the Constructors' Championship.

It had been steady progress for five years but now came the charge: the RB6 won nine times in 2010, with both Webber and Vettel title contenders throughout. The team secured its first Constructors' Title in Brazil at the penultimate race, and Sebastian took the Drivers' Championship at a thrilling showdown in Abu Dhabi.

That 2010 success provided the springboard for a dominant 2011. The RB7, with a dozen victories, propelled Red Bull Racing to a second team title and Vettel to another Drivers' crown and a host of new records, including a new standard of fifteen pole positions in a single season.

However, 2012 was more of a scrap. Six different teams won races in an incredibly competitive season, but the RB8's seven victories were enough to secure the Constructors' Title at the Circuit of the Americas while Vettel left it very late to take his third Drivers' title after a nail-biting season finale in Brazil.

The 2013 campaign was a lot more straightforward. Both titles were in the bag in India with three races to spare—but Vettel would go on to establish a new record of nine consecutive wins with the RB9, while the team would deliver the first competitive sub-two seconds pitstop.

And so, what of our tenth season? Change was in the air. Mark Webber had retired, replaced by Daniel Ricciardo, another product of the Red Bull Junior Team production line. Meanwhile the technical regulations of F1 were undergoing a dramatic reboot with new hybrid engines and a comprehensive redrafting of the rules governing aerodynamics.

Pre-season, things didn't look so good. While some elements of the RB10 suggested it was a worthy successor to its predecessors, other elements needed refining and in addition, all the downforce in the world wouldn't be enough as F1 turned back into a formula governed by horsepower and it became clear we didn't have enough.

Nevertheless, the RB10 and the team surprised many by qualifying and racing at the top of the order from the first race onwards. Having completed no more than twenty consecutive laps during pre-season testing, it was an awesome turnaround. Daniel's disqualification from a podium finish in Australia was a bitter blow but he and the team bounced back superbly with podiums in Spain and Monaco, followed by his maiden F1 victory in Canada. It was a marvellous day for the team and a definite highlight of the season—and the generosity of our rivals in applauding him home demonstrates how popular a victory it was.

Daniel followed that with two more wins: a finely judged tactical triumph in Hungary and a flat-out blast around the mighty Spa-Francorchamps. Along the way there were other drives just as good from both Daniel and Seb and great team efforts that, while perhaps not as eye-catching for the outside world, made us very proud indeed. As seasons go, it didn't quite scale the heights but it certainly provided some great memories. Of course, we're hoping for more from 2015 and the RB11.

2014
Three Wins
Nine additional podium finishes
405 Championship Points
Second in Constructors' Championship
Third in Drivers' Championship (DR)
Third: Malaysian Grand Prix (SV)
Third: Spanish Grand Prix (DR)
Third: Monaco Grand Prix (DR)

Win: Canadian Grand Prix (DR); Third: Canadian Grand Prix
   (SV)
Third: British Grand Prix (DR)
Win: Hungarian Grand Prix (DR)
Win: Belgian Grand Prix (DR)
Second: Singapore Grand Prix (SV); Third: Singapore Grand
   Prix (DR)
Third: Japanese Grand Prix (SV)
Third: US Grand Prix (DR)

2013
Drivers' and Constructors' Championships
Youngest-ever quadruple World Champion (SV)
Thirteen wins
Eleven pole positions
Eleven additional podium finishes
596 Championship Points
Win: Malaysian Grand Prix (SV); Second: Malaysian Grand
   Prix (MW)
Win: Bahrain Grand Prix (SV)
Third: Monaco Grand Prix (MW)
Win: Canadian Grand Prix (SV)
Second: British Grand Prix (MW)
Win: German Grand Prix (SV)
Win: Belgium Grand Prix (SV)
Win: Italian Grand Prix (SV); Second: Italian Grand Prix (MW)
Win: Singapore Grand Prix (SV)
Win: Korean Grand Prix (SV)
Win: Japanese Grand Prix (SV); Second: Japanese Grand Prix
   (MW)
Win: Indian Grand Prix (SV)
Win: Abu Dhabi Grand Prix (SV); Second: Abu Dhabi Grand
   Prix (MW)
Win: US Grand Prix (SV); Third: US Grand Prix (MW)
Win: Brazilian Grand Prix (SV); Second: Brazilian Grand Prix
   (MW)

2012
Drivers' And Constructors' Championships
One of four teams to win three consecutive Formula One titles
One of three drivers to have won three consecutive championships (SV)
Youngest-ever triple World Champion (SV)
Seven wins; eight pole positions; seven additional podium finishes
460 Championship Points
Win: Bahrain Grand Prix (SV)
Win: Monaco Grand Prix (MW)
Win: British Grand Prix (MW); Third: British Grand Prix (SV)
Win: Singapore Grand Prix (MW)
Win: Japan Grand Prix (SV)
Win: Korean Grand Prix (SV); Second Korean Grand Prix (MW)
Win: Indian Grand Prix (SV); Third Indian Grand Prix (MW)
Second: Australian Grand Prix (SV)
Second: Belgian Grand Prix (SV)
Third: Abu Dhabi Grand Prix (SV)
Second: US Grand Prix (SV)

2011
Drivers' and Constructors' World Champions—the team was on the podium at every race in 2011 except for one (Abu Dhabi)
Record: Most poles achieved by a team in a season
Record: Most poles achieved by a driver in a season (SV)
Record: Most laps led by a driver in a season
Twelve Wins; eighteen pole positions
Fifteen additional podium finishes
650 Championship Points
Win: Australia Grand Prix (SV)
Win: Malaysia Grand Prix (SV)
Second: China Grand Prix (SV)
Third: China Grand Prix (MW)
Win: Turkey Grand Prix (SV); Second: Turkey Grand Prix (MW)
Win: Spain Grand Prix (SV)
Win: Monaco Grand Prix (SV)

Second: Canada Grand Prix (SV); Third: Canada Grand Prix (MW)
Win: Europe Grand Prix (SV); Third: Europe Grand Prix (MW)
Second: British Grand Prix (SV); Third: British Grand Prix (MW)
Third: German Grand Prix (MW)
Second: Hungary Grand Prix (SV)
Win: Belgium Grand Prix (SV); Second: Belgium Grand Prix (MW)
Win: Italy Grand Prix (SV)
Win: Singapore Grand Prix (SV); Third: Singapore Grand Prix (MW)
Third: Japan Grand Prix (SV)
Win: Korea Grand Prix (SV); Third: Korean Grand Prix (MW)
Win: India Grand Prix (SV)
Win: Brazil Grand Prix (MW); Second: Brazil Grand Prix (SV)

2010
Drivers' And Constructors' World Championships
In only six seasons, youngest-ever F1 World Champion (SV)
Nine wins; 15 pole positions
Eleven additional podium places
498 Championship Points
Sebastian Vettel awarded Driver of the Year—fifteen pole positions in the year
Win: Malaysia Grand Prix (SV); Second: Malaysia Grand Prix (MW)
Win: Spain Grand Prix (MW); Third: Spain Grand Prix (SV)
Win: Monaco Grand Prix (MW); Second: Monaco Grand Prix (SV)
Third: Turkey Grand Prix (MW)
Win: Europe Grand Prix (SV)
Win: British Grand Prix (MW)
Third: Germany Grand Prix (SV)
Win: Hungary Grand Prix (MW); Third: Hungary Grand Prix (SV)
Second: Belgium Grand Prix (MW)
Second: Singapore Grand Prix (SV); Third: Singapore Grand Prix (MW)

Win: Japan Grand Prix (SV); Second: Japan Grand Prix (MW)
Win: Brazil Grand Prix (SV); Second: Brazil Grand Prix (MW)
Win: Abu Dhabi Grand Prix (SV)

2009
Six wins
Five pole positions
Ten additional podium places
153.5 Championship Points
Second in Constructors' and Drivers' Championships (SV)
World Championship first pole position: Chinese Grand Prix
    (SV)
Win: Chinese Grand Prix (SV); Second: Chinese Grand Prix
    (MW) (First team win)
Second: Bahrain Grand Prix (SV)
Third: Spain Grand Prix (MW)
Second: Turkey Grand Prix (MW); Third: Turkey Grand Prix
    (SV)
Win: British Grand Prix (SV); Second: British Grand Prix (MW)
Win: German Grand Prix (MW, first win and pole position);
    Second: German Grand Prix (SV)
Third: Hungary Grand Prix (MW); First fastest lap for Webber:
    Hüngaroring (1:21.931)
Third: Belgium Grand Prix (SV)
Win: Japan Grand Prix (SV)
Win: Brazil Grand Prix (MW)
Win: Abu Dhabi Grand Prix (SV); Second: Abu Dhabi Grand
    Prix (MW)

2008
Twenty-nine Championship Points
Seventh in Constructors' World Championship
First-ever front row on the Grid; P2 in qualifying at Silverstone
    (MW)
Third: Canada Grand Prix (DC)
David Coulthard retires

2007
Twenty-four Championship Points
Fifth in Constructors' World Championship
Third: Europe Grand Prix (MW) (Webber's first podium with
    the team)

2006
Third: Monaco Grand Prix (DC) (first podium)
Sixteen Championship Points
Seventh in Constructors' World Championship
Renault powered
Mark Webber signs for the team

2005
Thirty-four Championship Points
Seventh in Constructors' World Championship in first racing
    season
First F1 Points for the team: Australian Grand Prix (DC fourth;
    Christian Klien seventh)
Adrian Newey signs for the team

WILLIAMS: A RICH HERITAGE

1977

Williams Grand Prix Engineering Ltd is founded by Frank Williams and ambitious British engineer Patrick Head. The company is based in an empty carpet warehouse in Didcot, Oxfordshire, and enters a purchased March chassis in order to compete in F1 during the latter half of the season. It sets about designing a car to contest the 1978 FIA Formula One World Championship with a staff of just seventeen people. Frank Williams finds a consortium of Middle Eastern backers to support the team's efforts and the first car, the FW06, is shaken down at the end of the year with Australian ace Alan Jones behind the wheel.

1978

The team fields a one-car team during its first season of competition. Alan Jones immediately forms a close working relationship with Williams and Head, and the car is competitive. "AJ" finishes fourth in only the team's third race, in South Africa, and he bags a podium at the penultimate race of the year, the USA GP East. Staff numbers swell from seventeen to fifty and Frank buys an entry for a second car in 1979 from John Surtees, whose team is closing down.

DID YOU KNOW?

Alongside his driving duties for Williams in 1978, Alan Jones wins the Can-Am title in the United States.

1979

Williams finishes runner-up in the Constructors' Championship beaten by Ferrari, who also win the Drivers' Title with Jody Scheckter. Jones and Clay Regazzoni are forced to race FW06s until the Belgian Grand Prix when the ground effect FW07 makes its debut. Regazzoni wins the team's first race, at Silverstone, and Alan Jones is the man to beat during the second half of the year, winning four races. Off-track, Patrick Head understands the increasing importance of aerodynamics in F1, so the team buys a wind tunnel from Lola.

DID YOU KNOW?
Frank Williams has very fond memories of 1979. "The car was a beauty that year," he says. "Whenever it finished, it won. Suddenly Williams were the team to drive for."

1980
The FW07B is the car to beat. Jones dominates the season en route to winning the World Title and he is ably supported by his new teammate, Carlos Reutemann. The Constructors' crown falls to the team as well, leading Frank Williams to comment, "This is the best feeling in the world." The team takes six victories (five for Jones, one for Reutemann) and nineteen podiums during the year, much to the delight of new sponsor Leyland Vehicles.

DID YOU KNOW?
Alan Jones contracts pleurisy prior to the USA Grand Prix West, following poor weather at a Silverstone test session and a damp bed in South Africa. He still competes at Long Beach, but crashes out of the race.

1981
Carlos Reutemann and Alan Jones win four races and bag the team its second consecutive Constructors' Championship. Reutemann loses out on the Drivers' Title by one point at the final race in Las Vegas, leading him to announce his retirement. The team's meteoric rise is acknowledged by Her Majesty the Queen, when Frank Williams receives the Queen's Award for Export. The team's title sponsorship changes hands mid-season from Abilad Williams to TAG Williams.

DID YOU KNOW?
Alan Jones breaks his little finger in a fight in the middle of the season, following a traffic incident in London. He carries on driving despite four of his fingers having to be bandaged.

1982
Rosberg wins the world title by five points with only one victory. The team's other FW08 is shared by Reutemann, Derek Daly and Mario Andretti. Reutemann finishes second in the season-opener in South Africa, but he retires from driving after the second race and Daly takes over the car. Andretti makes a guest appearance at Long Beach. The team develops a four-wheel-drive, six-wheeled F1 car for 1983, but the technology is banned before the team has a chance to race it.

DID YOU KNOW?
There were eleven different winners from sixteen races in 1982—a record.

1983
With turbo engines fully entrenched in F1, the Williams FW08C struggles in 1983 with its normally-aspirated Cosworth V8. Rosberg does brilliantly to win at Monaco but there are few highlights for Jacque Lafitte in the team's least successful season since 1978. At the last race in South Africa the Honda-powered FW09 turbo makes its debut, with Rosberg coming home in an encouraging fifth place. Over the winter the team moves into new, bigger premises in Didcot.

DID YOU KNOW?
Williams gave Ayrton Senna his first Formula One test at Donington Park in 1983. As you might expect, the Brazilian was immediately on the pace.

1984
The season is blighted by unreliability as Williams and Honda get to grips with turbo power for the first time. When the engine holds together, the car is quick and Rosberg takes a brilliant victory in the USA Grand Prix in Dallas. Jacques Laffite endures a disappointing season. He scores just five points to Rosberg's twenty-point-five and the team announces in September that Nigel Mansell will replace him in 1985.

DID YOU KNOW?
The race morning warm-up was held so early in Dallas that
Jacques Laffite turned up in the pits wearing his pyjamas!

1985
The FW10 is Williams' first carbon fibre chassis and the improved
torsional stiffness of the car has an immediate impact on performance. The FW10 is the car to beat from mid-season, once reliability is established. Rosberg leaves tongues wagging at Silverstone
when he sets a pole position lap of 160.938 mph—the fastest lap
ever recorded by an F1 car at the time. At the Austrian Grand
Prix in August, Rosberg announces that he is to leave the team
after four seasons and switch to McLaren. Nelson Piquet will
partner Mansell from the start of 1986.

DID YOU KNOW?
One of Nigel Mansell's first jobs after joining the team was to take
Frank's children on the dodgems at a fairground near Zandvoort
in Holland!

1986
The year gets off to a devastating start, when Frank Williams is
seriously injured in a road accident in the south of France. He
crashes while driving to the airport after a pre-season test at Paul
Ricard and his injuries leave him confined to a wheelchair. The
dominance of the FW11 allows Williams to clinch its third
Constructors' Title with Nelson Piquet and Nigel Mansell. They
enter the final race in Adelaide as overwhelming favourites to
claim the Drivers' Title too, but they suffer tyre problems and are
beaten to the crown by Alain Prost in a McLaren.

DID YOU KNOW?
To cap his most successful season in F1, in which he scores five wins,
Mansell is named BBC Sports Personality of the Year in the UK.

1987
Nine months after his car accident, Frank Williams is back at the
helm of Williams. He is awarded a CBE in the Queen's New Year

Honours list and, fittingly, 1987 is a dominant season for the team in which it wins the Drivers' and Constructors' Titles. The team becomes unstoppable from mid-season. Nelson Picquet wraps up the Drivers' Title at Suzuka when Nigel Mansell's season comes to a premature end, following a heavy crash. At the Italian Grand Prix Honda announces that it will end its partnership with Williams at the end of the season, one year ahead of schedule.

DID YOU KNOW?
The team debuts its first active suspension car at the Italian Grand Prix and Nelson Piquet duly romps to an impressive victory.

1988
While it waits for the arrival of Renault power in 1989, John Judd steps into the engine breach with his normally-aspirated V8, but it proves a disappointing year for the team. The highlights are Nigel Mansell's two podiums, at Silverstone and Jerez. Mansell suffers a bout of chickenpox mid-season, forcing him to miss the Belgian and Italian Grands Prix. He is replaced by Martin Brundle and Jean-Loius Schlesser, the latter's race at Monza becoming infamous for a misunderstanding between him and race leader Ayrton Senna at the first chicane. Senna retires and Monza is the only race of the year that McLaren fails to win.

DID YOU KNOW?
Riccardo Patrese is fined a whopping $10,000 for impeding Julian Bailey during qualifying for the Spanish Grand Prix.

1989
Thierry Boutsen wins a couple of races (in Canada and Australia) and the team secure second place in the Constructors' Championship. Renault's V10 hits the ground running and, in the back of the FW13, which is introduced for the final four races of the year, it's a particularly potent force. The drivers score four podiums in four races, helping the team to leapfrog Ferrari in the Constructors' standings.

DID YOU KNOW?
At the season-opening Brazilian Grand Prix, Riccardo Patrese starts his 177th race—a record at the time for the most Grand Prix starts by a driver.

1990
The team starts the year with an updated FW13 and there are only two highlights: Riccardo's win at Imola and Thierry's tenacious drive to victory in Hungary, in which he leads the entire race and comes home just 0.2 seconds ahead of Senna. However, the team needs to up its game in 1991 so in September Mansell re-signs, joining new aerodynamicist Adrian Newey.

DID YOU KNOW?
Drug testing in Formula One was introduced at the 1990 British Grand Prix.

1991
This year marks the beginning of a very successful period for Williams-Renault. Riccardo Patrese in the FW14 takes the team's first win of the year in Mexico, and Nigel Mansell goes on to win five races to Patrese's two. Mansell's victory at Silverstone produces one of the most iconic images of the year when he gives Senna, whose McLaren has run out of fuel, a lift back to the pits after the chequered flag.

DID YOU KNOW?
Riccardo Patrese's victory in the Portuguese Grand Prix was the team's fiftieth win.

1992
The FW14B is still regarded as one of the most advanced racing cars ever built. It features a semi-automatic gearbox, active suspension, traction control and anti-lock brakes. Mansell wins the opening five races of the year and wraps up the title at the Hungarian Grand Prix in August. With Patrese's impressive consistency Williams are Constructors' Champions for a fifth time. Nigel and Riccardo leave the team at the end of the season.

DID YOU KNOW?
Nigel claimed Williams' fiftieth pole position at the Portuguese Grand Prix.

1993
Alain Prost joins the team and Damon Hill is promoted from test driver to race driver in January. Prost wins the opening race of the season in South Africa, and Hill quickly gets up to speed winning three consecutive races (Hungary, Belgium and Italy), earning him a new contract for 1994. In the Constructors' Championship Williams scores exactly double the number of points of second-placed McLaren. In September the team announces Ayrton Senna as its lead driver for 1994, in place of Prost. The Frenchman retires from the sport as statistically its most successful driver.

DID YOU KNOW?
David Coulthard had his first test for the team in September.

1994
The year is marred by tragedy when Ayrton Senna is killed at the third race of the season. He crashes at the Tamburello corner at Imola and succumbs to his injuries while in hospital. Since his death, all Williams cars have been branded with the Senna "S" on or around the nose cone.

Damon Hill runs Michael Schumacher close in the world title fight. He loses out at the last race following a controversial collision that takes both cars out of the race. The team's second FW16 is shared by David Coulthard and Mansell as the team retains the Constructors' Title.

DID YOU KNOW?
Hill and Coulthard's one-two finish in the Portuguese Grand Prix was the first British one-two in Formula One since Graham Hill and Piers Courage did the same at the Monaco Grand Prix in 1969.

1995
Damon Hill wins four races but he is still thirty-three points adrift of Schumacher come the end of the year. David Coulthard takes

his first pole in Argentina and his first win in Portugal to finish third in his debut year. However, in August the team signs Jacques Villeneuve as Damon Hill's teammate for 1996 and moves to its current premises in Grove.

DID YOU KNOW?
Damon Hill was the highest paid British sportsman in 1995, ahead of boxer Chris Eubank.

1996
Williams is utterly dominant. The team wins all but four races and scores more than double the number of points of its closest rival, Ferrari. Damon takes driver's the crown at the final race in Japan from Villeneuve who has an impressive debut season. The French-Canadian stays on for 1997, but Hill moves to pastures new.

DID YOU KNOW?
When the team moved into its new factory in Grove, a section of the A34 had to be closed in order to move the wind tunnel from its old factory in Didcot.

1997
The FW19 is immediately quick, winning three of the opening four races and Villeneuve goes on to pip Michael Schumacher to the World Championship. To celebrate, Jacques' mechanics wear blond wigs in deference to his new look. Jacques also claims Williams' 100th pole and the team's 100th victory. Off-track, it is another change of engine for Williams as Renault quits F1 to be followed by the return of BMW in 2000.

DID YOU KNOW?
Heinz-Harald Frentzen scores his first F1 victory at the San Marino Grand Prix.

1998
Williams commits to a two-year supply of customer engines based on the Renault French V10 rebadged as Mecachrome.

There is a major overhaul of the technical regulations aimed at improving the spectacle. The cars are narrower and they run on grooved tyres, much to the dismay of the drivers. "They feel so unstable compared to slicks," says Jacques Villeneuve. The team finishes third in the Constructors' Title and fails to win a race for the first time in ten years.

DID YOU KNOW?
The Williams FW20 did not suffer a single mechanical failure that was attributable to the team.

1999
Ralf Schumacher and Alex Zanardi debut for the team; Schumacher finishes on the podium in his first race but the FW21 lacks the pace of the Ferraris and McLarens to challenge consistently at the front. There is disappointment for the team at the European Grand Prix, which is held in atrocious conditions at the Nürburgring. Schumacher is leading the race and he looks set for his first win in F1 when a puncture forces him to make an extra pitstop, dropping him to fourth. The team also announces a tyre deal with Michelin for the following season.

DID YOU KNOW?
Frank Williams received a knighthood in the New Year honours list.

2000
Williams finishes a credible third in its first year with BMW. Schumacher takes three podiums and the FW22 demonstrates great consistency. Jenson Button is confirmed as Schumacher's teammate in January. Jenson retires from a points-scoring position in his first race and he finishes sixth in his second, the Brazilian Grand Prix. At the United States Grand Prix the team confirms that its test driver, Juan Pablo Montoya, will partner Schumacher in 2001. The season highlight was Ralf Schumacher's third place in Melbourne—just twenty seconds behind race winner Michael Schumacher.

DID YOU KNOW?
BMW Williams introduces a major new sponsorship portfolio for 2000. It includes global giants Compaq, Reuters and Allianz.

2001
The team scores more than double the number of World Championship Points of 2000 and it racks up four victories en route to third place in the Constructors' Championship. Schumacher takes his and BMW Williams' first victory at the San Marino Grand Prix and Montoya stands atop the podium for the first time at the Italian Grand Prix.

DID YOU KNOW?
When Ralf Schumacher wins the Canadian Grand Prix and Michael Schumacher finishes second, it's the first one-two for two siblings in the history of the sport.

2002
The team finishes second in the Constructors' Championship to Ferrari. The one-two finish in the Malaysian Grand Prix proves to be its only victory of the year. Juan Pablo Montoya is spectacular in qualifying, taking seven pole positions but Michael Schumacher wins eleven of the seventeen races and storms to his fifth world title. During qualifying for the Italian Grand Prix at Monza, Montoya records the fastest-ever lap by a Formula One car when he averages 161.449 mph en route to pole position.

DID YOU KNOW?
Despite being vehemently denied by team boss Frank Williams, there were rumours throughout the year that BMW was about to buy the team.

2003
For 2003, points are awarded down to eighth place and one-lap qualifying is implemented. Juan Pablo Montoya and Ralf Schumacher both score points in twelve out of the sixteen races, but Ferrari pips the team to the Constructors' Championship. At the Nürburgring BMW announces the continuation of its

partnership with Williams until 2009, while McLaren announces at the end of the year that Juan Pablo Montoya will join the team in 2005.

DID YOU KNOW?
Ralf Schumacher was fined $50,000 for causing a first-corner accident at the German Grand Prix.

2004
A late run of form is enough to ensure it beats long-standing rivals McLaren in the Constructors' Championship, finishing fourth. Montoya ends the year—and his Williams career—on a high in Brazil when he gives the team its only victory of the year. A mid-season technical reshuffle sees Sam Michael take the reins as technical director, while Patrick Head becomes the team's director of engineering. BMW, meanwhile, becomes the first manufacturer to exceed 19,000 rpm with its P83 engine.

DID YOU KNOW?
The threat of a typhoon over the Japanese Grand Prix weekend prompted organizers to cancel all of Saturday's on-track sessions, including qualifying.

2005
Mark Webber and Nick Heidfeld steer the team's fortunes to notch up a handful of podiums but no victories as the team drops to fifth place. The highlight of the season is a double podium at the Monaco Grand Prix, but, overall, it's a turbulent season for the team on and off the racetrack. There are several changes to the design team and BMW announces the termination of its partnership with Williams at the end of the season.

DID YOU KNOW?
The US Grand Prix at Indianapolis turns into a fiasco when Michelin withdraws from the race on safety grounds. As a result, neither Williams driver starts the race.

2006
The team joins forces with Cosworth for the new V8 era and it also has a new tyre supplier in Bridgestone. Nico Rosberg, son of 1982 world champion Keke, replaces Nick Heidfeld alongside Mark Webber. Rosberg hits the ground running at the season-opener in Bahrain, where he finishes seventh and sets the fastest lap of the race—the youngest driver in the history of the sport to do so. But it's an unrewarding year for the team, in which it finishes eighth in the championship. The team announces mid-season that Alex Wurz will replace Webber for 2007.

DID YOU KNOW?
Nico Rosberg was offered a place at Imperial College, London, to read aerodynamics before he entered Formula BMW in 2002.

2007
Williams joins forces with Toyota in 2007 and winds up fourth in the Constructors' standings. Alex Wurz takes the team's only podium of the year when he comes third in the Canadian Grand Prix. Nico Rosberg re-signs with the team in a deal that will keep him at Williams until the end of 2009.

DID YOU KNOW?
Nico Rosberg improved his average grid position from fourteenth in 2006 to ninth in 2007.

2008
A standard Engine Control Unit (ECU) is implemented by the FIA to ensure traction control and launch control are outlawed, and there's a new rule stipulating that each gearbox has to last for four races. Rosberg manages only two podiums as the team finish eighth in the Championship. Frank Williams becomes the sport's longest serving team principal (longer even than Enzo Ferrari). The team records its 50,000th racing lap in Turkey and it starts its 500th Grand Prix at Monza.

DID YOU KNOW?
In October Williams is listed as a 2008/09 cool brand joining others such as Tate Modern, Rolex and Aston Martin.

2009
The FW31—which has a double diffuser from the outset—lacks the punch of its rivals and the team comes home seventh in the Constructors' Championship. The return of slick tyres is celebrated by the drivers, as are the new extra wide front wings and narrow rear wings. Jenson Button of Brawn GP wins the world title. Austrian race driver and private equity investor Toto Wolff buys a minority shareholding in the team, and the Williams Technology Centre in Qatar is launched in October.

DID YOU KNOW?
At the end of the season the team heads to Losail in Qatar, where His Highness Sheikh Khalid bin Hamad Al-Thani becomes the first Qatari to drive an F1 car.

2010
Rubens Barrichello joins the team alongside Nico Hülkenberg. Williams returns to Cosworth for V8 power. Barrichello finishes fourth at Valencia and he's a regular points scorer for the remainder of the season. Hülkenberg gets up to speed quickly and at the penultimate race in Brazil gives Williams its first pole position since the Nürburgring in 2005 He comes home in eighth place as the team pips Force India by one point in the Constructors' Championship to bag sixth place overall.

DID YOU KNOW?
Frank Williams is honoured at the BBC's annual Sports Personality of the Year awards. He wins the Helen Rollason Award to add to his list of accolades.

2011
Pastor Maldonado lines up alongside Barrichello. The year proves to be difficult and the team finishes ninth in the Constructors' Championship. Technical Director Sam Michael and Chief

Aerodynamicist Jon Tomlinson make way for Mike Coughlan (Technical Director), Mark Gillan (Chief Operations Engineer) and Jason Somerville (Head of Aerodynamics). Williams Grand Prix Holdings Plc lists on the Frankfurt stock exchange in March.

DID YOU KNOW?
At the end of the season Patrick Head, co-founder of Williams and its director of engineering, steps down from his involvement with the race team in order to concentrate on Williams Hybrid Power.

2012
The FW34 is a much more competitive proposition as the car has a good balance and the driveability of Renault's V8 immediately goes down well with Maldonado and Bruno Senna. Maldonado wins the Spanish Grand Prix from pole position. The team finishes eighth in the Constructors' Championship but the raw pace of the car gives everyone confidence for 2013.

DID YOU KNOW?
Williams has won sixty-four races with Renault power—more than with any other engine manufacturer.

2013
The team retains Pastor Maldonado for 2013 but he is joined by 2011 GP3 Champion and longstanding Williams Test driver Valtteri Bottas. The FW35 remains Renault powered, meaning the team can build on the relationship formed in 2012.

Changes within top management saw Executive Director Toto Wolff leave while Claire Williams, Frank Williams' daughter, is appointed Deputy Team Principal and takes control of the day-to-day running of the team.

The car has a fully incorporated Coanda exhaust with similar rear bodywork to that of the 2012 Champions, Red Bull. This however doesn't work as desired in the early part of the season with handling being sporadic and difficult. Another mid-season overhaul of management sees Mike Coughlan depart and make way for Pat Symonds, formally of Benetton/Renault, in the role of Technical Chief.

A strong showing in a wet Canadian qualifying sees Bottas qualify third while Maldonado failed to repeat his victory in Spain from 2012. An abandoned blown exhaust towards the end of the season results in a change of form as Bottas scores his first-ever points, claiming an impressive eighth position at Austin, USA.

Maldonado announces he is to leave at the end of the season to be replaced by Ferrari's Felipe Massa, bringing with him bags of experience and a calm head under pressure. More changes ring out through the team with Dave Wheater coming in as Head of Aerodynamic Performance, Shaun Whitehead as Head of Aerodynamic Process, Jakob Andreasen as Head of Engineering Operations, Craig Wilson as Head of Vehicle Dynamics and Rod Nelson as Chief Test & Support Engineer.

DID YOU KNOW?
Mercedes is Williams' eighth different engine supplier since 1977.

2014
Widespread regulations changes for the 2014 Formula One season, with the introduction of new hybrid 1.6 litre turbo charged V6 power units, presented Williams with an ideal opportunity to improve its competitiveness on the racetrack compared to 2013. The decision to partner with Mercedes Benz as the team's new power unit supplier in a long-term deal proved pivotal, and the well powered FW36 showed strong early promise at the start of the season. A mixture of bad luck on the racetrack and difficulties in wet conditions hindered the team's ability to maximise its points haul in the first half of the 2014 season. But this was to change in the second half of the season as Williams out developed its rivals to emerge as the main challenger to eventual Championship winners Mercedes. The team picked up its first podium since 2012 at the 2014 Austrian Grand Prix, and this was to lead to another eight podium finishes to eventually claim third in the Constructors' Championship.

FERRARI—THE COMPLETE TIMELINE

1950

Ferrari takes part in the second race of the World Championship, the Monaco GP, held on 21 May on the city circuit. The drivers are Alberto Ascari, Luigi Villoresi and Raymond Sommer, with the 275 F1 and 340 F1 single-seaters.

1951

The year of the first win in Formula One for Ferrari with José Froilán Gonzáles at Silverstone. The 375 F1, with a naturally aspirated engine, consumes less, which enables the Scuderia to gain time during the pit stops.

1952

The first Formula One Championship is won with Ascari: seven wins in seven races for the Scuderia from Maranello and Alberto Ascari is the new World Champion.

1953

An important season with Ascari winning with the 500 F2: World Champion for the second time in a row. The Italian driver wins in Argentinia, in the Netherlands, in Belgium, England and in Switzerland.

1954

Formula One changes the rules: cylinder capacity is limited to 2.5 litres for naturally aspirated engines and to 750 cc for those with compressors. Ferrari chooses the first option and fights with Mercedes. The Germans calculate speed and gear changes for every single circuit, but in the end Ferrari wins the marques' title.

1955

The Drivers' Title is fought out with Mercedes, and Fangio wins. Ferrari works on materials, the chassis construction and an improved engine. The 500 series proceeds: after the 553 the Scuderia uses the 555 F1 aka "Supersqualo".

1956

A positive season for a new team of drivers, with World Champion Juan Manuel Fangio together with Luigi Musso, Eugenio Castellotti and Peter Collins, as well as in some races Olivier Gendebien, Alfonso de Portago and Maurice Trintignant. The single-seater, the D50 from Lancia and Ferrari, powers to the World Title with the Argentinian driver despite some problems with its reliability.

1957

The Scuderia Ferrari debuts with the new single-seater type 801, similar to the D50 for the V8 engine and De Dion rear axle. This was the last season with the Lancia V8 engines, but it wasn't as successful as previous years. Fangio did not race for Ferrari anymore, but a young Mike Hawthorn joined the team from Maranello.

1958

The new engine, the V6 from the Dino made its entry in this challenging season to fight against Stirling Moss with Cooper and later Vanwall. The Ferrari drivers Hawthorn, Musso and Collins raced with the 246 F1 and Fangio for Maserati. The Italian driver lost his life in Reims in July, while Collins died at the race on the Nürburgring. The second place gained by Hawthorn was worth the World Title, the third for the Scuderia Ferrari.

1959

While in 1958 the Scuderia could defend itself against Cooper, the team was less fortunate in 1959. The English team used a 2495 cc engine especially planned for Formula One. Brabham won the first race and the Title, while Ferrari replied with a bigger V6-engine with 2474.5 cc and 292 bhp. On the fast tracks the car was superior, but it wasn't enough this year due to problems with the braking system and parts of the chassis.

1960

A year of transition for the Scuderia Ferrari, competing with the last models with front engine and rear traction. Cooper dominated the races, but now Ferrari had Phil Hill, Wolfgang von Trips, Richie Ginther, Cliff Allison and Willy Mairesse, who won the first three places at the race in Monza, due to the new rear engine, starting point of the new 246 P and the 156 used in Formula Two.

1961

The year of the Championship won by the US driver Phil Hill: the mid-rear engine worked extremely well in the 156 F1. All Ferrari drivers gained excellent results in Belgium, England and in Monza, where Hill won the race, while Wolfgang von Trips died in a terrible accident. Both World Titles went to Maranello this year.

1962

The management and the technicians leaving, as well as the fierce competition, had an important impact on the performance of this season for the Scuderia Ferrari. Furthermore there were no compromises in terms of safety and the cars were heavier than the required minimum weight. Meanwhile the Scuderia had prepared a "mysterious" single-seater with eight cylinders and air cooling.

1963

New drivers for the Scuderia Ferrari: Lorenzo Bandini, Willy Mairesse and Ludovico Scarfiotti as well as the new-entry John Surtees, who was already the fastest on motorbikes. The single-seater (156 F1-63) was fitted with new suspension, bodywork and engine: a V6 with injection system and six gears. The competition was stronger on the track and Ferrari concluded the season with a fourth place.

1964

A completely new single-seater for the new motorsport season: the 158 F1 with a small and light eight-cylinder, a new concept. It was a difficult start with some retirements, but in Germany Surtees won the race, but in the classification the English driver was behind the competition of Lotus and BRM until the US Grand Prix. A second place in that event helped John Surtees to seal the titles for the Scuderia Ferrari in the Drivers' and the Constructors' Championships.

1965
A difficult season with two single-seaters racing for the Team from Maranello: the 158 F1 and the 512 F1 with V8 and V12 engines respectively. Surtees and Bandini behind the wheel with Scarfiotti and Pedro Rodriguez, experienced some problems with the cars, unforeseen events and accidents. The season ended with an unsatisfying fourth place.

1966
Engine capacity was doubled to three litres for this campaign. The 312 F1 debuted in Syracuse with a victory by Surtees in a race that didn't count for the Championship. The season was not a success and Surtees was dismissed midway through the season, while Bandini gained results that did not count for the Championship.

1967
The car's official debut was at the Champions' Race at Brands-Hatch with Bandini and Scarfiotti behind the wheel. It was the last Grand Prix for Lorenzo Bandini, who lost his life in Monte Carlo on 7 May. After that terrible loss the season was not very satisfying and the competitors showed up with more advanced technology.

1968
Franco Gozzi was nominated Head of the Scuderia. The new 312 F1-68 was driven by Chris Amon, Andrea de Adamich and Jacky Ickx with mixed results, but there were constant technological innovations. The first victory was gained by Ickx in France in the rain, but it was not enough to shake up the classification, in which Ferrari was outstripped by Lotus and Brabham.

1969
This was a dark year for the Scuderia in terms of problems with the organization, the technology and financing. Pedro Rodriguez

and Chris Amon (just after mid-season) gained only one place on the podium, in Holland.

1970
After a difficult start to the season Ferrari was again ruling in Formula One thanks to the competitive 312 B with its 180-degree V12 engine. Next to Jacky Ickx, Ignazio Giunti and Clay Regazzoni were racing for the Prancing Horse. Regazzoni gained his first win in F1 in Monza. With four successes and a pair of one-two wins Ferrari gained second place in the Constructors' Championship. The Drivers' Title went posthumously to Jochen Rindt, who had died in an accident during the qualifying sessions for the Italian Grand Prix.

1971
The Scuderia, with Peter Schetty now at the helm, opened the season with a victory at the South African Grand Prix with the Italian-American Mario Andretti behind the wheel of the 312 B. The Firestone slicks were used for the first time in Spain. At the third race, held in Monte Carlo, the new 312 B2 went onto the track. Its engine had a shorter stroke for higher revs and improved performance as well as innovative rear suspension. It was an exciting project, which helped Ferrari especially in qualifying, but the car's racing reliability only led to a third place in the Constructors' Championship.

1972
Problems with the engine and with performance below expectations this was a difficult year for the Scuderia with fierce competition from England. On the challenging Nürburgring Jacky Ickx, behind the wheel of the 312 B2 starting from pole position, garnered the only win of the season, which ended with a fourth place in the Constructors' Championship.

1973
Despite the intention to start with only one car in the F1 Championship, in the end two 312 B2s were used, with Jacky Ickx and Arturo Merzario behind the wheel. At the fourth race,

held in Spain, the 312 B3-73 made its first appearance with a
self-supporting chassis created by Englishman John Thompson.
It was a difficult year for the Scuderia with no F1 wins. In July
1973 Luca di Montezemolo was hired as an assistant to the
management.

1974

With Luca di Montezemolo as the head of the team and Clay
Regazzoni and Niki Lauda behind the wheel of the 312 B3-74
designed by Mauro Forghieri, the Scuderia quickly forgot 1973,
gaining ten pole positions and three victories. Lauda's win at the
Spanish Grand Prix was the Scuderia's fiftieth F1 Grand Prix
success. The Scuderia and Regazzoni were fighting for both titles
right until the last race, staged at Watkins Glen (USA). In the end
Emerson Fittipaldi—for the second time in his career—and his
McLaren-Ford secured both the Drivers' and Constructors'
Championship.

1975

After an eleven-year gap the Scuderia regained the Drivers' Title
with Niki Lauda, thanks to his five wins, eight podium places,
nine pole positions and securing points in twelve of the fourteen
races. The Constructors' Title was also captured thanks to the
contribution of Clay Regazzoni, who won the Italian Grand Prix.
This was the end of the domination by the V8s with the first
V12-cylinder winning the Championship with almost 500 bhp.

1976

On 1 August Lauda suffered an appalling accident on the
Nürburgring: he was saved by two other drivers from the flames of
his single-seater. The Austrian was severely hurt, and yet forty-two
days later he was back on the track in Monza, gaining a fourth-
place finish. The Championship was decided at the last race, held
in Fuji, Japan. The Austrian retired after the second lap because of
the torrential rain. James Hunt, who crossed the line third with
McLaren, took the Drivers' Title finishing one point ahead of
Lauda. The Constructors' Title went to the Prancing Horse.

1977

In 1977 Ferrari used the 312 T2, again driven by Niki Lauda and Carlos Reutemann. With three wins and six second places Lauda won the World Championship with two races to go, at which point the Austrian left the Prancing Horse. In the last two races of the Championship, which ended with four wins in total and the third Constructors' Title in a row, the Scuderia selected Canadian Gilles Villeneuve to race in the team with Reutemann.

1978

The Scuderia, under the new head of the team Marco Piccini, changed its tyre supplier from Goodyear to Michelin, introducing radial tyres for the first time in F1. With the 312 T3 Reutemann and Villeneuve—whose courage on the track thrilled the fans— were the main players in a year of ups and downs but a total of five wins. Mario Andretti won the Drivers' Title with Lotus 79, the first proper "wing car" in the history of Formula One.

1979

In Formula One everybody was talking about "ground effect" and single-seaters were planned to generate downforce for higher corner speed. The 312 T4 was a competitive and reliable car helping Jody Scheckter to win the title. Gilles Villeneuve—whose fierce duel with René Arnoux at the 1979 French Grand Prix entered history—gained second place in the Drivers' standings, while Ferrari, with 113 points and six victories, won their sixth Constructors' Title.

1980

The season was a negative one for Ferrari: the 312 T5 is mystifyingly uncompetitive despite being based on the previous year's World Champion car. The Scuderia only came home eighth in the Championship. Not even Gilles Villeneuve could tame this difficult single-seater and help Ferrari to gain places on the podium in a challenging season. At the end of the year Jody Scheckter retired from Formula One.

1981

For Ferrari 1981 was the start of the turbo era with the 1.5-litre turbo powertrain. Ultimately, however, the season turned out to be an extremely difficult one with loads of problems related to reliability, as was the case for the other teams. With Villeneuve and Didier Pironi the Scuderia concluded the Championship in fifth position in the Constructors' standings with two victories and a third place for the Canadian driver. Off the track, FISA and FOCA made peace thanks to Enzo Ferrari's intervention: this is how the first Concorde Agreement was born.

1982

A dramatic year for the Scuderia: Gilles Villeneuve died in qualifying at the Belgian Grand Prix while in Germany a terrible accident put an end to Didier Pironi's career. The Frenchman participated in ten out of the campaign's sixteen races, coming in just five points adrift of Keke Rosberg in the Drivers' standings. The 126 C2, used this season, had a lighter and more rigid chassis thanks to the composite material introduced in Maranello by British engineer Harvey Postlethwaite. The car was then raced by Patrick Tambay and, for the last two races of the season, by Mario Andretti. In the end the single-seater carried the Scuderia to the Constructors' Title with a total of eleven podium places and three victories.

1983

The Scuderia started the Championship with the 126 C2B, an evolution of the 126 C2 set up for the new rules and fitted with Goodyear tyres. The car was driven by the Frenchmen Patrick Tambay and René Arnoux. The car gained two victories before the debut of the 126 C3 at the British Grand Prix (ninth race out of fifteen). The victories with Arnoux at the German and the Dutch GPs confirmed the strength of the new single-seater and the Scuderia carried away the Constructors' Title again this year.

1984

To limit the engine capacity the Federation decided that the tank could only have a volume of 220 litres and the cars could not be refuelled. This led to incredible research in the area of petrol and lubricants and consequently to higher costs. In Maranello, eleven years after Arturo Merzario, there was an Italian back in the fold: Michele Alboreto, discovered by Ken Tyrrell. Alboreto came to race next to René Arnoux. It was a difficult season for the team, but also for the fans. The Scuderia concluded the 1984 Championship with a frustrating second place in the Constructors' standings behind McLaren.

1985
In the 1985 Championship Ferrari raced the 156-85. The 1.5l V6 car won six places on the podium and gained two victories in the first eleven races of the year. Alboreto—with his Swedish teammate Stefan Johansson, who took over from Arnoux as of the second race of the season—arrived at the Italian Grand Prix fighting for the Drivers' Title with Alain Prost. However, technical problems with the car at the end of the season meant that Alboreto failed to gain a single point in the last five races, effectively "handing over" the World Title.

1986
The F1-86 was the protagonist of a season to forget. Due to reliability issues Alboreto and Johansson completed the season without a victory and a lowly fourth place in the Constructors' standings, more than 100 points behind the winning Williams team. At the end of the season Enzo Ferrari announced that John Barnard would come to Maranello. Barnard had been the first to introduce a carbon fibre chassis in F1 in 1981 (with the McLaren MP4/1). He had won five Drivers' and Constructors' Titles between 1984 and 1986 as the father of the MP4/2 project.

1987
The rules now permitted naturally aspirated engines with a displacement of 3,500 cc and no tank limit. The Scuderia started the season with the F1-87 with a 90-degree V6 turbo engine. After a disappointing first part of the Championship that brought two third places by Michele Alboreto, the Scuderia improved with a second place in Portugal. Gerhard Berger

secured pole and won in Japan and in Australia, with Ferrari gaining a one-two victory.

1988
On 14 August 1988 Enzo Ferrari died; he was ninety years of age. The season was dominated by McLaren with Senna-Prost, winning fifteen out of sixteen races. The Drivers' Title went to Senna. Ferrari interrupted the McLaren procession with an unexpected one-two win in Monza about a month after the death of its founder; this helped the Scuderia gain second place in the Constructors' Championship.

1989
Ferrari started a new era with the revolutionary F1-89. With innovative aerodynamics and pushrod suspension it was fitted with a semi-automatic seven-speed gearbox, activated on the steering wheel. The clutch pedal was only used at the start. It was powered by a naturally aspirated 65-degree V12 cylinder with 600 bhp at 12,000 rpm. The new driver, Nigel Mansell, won in Brazil and Hungary. Gerhard Berger had a terrible accident at the San Marino Grand Prix, where the car went up in flames at the Tamburello corner. The Austrian had light burns on his hands and couldn't race in the next event, but still won in Portugal and gained second places at the Italian and the Spanish races.

1990
The Scuderia started with the F1-90, sporting the number one thanks to World Champion Alain Prost. The single-seater was especially competitive and Prost won five races, while Mansell won in Portugal. Prost came to Suzuka, the penultimate race of the season, to challenge for the title with his old rival Ayrton Senna but a collision between the two put an end to their race and the Frenchman's Championship challenge.

1991
Jean Alesi joined Prost behind the wheel of the F1-91 for what turned out to be a disappointing season. After the Monaco Grand Prix Cesare Fiorio left in anger, while Alain Prost heavily criticized

the single-seater. It was the beginning of the end of the relationship between the French driver and Ferrari. For the season's last race Gianni Morbidelli started for the Prancing Horse. The Championship, with Senna and McLaren winning the titles, ended without a win for the Scuderia. At the end of the year Luca di Montezemolo was back at Ferrari as Chairman and CEO.

1992

In 1992 the Scuderia drivers were Jean Alesi and new-entry Ivan Capelli. The single-seater F92 was a completely new approach but it wasn't fitted with active suspension and this allowed Williams to dominate the Championship. Nigel Mansell claimed the title with five races to go. For Ferrari it was a season to forget—no wins and a mere twenty-one points in the Constructors' Championship.

1993

In the summer of 1992 John Barnard returned to Ferrari. It was nevertheless another bitter season for the Scuderia. Alesi and Berger accumulated twenty-eight points, compared to the 168 from Williams-Renault, who secured the Constructors' and Drivers' World Titles (with Alain Prost). Prost ended his career at the end of the season having won four titles. In July of 1993 Jean Todt came to Maranello—after a successful motorsport career at Peugeot—to revitalize the team as Head of the Scuderia.

1994

A truly dramatic year for Formula One with both Roland Ratzenberger and Ayrton Senna losing their lives at the San Marino Grand Prix in Imola. Ferrari concluded the Championship with third spot in the Constructors' standings; the title was won by Williams. Meanwhile Michael Schumacher (Benetton-Ford) won his first Drivers' Title.

1995

The Scuderia's car for the 1995 season was the 412 T2, designed by Barnard in England. The engineer abandoned the fashion of the high nose and presented the clutch lever behind the steering wheel, plus an intelligent mechanical brake power distributor (which was a first in F1). Alesi and Berger were more competitive, securing eleven places on the podium. The World Titles went again to Michael Schumacher and Benetton-Renault, while the Scuderia came in third in the Constructors' Championship.

1996

In August 1995 Ferrari confirmed Michael Schumacher as an official driver alongside Eddie Irvine. They raced the F310 to four pole positions, three wins, three second and two third places. Michael Schumacher gained third place in the Drivers' Championship, behind Damon Hill (Williams-Renault) and his teammate Jacques Villeneuve. With 112 points Ferrari secured second place in the Constructors' Championship.

1997

In the fiftieth year of Ferrari Michael Schumacher won five races. At the last race of the season in Jerez he was one point ahead of Villeneuve. With twenty-one laps to go race leader Schumacher reacted decisively to an attack from the Williams driver and the two cars collided. Schumacher had to retire, while Villeneuve crossed the line third, winning the World Title. The FIA subsequently disqualified Schumacher from the standings at the end of the year, while confirming the points the German gained in the Constructors' Championship, thereby helping Ferrari to second place.

1998

Six wins for Schumacher, plus two second places and three third places, as well as three second places and five third places for Eddie Irvine, gave the Scuderia Ferrari 133 points and second spot in the Constructors' Championship. In 1998 slicks were replaced by grooved tyres and the F300 was a completely new design. The team from Maranello arrived at the last race in Suzuka with Schumacher fighting for the title with Mika Häkkinen and his McLaren-Mercedes. However, a tyre failure put paid to any Drivers' Title dreams. Häkkinen won the race, and with it the title by fourteen points.

1999
The Scuderia claimed the Constructors' title, thereby ending a barren spell stretching back to 1983. After Schumacher broke his leg at the British Grand Prix, Eddie Irvine challenged for the title. He arrived at the last race, in Suzuka, ahead of Häkkinen by four points in the Drivers' standings. But the Ferrarista only managed to cross the line third. Häkkinen won the race and the Drivers' Title for the second time in a row.

2000
Twenty-one years after Jody Scheckter's triumph the Drivers' World Title was back in Maranello thanks to the competitiveness of the F1-2000, which set the standard for the years to come. Michael Schumacher gained nine victories and won a season-long battle against Mika Häkkinen of McLaren at the penultimate race of the season in Suzuka. With a total of 170 points the team also reprised their 1999 Constructors' title success. For the Scuderia this was the start of one of the most successful cycles in the history of Formula One.

2001
Michelin joined Bridgestone in Formula One as an official tyre supplier and by the Hungarian Grand Prix, the thirteenth race of seventeen on the race calendar, Michael Schumacher had secured the Drivers' Title, the fourth of his career. At the end of the season the Scuderia had nine victories, fifteen places on the podium and a hat-trick of one-two wins for a total of 179 points to once again top the Constructors' standings. The main star of this incredible season was the F2011, nicknamed the "anteater" due to the shape of its nose. It sported a high chassis, with a concave lower section and relatively short and high sidepods.

2002
Ferrari dominated the 2002 Formula One campaign with fifteen victories in seventeen races. The eleven wins by Michael Schumacher and the four by Rubens Barrichello (including nine one-two successes) powered the Scuderia to the Constructors' Championship with 221 points—as many as the other ten teams

combined. The 2002 single-seater was an entirely new project with smaller aerodynamic sides, a titanium-fusion gearbox, a different chassis design and construction and a new engine with a lower centre of gravity producing 835 bhp at 17,800 rpm.

2003

The 2003 Championship saw a re-invigorated competition thanks to the rivalry between Bridgestone and Michelin, resulting in eight different winners in the sixteen-race calendar. In the end it was Michael Schumacher again, clinching the World Title in the last race, held in Suzuka. The German driver beat the record of five World Titles held by Juan Manuel Fangio, becoming the most successful driver in the history of Formula One. Thanks to the F2003 GA, named in homage to Gianni Agnelli who had died in January 2003, the Scuderia gained its fifth successive Constructors' Title.

2004

Out of eighteen Grands Prix the Scuderia won fifteen (thirteen for Schumacher, two for Barrichello), including eight one-two victories. Their total of 262 points secured the Scuderia's fourteenth Constructors' Championship. Schumacher gained his fifth title in a row, the seventh of his career, in Belgium on the same track where he debuted in 1991. The F2004, fitted with the new ten-cylinder 053 engine producing 865 bhp at 18,300 rpm, exceeded all expectations in terms of performance and reliability.

2005

In the end the F2005, debuting with Schumacher at the Bahrain Grand Prix, failed to live up to expectations. The single-seater had plenty of potential but failed to compete with its rivals on the track. The season ended with just one win, gained in Indianapolis at a Grand Prix with just six cars on the track caused by a boycott of the teams running on Michelin tyres. The team achieved third place in the Constructors' Championship, won by Renault, while Fernando Alonso, with Renault, gained the Drivers' Title.

2006
Ferrari made up for a weak 2005 campaign as V8 2.4-litre engines were introduced, as were tyre changes during the race. The seven victories by Schumacher and the three by new-entry Felipe Massa weren't enough to stop Alonso and the Enstone team repeating their successes of the previous season. At the end of the campaign, with his retirement announced after the win in Monza, the Schumacher era in Maranello came to an end. In 180 GPs with Ferrari the German driver had won seventy-two races, gained fifty-eight pole positions, recorded fifty-three fastest race laps and won five Drivers' and six Constructors' Titles.

2007
The year 2007 will enter the annals of Formula One as one of the most embattled and intense seasons in the history of the series: two teams, Ferrari and McLaren, and their drivers, Massa, Räikkönen, Alonso and the rookie Hamilton fought for the World Title until the last race. In the season where Bridgestone again became the sole tyre supplier, the Scuderia achieved nine victories (six by Räikkönen, three from Massa) winning the Constructors' Championship with 204 points and also the Drivers' title—at the last race of the season in Brazil Kimi Räikkönen gained the upper hand on the McLaren duo.

2008
The Scuderia, with Stefano Domenicali as the new Team Principal and Aldo Costa as the Technical Director, defended the World Title with the F2008 on the track. At the end of an intense season Ferrari won its sixteenth Constructors' Title with eight victories (six for Massa, two by Räikkönen). As for the Drivers' Title, it was a sad day for the Prancing Horse at the last race, staged in Brazil. Felipe Massa won the race and was World Champion for less than forty seconds, that was when Lewis Hamilton (McLaren-Mercedes) crossed the line. At the last corner the English driver had grabbed fifth position in the race and the four necessary points to win the title by one point from the Brazilian Ferrarista.

2009
This campaign saw a host of regulation changes that radically changed the look of the cars. The front wing was much wider than in the past, the rear wing much narrower and higher, the diffusor further back, the bodies were without air-outlets and the aerodynamic devices reduced. The drivers had the possibility to move the flaps' angles from the cockpit and there was the introduction of the KERS (kinetic energy recovery system). Slicks were back, too. The F60—whose name celebrated the sixtieth participation of the Scuderia in the F1 Championship—could not offer a realistic challenge to teams such as Red Bull and Brawn GP. Ferrari gained just one success, with Kimi Räikkönen in Spa-Francorchamps, while a terrible accident during the qualifying at the Hungarian Grand Prix put an end to Felipe Massa's season; he was replaced by Luca Badoer and then Giancarlo Fisichella. The 2009 Championship, won by Jenson Button and Brawn GP, ended for the Scuderia with a disappointing fourth place in the Constructors' Championship.

2010
Race refuelling is banned and drivers reaching Q3 must start the race with the tyres they last used in qualifying. At Ferrari, Fernando Alonso comes in to replace Kimi Räikkönen alongside Felipe Massa. Fighting closely for the title are the Spanish Ferrari driver, Lewis Hamilton for McLaren and the two Red Bulls of Mark Webber and Sebastian Vettel. The F10 has a perfect start to the season, taking a one-two in Bahrain. Fernando wins in Germany, Italy, Singapore and Korea with the title fight coming down to the last event. The Abu Dhabi race is a heartbreaker for Ferrari, though, as Fernando finds himself behind Russia's Vitaly Petrov in his Renault, who has a good top speed and so is difficult to overtake; this gives the title to Vettel.

2011
The Championship is dominated by Red Bull as DRS is introduced and Pirelli become sole tyre supplier. Sebastian Vettel is again Champion, this time clinching the title in Japan, no fewer than four races before the end. The Prancing Horse F150 Italia

car only wins one race, the British Grand Prix, with Alonso, who is very consistent, getting to the podium a further nine times. At the end of the season, the team is third in the Constructors' Championship with Fernando fourth in the Drivers' standings.

2012
The 2012 season is another exciting one for Scuderia Ferrari. The team, along with Fernando Alonso, are in the hunt for the title right down to the wire. The F2012 emerges from the Maranello factory and improves as the season goes on. Alonso scores some exceptional wins: in Malaysia the rain plays its part, as he wins from eighth on the grid, and he drives an incredible race in the European Grand Prix.

2013
The start of the season is encouraging, as there are two wins in the first five races—in China and Spain. However, from the mid-season onwards, Sebastian Vettel strings together a run of victories to take the title in India with three races in hand. Fernando Alonso is best of the rest, taking seven podiums. At the end of the season, with the team third in the Constructors' classification, Felipe Massa goes to Williams and Kimi Räikkönen returns to Maranello.

2014
The 2014 season was far from satisfactory for Scuderia Ferrari. For the first time since 1994, it failed to win a single race as the team dropped to fourth in the Constructors' classification. Alonso ended up sixth while Kimi Räikkönen could do no better than twelfth place, never seeing the podium. The Scuderia's car, named F14 T following an on-line poll among the fans, suffered in the face of the new engine regulations and because the aerodynamic package had limitations.

MCLAREN: IN THE BEGINNING . . .

Bruce McLaren was born in 1937 in Auckland, New Zealand, and in 1963 founded Bruce McLaren Motor Racing in order to develop and race sports cars alongside his commitment as lead driver in the Cooper Grand Prix team.

He had arrived in the UK in 1958 with the "Driver to Europe" scheme that was designed to encourage antipodean drivers to compete with the cream of the world's drivers. His mentor was Jack Brabham who introduced Bruce to Cooper Cars, the small Surbiton (London)-based team who were poised to create a revolution with compact, lightweight Grand Prix cars powered by an engine behind the driver. Following an auspicious start to his F2 career in 1958 he joined the F1 team for 1959 and stayed with Cooper for seven years.

Bruce made an impact almost immediately by winning the 1959 US Grand Prix aged just twenty-two years and eighty days, at that time the youngest Grand Prix winner. He went on to win three more Grands Prix and countless sports car victories. Yet Bruce was no ordinary driver. His upbringing was steeped in cars and practical engineering at his parents' service station and workshop. By the age of fourteen he had entered a local hill climb in an Austin 7 Ulster and shown promise both as a driver and an engineer.

Back in the 1960s Bruce raced, as did most Grand Prix drivers of this time, in sports cars, Grand Touring cars and more humble saloon cars alongside his commitments to Cooper in Formula 1. He drove for Jaguar, Aston Martin and Ford with whom he won the 24 Hours of Le Mans in 1966.

He was a true competitor who excelled at innovating and developing racing cars. It was this passion that led Bruce to start his own company, first to develop and race a Cooper with a rear-mounted Oldsmobile engine that helped to kick start the "big banger" sports car era. In a show of loyalty to Cooper cars Bruce engineered two 2.5-litre Coopers for the 1964 Tasman series which he won.

MCLAREN BECOMES A CONSTRUCTOR

In 1964 Bruce and his small team built the first true McLaren sports car—the M1A—which became a top contender in sports car racing both in Europe and America. After proving its credentials the orders rolled in and twenty-four examples were built. Its successor, the M1B, was quicker still and carried Bruce's nascent team into the inaugural Can-Am (Canadian–American Challenge Cup) championship. These cars were faster than the Formula 1 cars of the age, providing a spectacle of colour accompanied by the deep rumble of highly-tuned, large American V8 engines. The inaugural year of this championship did not yield a victory for McLaren but Bruce came third in the series.

The following year, 1967, saw the start of one of the most dominant episodes in motor sport history.

Now in its trademark papaya orange livery, Bruce and fellow Kiwi Denny Hulme's Can-Am cars won five of the six races, with Bruce taking McLaren's first title. In the following five seasons what became known as the "Bruce and Denny show" rolled on with Hulme winning the title in 1968 and 1970, while Bruce claimed his second crown in 1969. Peter Revson won for McLaren in 1971. Between 1967 and 1971 the works McLarens won thirty-seven of the forty-three races, including nineteen one-twos. Such dominance won many admirers and many sales of racing sports cars and, just occasionally, a customer car won too. Over the duration of the Can-Am series McLaren was the dominant victor with forty-three victories, almost three times more than its closest competitor, Porsche.

McLAREN MOVES INTO FORMULA ONE

Back in 1965 Bruce had already decided to leave Cooper and build his own Formula One car for the first season of the new three-litre formula. Having built a "mule" chassis for testing in 1965, the first McLaren F1 car, the M2B, made its bow at the Monaco GP. Although saddled with underpowered and unreliable engines, Bruce scored a point for sixth place in only its third race, at Brands Hatch, with a further two points later in the season. It was a respectable start but the real mark left by McLaren's first F1 car was the innovation it featured.

Establishing a tradition that has long guided McLaren, the car's designer, Robin Herd, was recruited from the aerospace industry at Farnborough. Herd had worked with a material called Mallite—endgrain balsa wood sandwiched between two sheets of aluminium in a honeycomb, from which he constructed the entire inner and outer monocoque of the M2B. It was strong and light—a watchword for the aviation industry and a prescient and enduring quest for McLaren to this day.

It took only another season for the McLaren F1 team to make it to the top step of the podium, a feat achieved, appropriately, by Bruce himself at the 1968 Belgian Grand Prix. The Cosworth-powered M7A was among the fastest cars of the season and was liveried in McLaren orange for the first time. Denny Hulme won a further two races in 1968, the latter in Canada yielding the team's first one-two. Hulme went on to win four more Grands Prix in the following years.

Bruce's tragic death while testing at Goodwood in 1970 would have thrown lesser teams into disarray, but under the guidance of Teddy Mayer and with the support of Denny Hulme, who stayed loyal to his compatriot's team until he stepped down from Formula One and McLaren at the end of 1974, McLaren was on the cusp of achieving the ultimate success. The team's first Drivers' and Constructors' Championships came in 1974 when Brazilian Emerson Fittipaldi won three races and took the crown in the McLaren M23. The same model, now in its fourth season, also powered James Hunt to the 1976 Drivers' Championship after a season-long, and enthralling, battle with Niki Lauda and Ferrari.

McLAREN AT INDIANAPOLIS

In the 1970s McLaren had also been very active in the USA. Not only had McLaren created history with its Can-Am success but the team also coveted glory at the prestigious Indianapolis 500.

Following an unlucky accident that precluded Hulme from competing with the M15 in 1970, McLaren bounced back in 1971 with the F1-inspired M16. Powered by the ubiquitous turbo-charged Offenhauser engine and presenting the first wedge-shaped car at the Brickyard, Peter Revson and Mark Donohue were both super-quick but failed to take the win, Revson finishing second.

The following year Donohue won the Indy 500, and several other USAC races, in Roger Penske's M16B. It wasn't the orange car that won, but it was a McLaren. Two years later Texan Johnny Rutherford took the flag at the Brickyard in a McLaren M16 C/D and this time it was orange. The now venerable M16 line of cars progressed into D and E specifications with Rutherford finishing second in 1975 and winning again in 1976. Although this marked the end of McLaren's active involvement at Indy, customer examples of the M16 continued at the 500-mile race until 1981 when Vern Schuppan's example still managed third place.

THE IMPORTANCE OF CUSTOMERS

The 1980s were to see major upheaval at McLaren, setting the tone for McLaren's Formula 1 successes. Before moving on to this significant chapter in the history of McLaren there were some important learning points from the company's first seventeen years. First McLaren learned that success breeds success: with each new Can-Am car it experienced strong demand from customers who wanted the fastest cars available. The company also realized that it needed a production partner in order for it to be able to focus the efforts of the young company on developing its products. Accordingly McLaren established a partnership with Trojan to build customer cars. Between 1965 and 1976 Trojan built around 160 customer Group 7 Can-Am cars, fifty-two Formula 500/A cars and twenty-five Formula 2/B cars. In addition McLaren made no fewer than twenty-four cars for USAC racing in America.

The company's fame in the United States led it to form McLaren Engines based in Livonia, Michigan, in order to be close to its racing centres and provide on the spot support. Its experience of IndyCar racing delivered tremendous experience in aerodynamics due to the high speeds generated on the oval circuits—average speeds came close to 200 mph—and in the use of turbocharged engines at a time when almost all European racing was with normally-aspirated units. All this experience would prove valuable in the new era of the 1980s with Ron Dennis at the helm.

RON DENNIS STARTS A NEW ERA AT McLAREN AND REVIVES ITS SUCCESS

After McLaren's purple patch in the mid-1970s, the team's performance went downhill in 1978, '79 and '80. It was a time that saw the emergence of hugely powerful turbo cars from the big manufacturers competing against the small teams equipped with the normally-aspirated Cosworth engine that made its debut back in 1967. So in 1980 McLaren merged with Ron Dennis' Project 4 Racing team.

Ron's arrival was timely. He had worked in Formula 1 since 1966, joining Cooper Cars soon after Bruce McLaren departed, then started his own F2 team in 1971.

Not only did he bring a new drive and ambition to the famous team but he also brought back a skilled designer, John Barnard. Barnard was working in America where he designed the Chaparral 2K that won Indianapolis in 1980, but he had been at McLaren earlier in the 1970s where he worked on the M23 car that delivered two Formula 1 Drivers' Championships (Fittipaldi and Hunt) and McLaren's first Constructors' title (1974).

More significantly, Barnard was interested in a material new to racing car design, carbon fibre composite. This material was used in aerospace applications but had never been applied to a complete racing car monocoque. McLaren pioneered the use of carbon fibre in motor racing with its new car, the MP4/1, and revolutionized racing car construction. The carbon fibre chassis was built by Hercules Aerospace and brought new levels of rigidity and driver safety to Formula 1.

The MP4/1 series of cars raced for three years, delivering one victory in 1981, four in 1982 and another in 1983, by which time the turbo cars were outgunning the more nimble Cosworth-powered teams. Towards the end of 1983 McLaren's long-awaited turbo engine arrived in the form of a Porsche-designed V6 named TAG (Techniques d'Avant Garde). TAG principal Mansour Ojjeh became a shareholder in McLaren and shared in a period of rewarding success for the company. Ron attracted Niki Lauda out of retirement to join John Watson in 1982/3 on driving duties and both were to win races.

The 1984 season saw race wins turn into championships. Frenchman Alain Prost replaced Watson, but it was Lauda who took his third title, despite Prost winning seven Grands Prix to the Austrian's five. Guile and experience won over youth and pace, but McLaren had won its second Constructors' Championship and celebrated its most successful season so far with twelve victories from sixteen events. The MP4/2 B repeated its championship victories in 1985, Prost lifting the driver's trophy with five wins while Lauda managed just one before retiring. Prost went on to win the Drivers' Championship in 1986 and 1989 for McLaren.

THE SENNA AND PROST ERA

For 1988 McLaren entered what would be an enormously fruitful relationship with Honda, firstly with the Japanese company's 1.5-litre turbo engine then, when turbos were banned, with 3.5-litre V10 and V12 power plants. Also new for 1988 was Ayrton Senna. The explosive combination of Senna, the fastest driver in the world, and the master tactician and strategist Prost would yield two Championships (Constructors' and Drivers'). The first season of this partnership yielded almost the perfect score with Senna winning eight races and Prost seven, leaving just one for the other teams, and McLaren scored no fewer than ten one-twos. In 1989 the score was Senna six and Prost four, but the latter won the title.

Into the new decade, Senna had a new teammate in Gerhard Berger and he continued to dominate, winning Drivers' Championships in 1990 and 1991 with Berger's consistency helping the team to two more Constructors' Championships. The last year for McLaren and Honda was 1992 and although the team could not celebrate five Championships on the bounce, they won five races and finished second. Honda withdrew from Formula 1 leaving McLaren to use Ford and then Peugeot engines before linking up with Mercedes-Benz in 1995, a relationship that endured until 2014.

McLAREN'S FIRST TRIP TO LE MANS

In 1995 McLaren also entered Le Mans for the first time in its thirty-year history. The company's decision to build the F1, the ultimate super sports car was never intended to spawn a racing car. However, the burgeoning interest amongst racing teams for a GT series using road-derived cars, and the eagerness of some McLaren owners to compete, stirred the competitive spirit at Woking. McLaren set about strengthening the iconic F1 road car for the parts that might not stand the punishment meted out in endurance racing.

The basis of the car was good—a carbon fibre tub for strength, high torsional rigidity and light weight and a 6.1-litre BMW V12 engine that issued 627 bhp. It was almost a modern Can-Am car for the road. The resultant racing version was named the McLaren F1 GTR and from an intended production of three, nine were produced in 1995 alone.

Weight was reduced by 90 kilos, bigger brakes and wheels, a roll cage, a faster steering rack, a reinforced gearbox, and a rear wing were added. Engine power was reduced over the standard car to 600 bhp in order to comply with Le Mans regulations. It must be the only car in the world that went to the track with less power than its road-going sibling.

The GTR's first outing was the first race in the BPR Global GT Championship at Monza, a series for professional racers and gentlemen drivers. Three GTRs entered and owner Ray Bellm had the honour of giving his car a debut victory with Maurizio Sandro Sala as co-driver. The GTR won its first six races and then headed for the big one—the 24 Hours of Le Mans. Six McLarens entered the race with drivers ranging from ex-GP aces JJ Lehto, Mark Blundell and Yannick Dalmas to long-distance specialists Derek Bell and Andy Wallace.

It was a wet race that placed a premium on delicacy of touch and, although the lighter prototypes were expected to be faster, the conditions played into the hands of the McLarens with Lehto in particular driving spectacularly well. At one stage in the night he was lapping ten seconds faster than any other car on the track. In the end, the black GTR of Lehto, Dalmas and Masanori Sekiya came home first by a single lap. The other McLaren GTRs finished third, fourth, fifth and thirteenth, with only one retiring due to a crash.

It was a remarkable achievement in that a real road car with only minimal modifications had taken on and beaten the best in the world's most gruelling race, at its first attempt and in the first year of production. The result also guaranteed that the McLaren F1 would claim an iconic status in the eyes of aficionados the world over.

It also secured for McLaren a first in that it is the only manufacturer to win the triple crown—The Formula 1 World Championship, the Indianapolis 500 and the 24 Hours of Le Mans. It remains a unique achievement.

The F1 story continues with a resounding victory in the BPR GT series in both 1995 and 1996 and the All-Japan GT Championship. The F1 revisited Le Mans in 1996, finishing fourth, fifth, sixth, eighth, ninth and eleventh and again in 1997 with a revised long-tailed GTR finishing second and third.

FORMULA ONE IN THE MODERN ERA

Back in the Formula One arena the relationship between McLaren and Mercedes Benz gelled and the driver team of Mika Häkkinen and David Coulthard complemented each other for six seasons of positive results. In this period the Finn won twenty Grands Prix and took the Drivers' title in 1998 and 1999, while the Scot won ten Grands Prix with another two to come after Häkkinen had retired. In 1998 McLaren won its eighth Constructors' Championship.

Another Finn, Kimi Räikkönen, replaced Häkkinen as Coulthard's partner and finished second in the championship in both 2003 and 2005, taking nine victories. He was joined by another exciting driver, Juan Pablo Montoya, for 2005 and 2006 who took three wins for McLaren before going back to America.

In 2007, reigning double World Champion Fernando Alonso arrived at McLaren to challenge for a third title while the gifted protégé of the McLaren and Mercedes-Benz team, Lewis Hamilton, would start his rookie year alongside an established master. Hamilton was quick "out of the box" and went on to win five Grands Prix and take a close second in the title race. Alonso also took four wins to finish third. Alonso was to leave after just a single season, but Hamilton went on to take five more victories in 2008 and secured the Drivers' Championship at his second attempt—the youngest driver ever to do so.

McLaren now has a heritage of fifty-one years, in forty-nine of which it has been represented at the pinnacle of the sport.

McLAREN F1 STATISTICS

In Formula 1, it has won no fewer than 182 of the 761 Grands Prix in which the team has competed. It has achieved 485 podium positions, 155 pole positions and 152 fastest laps.

It has delivered twelve World Championships for its drivers, Emerson Fittipaldi (1974), James Hunt (1976), Niki Lauda (1984), Alain Prost (1985, '86, '89), Ayrton Senna (1988, '90, '91), Mika Häkkinen (1998, '99) and, most recently, Lewis Hamilton in 2008.

It has won eight Constructors' Championships: 1974, '84, '85, '88, '89, '90, '91 and '98.

In addition McLaren has won five Can-Am titles with forty-three race wins, three Indy 500 victories, the24 Hours of Le Mans and a host of F5000 and Formula A races.

FORCE INDIA

Sahara Force India, as the team is now known, was born in late 2007 after Dr Vijay Mallya joined forces with the Mol family, co-owners of the then Spyker team.

The 2008 season was about establishing a base from which to proceed, with veteran Giancarlo Fisichella joined by promising young talent Adrian Sutil. Indeed, Adrian provided the highlight of the season, running a brilliant fourth in the closing stages of a wet Monaco Grand Prix until he was pushed off the road by Kimi Räikkönen's Ferrari. It was a difficult building year that saw the team finish tenth in the championship.

Meanwhile, Dr Mallya was putting the pieces in place for future growth. Key among them was a switch for 2009 to Mercedes power and gearboxes from McLaren Applied Technologies. The change provided the team with a solid starting point from which to develop the VJM02. The improvement in form was immediately obvious, as Fisichella and Sutil moved up the grid and began to challenge for points.

The car had traditionally favoured faster circuits and at Spa, Fisichella took a shock pole position and then finished a brilliant second to Räikkönen in the race. By the next event at Monza, the Italian was himself driving for Ferrari and third driver Tonio Liuzzi was promoted to a race seat. At that race Adrian underlined the speed of the car by qualifying second, setting the fastest lap, and finishing fourth. At the season's end the team had moved up to ninth in the World Championship.

Sutil and Liuzzi continued the team's improving form into 2010. Adrian became a regular points-scorer, taking a best of fifth place in both Malaysia and Belgium, while Tonio had less luck but earned a good sixth in Korea. The progress was emphasised by a seventh place in a championship which now included three new teams, while Adrian earned 11th in the drivers' standings.

In 2011 the team continued to move forward as reserve driver Paul Di Resta was promoted to a race seat alongside Sutil. Both men challenged for points at nearly every race and the team ultimately earned sixth place, narrowly missing out on fifth. Paul was the best placed 2011 rookie in 13th, while Adrian was in ninth, beaten only by the drivers from Red Bull, McLaren, Ferrari and Mercedes. October 2011 also saw the signing of a historic partnership with Sahara India Pariwar, who became co-owners of the team.

2012 saw the team build on the strong foundations of 2011 as Nico Hülkenberg was promoted from third driver duties to a race seat alongside Paul. Recognised as one of the most exciting driver line-ups on the grid, the pair was closely matched all season with both men securing career-best fourth place finishes. Despite a challenging start to the campaign, the team scored points in 80 per cent of the races, earning 109 points and seventh place in the standings. The season ended on a high note as Nico led 30 laps of the Brazilian Grand Prix.

A wisely planned pre-season testing heralded a brilliant start of 2013 for the team. Adrian Sutil, returning to the team to partner Di Resta, led the Australian Grand Prix on the way to the first of many double points finishes for the team. A fourth place by Di Resta in Bahrain and a fifth place for Sutil on the streets of Monaco saw the team climb up to fifth in the Constructors' Championship before a mid-season change of tyre construction by suppliers Pirelli hurt the competitiveness of the VJM06. A strong end of the season resulted in the team reclaiming sixth position in the standings.

The 2014 season brought many changes, with the team's colour scheme morphing into an aggressive, sophisticated black and a new driver line-up featuring returning Nico Hülkenberg alongside young ace Sergio 'Checo' Pérez. The new pairing, together with the competitiveness of the VJM07, brought immediate results with Pérez ending the team's five-year wait for a podium in only his third race with Sahara Force India. An impressive ten-race run of consecutive points placements ended mid-season, but the year progressed strongly and resulted in ten double-point finishes: the team shattered the previous points record, 109, amassing an incredible 155 points and knocking at the doors of the top five in the Constructors' Championship.

The shift to a new wind tunnel and the consistency brought by the retained line-up of Hülkenberg and Pérez would hold great promise for the 2015 season. Sahara Force India is ready to claim a spot among the sports' greats.

2007
The Force India Formula One Team is created as the Orange India Holdings group, led by Dr Vijay Mallya and Jan and Michiel Mol, purchases the Silverstone-based outfit from Spyker. The deal is formally announced in October.

2008
Force India makes history in Australia as the first Indian team to contest a round of the FIA Formula One World Championship. Giancarlo Fisichella partners Adrian Sutil, aided by Tonio Liuzzi as test and reserve driver. Adrian runs in a brilliant fourth place in the closing stages of the Monaco event, only to be hit by Kimi Räikkönen's Ferrari. Fisichella earns a season best tenth in Spain, but despite making strong progress the team finishes the year with no points. It also agrees a partnership with McLaren Applied Technologies and Mercedes-Benz for the supply a complete drivetrain for the 2009 season.

2009
Sutil and Fisichella start the season, and solid car development soon shows Force India's potential. The team comes close several times before finally securing its first points in style as Fisichella takes a brilliant pole position and finishes second at the Belgian Grand Prix. For the Italian Grand Prix Fisichella has the chance to join Scuderia Ferrari, so Liuzzi steps up from the third driver role. He qualifies an encouraging seventh in his first outing at Monza, where Sutil starts a superb second and grabs his first points with a strong drive to fourth. The German also posts the fastest lap of the race. He goes on to qualify a brilliant third on a damp track in Brazil. The team finishes the

year in ninth place in the World Championship, having scored thirteen points.

2010
Sutil and Liuzzi continue as Force India's race drivers, while talented Scot, Paul Di Resta, joins as test and reserve driver. From the start of the season the cars are both regular top ten qualifiers and points-scorers. Sutil finishes in the points nine times, with a best of fifth place in both Malaysia and Belgium. He finishes eleventh in the World Championship. Liuzzi makes the top ten on five occasions, highlighted by sixth in Korea. Under the new system the team scores sixty-eight points and moves up to seventh in the World Championship, just a point behind Williams in sixth. Between his DTM commitments with Mercedes, Di Resta gains valuable F1 experience by taking part in eight Friday practice sessions on Grand Prix weekends.

2011
Di Resta is confirmed as Adrian Sutil's teammate for 2011, becoming the third British driver on the grid. Nico Hülkenberg joins as test and reserve driver, having previously raced for Williams. Together they form one of the most exciting line-ups on the grid. The season starts well with points for both cars in Australia, but it is not until the second half of the year that the team really finds its stride. The highlights include a superb sixth place for Sutil in Germany and a double-points finish for the team in Singapore. The team finishes a comfortable sixth in the standings, just four points behind Renault, to sign off Force India's most successful season to date.

In October 2011, Indian company Sahara India Pariwar, becomes a co-owner of the team with an investment of $100 million. The team is renamed Sahara Force India. In December, Nico Hülkenberg and Paul Di Resta are confirmed as the team's drivers for the 2012 season.

2012

Nico Hülkenberg gets promoted from his role as reserve driver to join Paul Di Resta, with Formula Renault 3.5 ace Jules Bianchi nominated as team tester.

The season starts positively, Di Resta opening the team's points tally in Australia and both drivers scoring in the following race in Malaysia; it is a prelude to a positive European season that culminates in double-scoring results in Monaco and Valencia. Hülkenberg is on song in Spa-Francorchamps and records a career-best fourth place, with Di Resta equalling this results two races later in Singapore. An inspired performance by Nico in treacherous conditions in the final race of the season sees the Sahara Force India VJM05 in the lead for thirty laps before finally settling down in fifth. The team finishes seventh in the standings, but with an impressive points haul and the confidence of being able to challenge the front-runners on merit.

2013

Adrian Sutil returns to the team to partner Paul Di Resta, recreating the pairing that achieved the team's best-ever Championship position of sixth in 2011. Halfway through the season, James Calado is announced as third driver.

The new VJM06 proves competitive from the start, with Sutil leading the opening race in Australia and the team scoring a double point finish; Di Resta is on form, matching his best-ever result with fourth in Bahrain and scoring in seven in the first eight races, while Sutil embarks on a giantkilling mission in Monaco to finish fifth in a race peppered with overtakes. The mid-season point sees dramatic rule changes that affect the car's performance, but the team bounces back in the closing races of the season, scoring crucial points at home in India and in Abu Dhabi to secure sixth place in the Constructors' Championship.

2014

The Mercedes-powered VJM07 combined with the new driver pairing of Hülkenberg and Pérez gave the team its biggest ever points haul of 155, which enabled it to match its highest ever finish of sixth in the World Constructors' Championship.

Nico's ninety-six points meant he secured ninth in the World Drivers' Championship with Sergio's fifty-nine points earning him tenth position—his third place in Bahrain was the highlight of the racing season and only the second podium finish in the team's history.

KEY PERSONNEL
Subrata Roy—Managing Director and Chairman,
Sahara India Pariwar

The Sahara India Pariwar is one of the country's major business conglomorates and has operations in financial services, life insurance, mutual funds, housing finance, city development, infrastructure and housing, print and television news media, entertainment channels, cinema production, consumer merchandise retail, healthcare, hospitality, manufacturing, sports, and information technology sectors.

Subrata Roy, affectionately called "Saharasri", founded Sahara India Pariwar in 1978 with 2,000 rupees, one small office room and two assistants. Today, Sahara has gross assets with a market worth of $25.94 billion, and 4,163 establishments that serve well over seventy million customers across India. With a million workers, Sahara India Pariwar is the second largest employer in India after the Indian Railways.

Sahara is based upon the Chairman's belief in the ethos of a family (the reason why the group is called a "Pariwar"—meaning family, in Hindi) and runs on a unique corporate philosophy of collective materialism—propounded by him, that advocates collective efforts for collective growth and finally collective sharing and caring. This has given immense opportunity to the employees and customers to grow together.

Roy's philosophies and beliefs have received acclaim across the world, through such occasions as his address to Harvard

University students, IIT (Indian Institute of Technology), IIM (Indian Institute of Management) and via books penned by him.

It is a firm conviction of "Saharasri" that every organization should contribute towards the cause of the society. Today, Sahara India Pariwar's universe of concern encompasses such diverse areas as providing financial assistance to 320 families of martyrs of Kargil war, the Mumbai terror attack and Dantewada Massacre victims, relocation and rehabilitation to victims of natural calamity, taking healthcare to remote corners of the country via a fleet of fifty-two mobile health vans, adult literacy and vocational training initiatives, rehabilitation of physically challenged people and annual mass marriage ceremonies of underprivileged girls from all religions.

Saharasri believes sports to be the most healthy, energetic and spirited facet of society, and promotes Indian sports on international, national and regional levels. Apart from sponsoring the Indian cricket and hockey teams for more than a decade, Sahara has also adopted Indian boxing, wrestling, archery, shooting, track and field as well as tennis, extending support to a total of 101 sportsmen across six sporting disciplines. It has very recently, extended support to Indian volleyball by sponsoring the men's national teams (senior and junior) for four years.

Under Saharasri's strong leadership, Sahara had played a major corporate role in the preparation and organizing of XIX Commonwealth Games 2010. The Group, under its company Sahara Adventure Sports Limited, owns the Pune franchise of Indian Cricket's coveted IPL League, and is an equal majority partner in the Sahara Force India Formula One Team. The Maharashtra Cricket Association's international Cricket stadium at Pune is named the Subrata Roy Sahara Stadium.

Dr Vijay Mallya—Team Principal and Managing Director
Dr Vijay Mallya is the charismatic businessman heading up one of India's largest and most significant business empires, the UB Group. With annual sales of over $4 billion and a

market capitalisation of approximately $12 billion, the Group has diverse interests in brewing, distilling, real estate, engineering, fertilisers, biotechnology, information technology and aviation. It is also the largest Indian manufacturer of beer and spirits.

In the late 1980s, Dr Mallya pursued global opportunities to transform the UB Group into India's first multinational company and in 1988, in a leveraged buyout, acquired the global Berger Paints Group with operating companies across four continents. Continuing the global growth in his businesses through the 1990s, Dr Mallya decided that the UB Group would only retain interests in businesses that were globally competitive and areas of core competence, thus transforming the vastly diversified UB conglomerate into a handful of key operating businesses including spirits, beer, fertilizers, engineering and aviation.

Dr Mallya has received several professional awards both in India and overseas. He was conferred an honorary Doctorate of Philosophy in Business Administration by the Southern California University and has also been nominated as a Global Leader for Tomorrow by the World Economic Forum. He has also received France's highest civilian award, the Legion of Honour.

One of Dr Mallya's major passions has always been cars, and motorsport in particular. He has had a longstanding presence in motorsport dating back to the seventies and eighties when he participated in several club races and non-championship Grands Prix, including the Madras and Calcutta Grands Prix, in Mo Nunn-owned Ensigns. When Dr Mallya was elected by shareholders as Chairman of UB in 1983 at the age of twenty-eight, and business interests began to take considerably more of his time, he took to collecting and racing historic and vintage cars in the UK.

In the late 1990s Dr Mallya returned to Formula One, this time as a sponsor. Kingfisher beer logos appeared on the Benetton Formula One Team in 1996 and 1997, while in 2007 Kingfisher Airlines was a sponsor of the Panasonic Toyota Formula One Team. During this time Dr Mallya's imagination was fired again for Formula One and, spotting a unique business opportunity, he formed a joint venture company with Jan and Michiel Mol to purchase the Spyker Formula One Team at the end of 2007. Force India was officially launched in January 2008 and has since gone from strength to strength.

SCUDERIA TORO ROSSO

Scuderia Toro Rosso has been competing in the Formula One World Championship since 2006. The team was created with a view to finding two extra cockpits for the stars of the future coming through the ranks of the Red Bull Junior Driver Programme. Nine years down the road, that was still very much the team's *raison d'être* in 2014 with a graduate of that programme, Jean-Eric Vergne competing in his third full season of Formula One. The French driver was teamed with Russian rookie, Daniil Kvyat, another graduate of the Red Bull Junior "finishing school" and winner of the 2013 GP3 Championship.

When the team was first established it operated partly as a satellite to Red Bull Racing running a car designed mainly by Red Bull Technology. However, for several years now Scuderia Toro Rosso has run completely independently, doing all the car design and manufacturing work in-house in Faenza. This necessitated a major expansion programme for the factory, still on-going, and the Italian side of the operation is supported by the team's wind tunnel facility in Bicester, England. A recent addition was a new building housing all the composites side of the operation, meaning the team produces virtually every component in house with the obvious exceptions of parts such as the engine.

On the engine front, after seven years being powered by Ferrari, for 2014 Scuderia Toro Rosso has switched to Renault, just as the technical rules went through a major change, featuring 1.6-litre turbocharged Power Units equipped with potent energy recovery systems.

Team Principal Franz Tost has been at the helm since 2005, while the technical side is managed by Technical Director James Key. Currently the team has one win and one pole position to its name, both courtesy of Sebastian Vettel, who produced the fairy-tale result at the team's home race, the Italian Grand Prix in Monza, back in 2008.

RACING RECORD
2014
Drivers: Jean-Eric Vergne and Daniil Kvyat
Car: STR9
Engine: Renault Energy-F1 2014
Constructors' Championship Points: 30
Final Position: Seventh

2013
Drivers: Daniel Ricciardo and Jean-Eric Vergne
Car: STR8
Engine: Ferrari V8
Constructors' Championship Points: 33
Final Position: Eighth

2012
Drivers: Daniel Ricciardo and Jean Eric Vergne
Car: STR7
Engine: Ferrari V8
Constructors' Championship Points: 26
Final Position: Ninth

2011
Drivers: Sébastien Buemi and Jaime Alguersuari
Car: STR6
Engine: Ferrari V8
Constructors' Championship Points: 41
Final Position: Eighth

2010
Drivers: Sébastien Buemi and Jaime Alguersuari
Car: STR5
Engine: Ferrari V8
Constructors' Championship Points: 13
Final Position: Ninth

2009
Drivers: Sébastien Buemi and Sébastien Bourdais (replaced by
    Jaime Alguersuari from Hungarian GP)
Car: STR4
Engine: Ferrari V8
Constructors' Championship Points: 8
Final Position: Tenth

2008
Drivers: Sebastian Vettel and Sébastien Bourdais
Car: STR2B—STR3
Engine: Ferrari V8
First Win and first Podium: Gran Premio d'Italia, Monza
    (Sebastian Vettel)
Constructors' Championship Points: 39
(Best) Final Position: Sixth

2007
Drivers: Vitantonio Liuzzi and Scott Speed (replaced by Sebastian
    Vettel from Hungarian Grand Pric)
Car: STR2
Engine: Ferrari V8
Constructors' Championship Points: 8
Final Position: Seventh

2006
Drivers: Vitantonio Liuzzi and Scott Speed
Car: STR1
Engine: Cosworth V10
First Championship point: US Grand Prix, Indianapolis (Vitantonio
    Liuzzi)
Constructors' Championship Points: 1
Final Position: Ninth

STR FACTORY—A WORK IN PROGRESS
While the race team is the visible high-profile side of the operation, seen at all the iconic Grand Prix tracks around the world, the
bulk of the work takes place on a bustling industrial estate on the

outskirts of the town of Faenza in northern Italy at the ever developing STR factory.

The original building was home to a Formula 1 team (Minardi F1) long before Scuderia Toro Rosso inherited it and you can get an idea of how much the sport has grown in complexity by the fact that the factory has now expanded to three buildings with a fourth on the way. There are also various other sites on the industrial estate, one used for the Machine Shop department, others for storage.

In fact, the fourth building will eventually replace the oldest part of the facility and provide a home to the team's very own simulator department. The reason why so much space is needed is that even if modern technology has replaced manpower in many areas, more and more people are needed to deal with all the specialist fields that go into a modern F1 car. In the old days, manufacturing and assembly employed the most staff, but today, the facility is packed with engineers, computer experts and specialists.

A large number work in the Drawing Office, with staff focusing on composites, suspension, hydraulics, transmission and system designs. In addition there are groups specializing in Aero Performance, Vehicle Dynamics, Computational Fluid Dynamics (CFD) and Stress Analysis. One thing modern technology has not erased is the need for face-to-face communication between everyone as all aspects of a Formula 1 car are linked together, which is why the designers of all components all work in the same space.

The main building is home to the Assembly Area, the workshop where the cars are actually built up prior to being sent off to races, and this houses the hydraulics, electronics and gearbox departments.

Building 2 is home to Research and Development and Quality Control where parts are tested under load, often to breaking point. The outfit might only produce a handful of cars each year, but there is still a need for the Stores to keep all the parts and, even though each component is extremely valuable, this area still looks like a typical car company's parts department, albeit rather more sophisticated.

Carbon fibre is the biggest single element used in the construction of Formula 1 cars and parts, which is why the Toro Rosso composites department takes up much of the newest structure, building 3, along with the autoclaves, or giant ovens that "cook" the parts once the layers of carbon fibre have been put together.

Typical of the sort of work that just did not exist when the factory was first built is the Operations Room, also housed in this building. In simple terms it replicates the pit wall at the track and for every practice and qualifying session, as well as the race itself, a team of engineers sit in the Operations Room seeing exactly the same data as their colleagues at the race track, in real time. With the number of team members allowed at the track limited by the regulations, this is an effective way of boosting the brainpower available to the team, with the guys at the factory working in ideal conditions to suggest ways of improving the car's performance at the track.

TEAM PRINCIPAL FRANZ TOST ON THE 2014 SEASON

Pre-season testing was extremely disappointing. Reliability problems meant the STR9 only did half the mileage of that covered by the Mercedes-powered cars and that put us at a big disadvantage for the start of the season.

Nevertheless, once the races began, we performed fairly well, with a particularly encouraging opening round in Melbourne, where Jev finished eighth and Daniil ninth, making him the youngest driver to ever score points in Formula 1.

The major change to the regulations meant that a high retirement rate was expected; however, such is the high level of technology in Formula 1, that the actual numbers across the entire field were fewer than predicted. At Toro Rosso, we actually made progress on the reliability front over the course of the year, as can be seen from the figures: from thirty-eight race starts, we posted ten DNFs, but nine of them came in the first ten races, up to the German Grand Prix. One was down to a collision, five were due to a Toro Rosso problem, with the remaining four being related to the power unit. Jean-Eric actually completed all races from Austria onwards. This means the general trend in terms of reliability was positive and it is something we must carry forward into 2015.

If we look at our targets for the year, we had intended to finish at least sixth in the Constructors' Championship and we did not hit this one, ending up seventh—but we can be reasonably pleased with having made a big step forward compared to previous years.

We also made progress with our working practices: the STR9 was the best car designed and manufactured by Toro Rosso so far. We made a big step forward. All the upgrades we introduced, such as those in Melbourne, Austria, Singapore and Japan, worked well. Also, we made progress on the fabrication side. The parts were of better quality and they fitted together much better. So well done to the different departments, we are working in the right direction.

As for the driver line-up, we have said goodbye to both our 2014 drivers, with Daniil being promoted to replace another former Toro Rosso "pupil", Sebastian Vettel at Red Bull Racing. Being a rookie, Daniil has done an excellent job and has shown in this first year in F1 to be a very fast driver and to have the potential to become one of the best. His strong motivation to learn and to achieve, his focus on how to improve his performance during the season, his great talent and his fighting attitude have given him the chance to step up quite quickly and to get the chance to move to RBR next year.

With Jean-Eric, it was a difficult decision to let him go because there is no doubt he is a very talented driver, but our team's remit is to bring on the youngsters in the Red Bull Junior Driver programme and Jev had completed three seasons with us. Our French driver had more than his fair share of bad luck this year and could have finished an impressive fifth in Monaco and he delivered another strong street circuit performance to come home sixth in Singapore. Now everyone within in the team is focused on 2015, applying the lessons learned this year.

FRANZ TOST, TEAM PRINCIPAL

Nationality	Austrian
Hometown	Trins, Austria
Date of birth	20 January 1956

As a young lad, Franz Tost's big hero was fellow countryman Jochen Rindt: his bedroom walls were covered with posters of the Austrian ace and when it was dissertation time at school, Franz's classmates would all groan, as they knew what was coming—another bloody eulogy to Rindt! Inevitably, Tost found himself behind the wheel, racing a Formula Ford. He was quick enough to win the 1983 Austrian FF Championship, but he felt he would not make it to the top as a driver so a degree in Sports Management from Innsbruck University was next on the agenda. This led to a job at the Walter Lechner Racing School at the Zeltweg circuit.

From there Tost moved to a team management role with EUFRA Racing and, at the end of 1993, he took the post of team manager with Willi Weber's Formula 3 team. It was here that he crossed paths with Ralf Schumacher, and Weber asked Tost to accompany the youngster to Japan. This led to looking after Ralf's interests at Jordan and then Williams, prior to taking on the role of Operations Manager with BMW's Formula 1 programme. From there, he took on the role of Team Principal with the newly formed Scuderia Toro Rosso in 2005.

CAREER HIGHLIGHTS

1978–82: Degree in Sports Sciences, University of Innsbruck

1982–5: Degree in Sports Management, University of Vienna (Magister Rerus Naturale—Master of Science)

1979–85: Formula Ford 1600, Formula Ford 2000 and Formula 3. Won the Austrian Formula Ford 1600 Championship in 1983.

1985–88: Walter Lechner Racing School—Director and Instructor, Walter Lechner Racing Team; Team Manager in Formula Ford, Formula 3

1989–91: Eufra Racing Team—Team Manager

1992–93: Vienna Racing Team—Deputy Manager

1993–95: WTS Formula 3 Team—Team Manager

1995–2000: WEBER Management—Marketing Manager

2000–05: BMW Williams F1—Race Track Operation Manager

Since 2006: Scuderia Toro Rosso—Team Principal

LOTUS AND RENAULT: THE ENSTONE ERAS

Lotus F1 Team entered a new era on 1 January 2015 as Mercedes power took over from Renault Sport F1, ending a twenty-year partnership with the French company.

The team's Formula One history started in 1981 as the Toleman Motorsport team, based in Witney, a few miles from its current base, Enstone in Oxfordshire (UK). Starting out in F1 in late 1980, the team struggled with their first car, the Toleman TG181 and it took until September's Italian Grand Prix for the car to qualify.

It was a further struggle in 1982 with the upgrades failing to make an impact before it was replaced. The following year saw the team take its first points finish in the Dutch Grand Prix.

A soon-to-be legend of F1 joined in 1984; Ayrton Senna quickly showed his burgeoning talent by taking the team's best-ever finish, second, in the Monaco Grand Prix, before adding two third-place finishes in Toleman's golden era.

After a tough 1985 for Toleman, Benetton took over and entered the 1986 F1 season. There was no Renault connection back then, as the team started with BMW and Ford. Berger clinched Benetton's inaugural win in the 1986 Mexican Grand Prix before before the team oversaw the 1994 Driver's title when Michael Schumacher edged ahead of Damon Hill by a single point.

A new era began in 1995 with Renault becoming the engine supplier, propelling the team to instant success. The first year with the new engine saw Benetton secure the double with Michael Schumacher retaining his World Drivers' Championship while the German and his British teammate Johnny Herbert combined for the Constructor's Championship, getting the new deal off to a dream start.

That was the peak of the team's existence, but they did however continue to attract top drivers. Jean Alesi joined in 1996, along with Gerhard Berger. And yet the defending Constructors' Champions struggled the following year. Alesi in particular found it tough in his two-year stint, with the Australian Grand Prix in 1997 proving to be the lowest point after confusion between radio and driver saw the car retire.

In contrast, Berger was able to get the most from the car, including victory in the 1997 German Grand Prix after storming through the field—thereby winning his first and last race with the team.

Alesi then left in 1998, with Giancarlo Fisichella the next Italian to turn out for the Benetton team, later to be joined by Austrian Alex Wurz when Berger departed. Drivers were not the headlining news, though, as Renault's withdrawal from Formula 1 in 1998 was to prove costly for the team.

Left to rely on the 1997 versions of the Renault engines, Benetton were at a disadvantage before racing began. It spoke volumes of the engines and drivers that Benetton were able to take second-place finishes in Montreal and Monaco, thanks to Fisichella. The Italian would later add a pole position, his first, in Austria.

Wurz also impressed in his first year, to the point that Ferrari supposedly became interested in his signature. In the second half of the year, the strong results dipped as Benetton lost traction to the teams around them with their latest engines.

Unfortunately, 1999 continued where the second half of 1998 left off, with inconsistency proving to be the season's pattern. Fisichella's best result was once again a second in Montreal, but generally it proved to be a challenging season for both the Italian and Wurz.

The new millennium saw a better start, with Fischella taking fifth and second in the first two races at Austria and Brazil while teammate Wurz took seventh in Austria. For 2000, the first half of the season proved a huge success for Benetton's Italian, Fisichella grabbing another two podiums in Monaco and Montreal with two third-place finishes.

Wurz struggled for consistency across the season and his teammate would soon join him, as the second half of the campaign became a tale of retirements and low finishes.

It was another false dawn, but Renault's return as owner of the team in 2001, along with promising Briton Jenson Button, resulted in a new-look team that looked set for better things in what turned out to be the final year of Benetton's existence.

Again, it was to be Fisichella who impressed in the B201 as a sixth place in Brazil showed the power of Renault's return. Reliability of the car improved, but the finishes proved to be outside the top ten once again.

There was to be an arrest to the second half of the season slump, as the Italian driver grabbed a fourth in Germany and then Benetton's last podium in Belgium thanks to a third-place finish. It was a steep learning curve for Button, but fifth in Germany showed he had the potential.

The duo would continue into the new era as Renault F1 Team in 2002, marking the next step in the long-running relationship between the team and engine supplier with Renault replacing Benetton as the constructor.

The eight-year spell proved to be highly successful with twenty wins, twenty pole positions and fifty-seven podiums combining for a total of 933 points. In addition, there were two World Drivers' Championship titles for Fernando Alonso in 2005 and 2006, with two World Constructors' Championships won in the same years.

There was a lot of work to do before any title celebrations, however, as the team started a challenging first year with Jenson Button and Jarno Trulli, who had replaced Fisichella in the off-season, the first to drive for the Renault F1 Team.

While there were some reliability issues, particularly for Trulli, there were also some significant points scored in 2002, with Button grabbing two fourth-place finishes, three fifth places and one sixth. His teammate weighed in with seven top ten finishes.

The two drivers combined to hand Renault a fourth-place finish in the Constructors' Championship, a commendable start for the team under its new identity.

The following year saw only one of the drivers retained, Trulli. Button had outscored his teammate in the previous season, but Renault replaced him with the class of Fernando Alonso. The rewards were instant, with Alonso taking a podium in only his second race following his first-ever pole position, the youngest driver in Formula 1 to do so, and his teammate followed with fifth.

Alonso followed this with his second podium in consecutive races at the Brazilian Grand Prix. He would go two better at Hungary, becoming the youngest race winner in Formula 1.

Trulli grabbed his best result of the season in the round prior, a third-place finish in the German Grand Prix. The duo had done enough to secure fourth for the team in 2003, while Alonso was sixth in the standings and Trulli eighth.

Fortunes reversed in 2004 as Trulli became the race winner, impressing in the Monaco Grand Prix in a season where top ten finishes were regular. Alonso, on the other hand, did not win a race, but did grab four podiums. He was regularly in contention and his strong results saw him pick up a respectable fourth in the drivers' standings.

The following year saw the team begin a two-year spell of dominance. Trulli left for pastures new with Giancarlo Fisichella returning, and Fisichella's campaign got off to a perfect start, taking victory in the Australian Grand Prix, kickstarting a run of four Renault victories to begin the season.

The team continued to impress and out of the nineteen Grands Prix that year, Alonso took seven victories, six pole positions and two fastest laps on his way to the title.

The Spaniard defended his title the following year, putting down a marker with another great opening run to the season. Alonso won the opener in Bahrain, before Fisichella won in Malaysia. The run continued into the Australian Grand Prix, where Alonso returned to the top step of the podium.

The reigning champion took a huge leap towards the title as the European leg of the season came to a close with three victories in a row in Spain, Monaco and Britain. He went on to win the Canadian Grand Prix before taking the chequered flag in Japan, the penultimate race of the season.

The team were brought back down to earth in 2007 when there was a new look to the team as Heikki Kovalainen replaced two-time World Champion Alonso to partner Fisichella.

Renault's new Finn managed a podium at Japan, although podiums proved hard to come by for both drivers. Even so, consistent top ten finishes meant that the team finished third in the standings.

In a bid to return to the top in 2008, Renault brought Fernando Alonso back to the team and partnered him with Nelson Piquet Jr. The move prompted two victories for the Spaniard, while

Piquet Jr took a best finish of second in Germany; however, the team struggled and only managed to take fourth in the standings.

The following season was notable for current driver Romain Grosjean joining the ranks for the first time, replacing Nelson Piquet Jr from the second half of the season, and partnering Fernando Alonso in what was a challenging year for the team.

The Renault F1 Team's final campaign in 2010 saw another new line-up, this time with the immensely talented Robert Kubica alongside Vitaly Petrov. Podiums in Montreal and Belgium saw Renault back on the up thanks to Kubica's strong debut season, while Petrov took a best result of sixth in Abu Dhabi to end the Renault F1 Team era on a good note.

While only lasting nineteen Grands Prix, the short-lived Lotus Renault Grand Prix team signalled Renault's return in 2011 to a full focus as an engine supplier rather than as a constructor. With the Lotus name back in use, Team Principal Eric Boullier announced a return to a British licence, moving the team closer to the British identity it has today.

However, there was a twist in the plan. Robert Kubica, the first Polish F1 driver, was unable to race for the team after a rally crash in February 2011. It meant that the first two drivers under the new team identity were Nick Heidfeld and Vitaly Petrov. The pair got off to a flying start with Petrov taking a podium in the team's first race in Australia before his teammate took another in the following event, in Malaysia.

Petrov continued to impress taking two top ten finishes in China and Turkey while Heidfeld was also consistent in securing four top ten finishes out of five between Turkey and Great Britain.

It was not all plain sailing, though, as Heidfeld had consecutive retirements in Germany and then Hungary, which prompted the appointment of Bruno Senna, nephew of Ayrton Senna, to take over driving duties alongside Petrov. The Brazilian returned the team to the top ten, taking ninth in his second race, in Italy. From there results dried up with only two more top ten finishes, both falling to Petrov in Japan and Brazil.

The team's current incarnation got the green light in 2012 and with a new driving duo behind the wheel. Previous World

Champion Kimi Räikkönen returned from his two-year sabbatical to join up with GP2 Series Champion and Enstone returnee, Romain Grosjean.

With a new name came fresh impetus. New-found pace was apparent from the off although the results took a little longer to arrive. Not too long though, as Bahrain witnessed Räikkönen coming close to his first victory on his return, settling for second as Grosjean grabbed third.

It set the team in a groove with Grosjean returning to the podium in Canada while he seemed set to do so again in Valencia before a mechanical problem ended his race. Thankfully, his teammate was just behind and the Finn was able secure a podium spot by finishing second.

Räikkönen then settled into a consistent run of form, taking consecutive podiums in Germany, Hungary and Belgium to make it three in a row. The Finn later made sure of a perfect end to the season, taking the Lotus F1 Team's first victory, as he stormed to victory in Abu Dhabi.

The pair returned for 2013 and in the E21 Kimi Räikkönen won the season opener, the Australian Grand Prix. It was no flash in the pan, as the Chinese Grand Prix saw the Finn take second place despite a difficult start to the race, before making it three second-place finishes in a row at Bahrain and Spain. In the former, Grosjean joined him on the podium as he finished third.

Some difficult race weekends followed but it was business as usual as Räikkönen took consecutive second place finishes at Germany and Hungary, later adding another at Korea.

Grosjean stepped up too, with three third-place finishes deservedly taken at Korea, Japan and India. He went one better in the US, taking second on his way to a final position of seventh in the standings. Heikki Kovalainen joined Grosjean for the final few races of the season as Räikkönen underwent back surgery.

In 2014, the outfit's final year with Renault, the supplier was up against it with the new engine regulations. The 2.4-litre normally-aspirated V8 engines were replaced by new 1.6-litre turbocharged V6 engines. It was a revolutionary change for Renault to combat in a challenging year.

Pastor Maldonado lined up alongside Romain Grosjean in

what turned out to be a difficult final year that showed pace, but not results; even so, both drivers were able to score points.

Romain impressed in the best race of the season for the team, qualifying fifth and turning it into an impressive eighth place in the Spanish Grand Prix. On the other side of the garage, Pastor drove impeccably in the United States, qualifying a strong eleventh before finishing in ninth spot, as Romain narrowly missed out on the top ten, coming home eleventh.

While the 2014 Abu Dhabi Grand Prix marked the end of a memorable relationship with Renault, the long-running association provided plenty of unforgettable highs that saw the team in its various incarnations win titles, take podiums and set fastest laps in what will always be known as "the Renault era".

SAUBER

TALES OF THE UNEXPECTED

Peter Sauber had never been particularly interested in cars, and motor racing did not do anything for him at all, not then at least. The fact that, in 2010, Sauber was able to celebrate the fortieth anniversary of Sauber Motorsport had a lot to do with chance in the early days, but afterwards it was down to sheer perseverance and, later on, a good deal of hard graft and skill.

Sauber's father owned an electrical company that employed around 200 staff and had premises in Zurich. Sauber's career path seemed to be mapped out. He trained as an electrical fitter with the aim of gaining further qualifications and following in his father's footsteps. But it would all turn out rather differently.

In 1967 Sauber used to drive to work every day in a VW Beetle—until a friend persuaded him to have some tuning work done. For a bit of fun he then entered it in a few club races in 1967. Far more significantly, it sparked his passion for tinkering with cars. He modified his Beetle to such an extent that eventually it was no longer fit for road use. This led to the next stage in Sauber's career: in 1970 he decided to set himself up as an independent builder of open two-seater racing sports cars. Out of the cellar of his parents' home in Zurich emerged the Sauber C1. He used the first name of his wife Christiane as the model designation.

That same year, he set up PP Sauber AG and moved into a specially built workshop on the premises of his father's company in Wildbachstrasse. With the C1 he won the 1970 Swiss sports car championship, but soon whittled things down to the occasional appearance as a racing driver. In 1974 he donned his helmet for the last time before turning his full attention to car construction. The "C" was retained as a trademark.

Sauber had set himself a difficult task: surviving on constructing racing sports cars in Switzerland seemed a doomed prospect. But he would not be deterred and was determined to battle on. The working day often stretched deep into the night, while money was in short supply.

SPORTS CAR SUCCESSES

Sauber achieved international prominence with the C5 in which Herbert Müller won the then acclaimed Interserie Championship in 1976. That was followed by his first forays at Le Mans. By this time Sauber Motorsport had four employees on the payroll. In 1981 Hans-Joachim Stuck and Nelson Piquet won the Nürburgring 1000 Kilometre race in a Sauber-built Group 5 BMW M1.

The following year was a decisive one for Sauber. He was commissioned by Swiss composite materials manufacturer Seger & Hoffmann to build a car for the Group C World Sports Car Championship: it was to become the Sauber C6. During this time he made contact with engineers at Mercedes who expressed an interest in motorsport—though all very much at a private level, as international motorsport had been an unmentionable subject for the Stuttgart carmaker since the tragic accident at Le Mans in 1955.

In 1985 Sauber began fitting Mercedes engines into his racing sports cars, moving that bit closer to the Stuttgart company. Just a year later, Henri Pescarolo and Mike Thackwell won the Nürburgring 1000 Kilometres in a Sauber C8. Further triumphs were to follow, ultimately prompting Mercedes' comeback to international motor racing. From 1988, Sauber and his crew acted as the Mercedes official works team.

Professor Werner Niefer, Chairman of Mercedes at the time, decided the cars should be painted silver, marking the revival of the famed "Silver Arrows". The highlight of this partnership was the year 1989, which brought not only the drivers' and manufacturers' titles in the World Sports Car Championship, but a one-two result in the legendary 24 Hours of Le Mans as well. The following year saw a repeat win of the World Championship title. Sauber Motorsport had grown to a workforce of fifty.

It was also during this time that the junior team was set up, based on an idea of Sauber's business partner of the time, Jochen Neerpasch. The drivers selected were Michael Schumacher, Heinz-Harald Frentzen and Karl Wendlinger. Peter Sauber paved the way for all three to enter Formula One.

FORMULA ONE

With the lustre of the World Sports Car Championship beginning to fade, Mercedes now looked to Formula One. In the summer of 1991 it was declared a joint project, and preparations went into full swing. Sauber set about building a new factory on the company site in Hinwil.

However, that November brought with it bad news: because of the straitened economic climate, the Mercedes board had decided against sending a works team into Formula One. Sauber had two options, one, to accept a financial settlement and withdraw, or alternatively to use the money as start-up capital for his own Formula One involvement. In January 1992 he took the plunge, and by autumn the first tests in the C12 were under way, with an Ilmor engine providing power. The company was then employing just under seventy staff.

On 14 March 1993, according to plan, two Sauber C12 cars driven by Karl Wendlinger and JJ Lehto lined up for the South African Grand Prix. With two World Championship points for fifth place claimed by the Finnish driver, this debut turned out an acclaimed success.

Contracts signed with Red Bull and Petronas in 1995 provided a solid foundation and enabled the Swiss team to establish itself as a firm fixture in Formula One. In 1995 and 1996 Sauber served as the works team for Ford, and from 1997 onwards the cars were powered by Ferrari engines bearing the name of the title sponsor Petronas.

The breakthrough was some time in coming, however. Finally, in 2001, three high points in the team's history arrived in rapid succession: the partnership with major Swiss bank Credit Suisse, fourth place in the Constructors' World Championship secured in mid-October and, just a few days later, the ground-breaking ceremony for the team's very own wind tunnel.

Sauber also decided to introduce some fresh blood into Formula One at this time, signing up Kimi Räikkönen and Felipe Massa to his team and later recommending Robert Kubica to the decision-makers at BMW.

BMW

Peter Sauber was on the lookout for a new engine partner in 2005. Now in his sixties, he was not disinclined to pass his life's work on into capable hands. An offer from BMW seemed like a good solution. The car manufacturer, which had been involved in Formula One with Williams since 2000, was keen to set up its own works team. On 22 June 2005, BMW announced its acquisition of a majority stake in the Swiss team.

The 2008 season—the third year of the BMW Sauber F1 Team—would mark the next milestone in the history of the team. The extension at Hinwil had in the meantime been completed and the workforce had grown to more than 400. The team's target for that year was to achieve its maiden victory—which turned out to be a one-two, with Robert Kubica winning in Canada ahead of Nick Heidfeld. In all, the BMW Sauber F1 Team notched up eleven podium places in 2008. Kubica claimed the team's first pole position in Bahrain and Heidfeld boosted the statistics with the first two fastest race laps. The team ended the World Championship in third place with 135 points.

Following a challenging start to the 2009 season, shock news broke on 29 July: at a press conference in Munich, BMW announced it was withdrawing from Formula One at the end of the season. The company bowed out with thirty-six points and sixth place in the World Championship.

STARTING OVER

The next press conference would be held on 27 November 2009, this time in Hinwil. Peter Sauber had reached an agreement with BMW and bought back his life's work. But the joy was tempered by disappointment as BMW had already decided to reduce the workforce. Employee numbers were whittled down from 388 to 260. It was with this pared-down workforce, with Ferrari as engine partner and drivers Kamui Kobayashi and Pedro de la Rosa, that the Hinwil team embarked on the 2010 race campaign.

The first half of the season was marred by numerous retirements for technical reasons, unprecedented in the team's history. After the first eight races, the team had a single World Championship point to its name. By the end of the season this had risen to forty-four, of which Kobayashi had picked up thirty-two, with De la Rosa and Heidfeld—who replaced the Spaniard for the last five Grands Prix—each contributing six points.

THE 2011 SEASON

The team hired another rookie, Sergio Pérez, for the 2011 season. The Mexican's arrival meant Kobayashi would have to take on leadership responsibilities in only his second full season on the F1 grid. The year began with the team getting to grips with the tyres developed by the new sole F1 supplier Pirelli, completing a promising programme of winter testing and jetting off for an opening race in which a strong team performance ultimately gave way to frustration.

Pérez and Kobayashi crossed the finish line seventh and eighth in Melbourne, only to be subsequently disqualified after a rear wing element was deemed to have contravened the rules. The team lost the ten points its performance had earned, but consolation arrived in the knowledge that the necessary speed was there.

Strong showings duly followed in the next few races. In Monaco, for example, Pérez had just made it through to the top-ten qualifying shootout for the first time when he lost control of the C30 on the exit from the high-speed tunnel section and slammed into the barriers with devastating force. The Mexican youngster was initially motionless in the car. After what felt like an eternity the news came through that he had got away with severe concussion.

Kobayashi went on to show great mental strength to finish fifth in the race, the best result of the season for the Sauber F1 Team. Pérez also had to sit out the next race in Canada, with De la Rosa taking his place at short notice.

After a good first half to the season, which saw the team occupying what looked like a safe sixth place in the Constructors' World Championship, the team endured a drop in form. The cause of the downturn was rooted in a controversial technology: diffusers fed by the car's exhaust flow, even—thanks to sophisticated engine mapping—when the driver is off the throttle. The FIA announced a ban on the practice, only to subsequently reverse its decision.

In the meantime, the team had stopped development of an

"outboard blown" diffuser for the C30, which put it at a disadvantage against rival teams still running the technology. Despite this handicap of well over a second per lap, the young drivers still managed to add to the team's World Championship Points haul. The Sauber F1 Team eventually finished seventh in the Constructors' Championship on forty-four points. Kobayashi was responsible for thirty of those, with Pérez recording fourteen points. Both Kobayashi and Pérez, together with Mexican reserve driver, Esteban Gutiérrez, were confirmed for the 2012 season as early as the summer.

THE 2012 SEASON

The Sauber F1 Team lined up for 2012 with the unchanged pairing of Pérez and Kobayashi in the race seats. And the season began strongly, Pérez coming home eighth and Kobayashi sixth at the opening race in Melbourne.

But that was only the start; even greater excitement was to follow in Malaysia, where Pérez delivered a sensational performance in fluctuating weather conditions. A clever tactical move in the early stages saw him make up a number of places, and the Mexican driver was subsequently the fastest man on a wet, then merely damp and finally drying track. Moving up into second place, he even put the race leader—Ferrari's Fernando Alonso—under pressure before briefly running wide and losing critical seconds. However, second was still an outstanding result and, most of all, it underlined what an excellent car the team had developed in the Sauber C31-Ferrari.

The next highlight of the season was not long in coming. Pérez qualified fifteenth for the Canadian GP, but a well thought-out strategy and the Mexican's ability to look after his tyres allowed him to work his way up to third—giving him and the team their second podium of the season at this still early stage.

The low point came at Spa. The weekend had begun perfectly; Kobayashi secured second place on the grid, with Pérez starting immediately behind him. However, the race had barely begun when both the Sauber cars were involved in the same collision caused by a rival driver. Their race was ruined and the disappointment was immense.

However, compensation for the Sauber F1 Team arrived just a week later in Monza, Pérez providing further evidence of his tyre-preserving prowess. The Mexican cut through the field like a hot knife through butter—most notably in the latter stages of the race after taking on more fresh rubber—to wrap up another second place.

The final highlight of the season came courtesy of Kobayashi in his home Grand Prix at Suzuka. The Japanese star had already qualified third to send his compatriots into raptures. Then he also made a fine start to the race, cementing his position at the business end of the field. Going into the final quarter of the race he came under increasing pressure from the ever-closing Jenson Button, but the local hero held firm to set the seal on his first podium finish in Formula One. For many in the team, the podium ceremony provided the season with its most emotional moment.

It was a very good year for the Sauber F1 Team, headlined by four podium finishes, 126 World Championship Points and sixth place in the Constructors' standings—a position higher once again than the previous year and an achievement that earned the praise of many outside observers.

HANDING OVER THE REINS

There was a major announcement on 11 October 2012—and with it a milestone in the team's history—it was the day that Sauber stepped down as Team Principal and passed on the baton to Monisha Kaltenborn.

THE 2013 SEASON

The Sauber F1 Team had a new pairing for the 2013 season. Nico Hülkenberg joined the Swiss squad for his third Formula One season. Esteban Gutiérrez was promoted from test and reserve driver to a race seat.

The season opener in Melbourne was chastening. Nico Hülkenberg was not even able to start the race due to a leak in the fuel tank of the C32; Gutiérrez finished his debut race in thirteenth spot.

However, the first points were not as far away as the team might have thought. Nico Hülkenberg collected the first four points of the season in Malaysia and followed up with a tenth place in China. After this, though, came a number of races that gleaned no points.

Scoring seven points and qualifying only twice for Q3 in the first half of the season was not the expected return of the Sauber F1 Team. In Hungary the team introduced an update package for the Sauber C32-Ferrari, and was confident it would improve performance during the second half of the campaign.

Finally, things turned round in Monza: Hülkenberg qualified third and finished fifth in the race, adding ten points to the team's tally. The remainder of the season saw at least one car qualifying in Q3.

In Singapore Gutiérrez made the top ten for the first time and delivered one of his best weekends of the season. In Korea both drivers qualified in the top ten with the Mexican rookie just missing out on his first point by finishing eleventh. Hülkenberg drove one of his best races in the season finishing fourth. The Sauber F1 Team advanced to seventh place in the Constructors'World Championship.

Only one weekend later, another highlight took place in Japan. Hülkenberg and Gutiérrez both finished in the top ten, pulling in another fourteen points for the team.

Disappointment struck in India, where Hülkenberg was forced to withdraw from the race early due to a broken brake disc. Gutiérrez missed the points because of a drive-through penalty. And it didn't look any better as the team left Abu Dhabi empty-handed as well. This time Hülkenberg got the drive-through penalty. In the last two races Nico Hülkenberg netted a further twelve points for the team.

With fifty-seven points, the Sauber F1 Team finished the 2013 season seventh in the Constructors' World Championship. Given that the team finished sixth twelve months earlier, this result was hardly satisfying. However, there were also positive aspects: the C32 was barely competitive in the beginning of the season, but the team improved significantly during the second half of the campaign by scoring fifty points. Nico Hülkenberg finished the Drivers' Championship in tenth spot, his best result in his Formula One career. Esteban Gutiérrez scored six points and was the best-placed rookie of the season.

THE 2014 SEASON

Failing to secure a single point and finishing in tenth position in the Constructors' World Championship, the team endured a frustrating campaign in 2014. This was the first time in their Formula One history that the team had failed to score even one point.

The team was dogged by reliability issues throughout the season, which resulted in a combined fifteen race retirements. Not surprisingly both drivers also struggled with power issues in qualifying to such an extent that the only time the team progressed into Q3 was when Adrian Sutil made it into the top ten shootout at the US Grand Prix.

In November it was announced that Marcus Ericsson and Felipe Nasr would replace Sutil and Gutierrez in the 2015 driver line-up. Only three teams on this year's grid—Ferrari, McLaren and Williams—have been in Formula One longer than Sauber. Between 1993 and 2014 a total of twenty-five drivers have lined up for the Swiss team in 384 Grands Prix.

MANOR (MARUSSIA) F1

THEY'RE BACK: AND MANOR (MARUSSIA) F1 MAKE TEN

Formula One is the most elite team sport in the world and over the years a staggering 158 teams have competed since its inception in 1950. Just in the nick of time, in terms of the FIA Championship entry deadline, not to mention the print deadline for this book, Manor (Marussia) F1 became number 159 as they literally rose from the ashes of 2014 to become the tenth team on the 2015 grid.

Speaking at the press conference before the start of The Australian Grand Prix in Melbourne back in March Graeme Lowdon, Team President and Sporting Director, had this to say:

"Just to be absolutely clear about the process, the team that competed last year, the entrant was Manor Grand Prix Racing Ltd. That company suffered financial issues and sought protection through administration. The process to settle the arrangement with its creditors, with unsecured creditors, was done using a Company Voluntary arrangement, very standard practice; the CVA and that mechanism is used to take care of the outstanding debts to those unsecured creditors. That allows the company then to return to trading with a fresh start if you like. So we have a new contract with Ferrari, one that we're very happy with. Hopefully it's one that [Ferrari] is very happy with and it allows us to go racing, which is what we want to do."

Having exited administration via the aforementioned CVA as late as February of 2015 it is a credit to the determination and organization of the team and its leadership that they have managed to successfully enter this year's championship. The long search for the necessary investment was finally complete when it was announced in March that Stephen Fitzpatrick, founder of energy company OVO, was to be their new investor.

He had this to say of his new venture: "It's personal funding for the time being, although we've been approached by several individuals and consortiums already who looked at the restructuring last year and walked away from it. So we're already discussing with quite a few different parties and that's something we're very open to. I'm certainly open to other investors that come in over time, but for the time being the investment so far has been personal.

"I know that Formula 1 is seen to be a rich man's sport. It's certainly an expensive sport. I don't think it has to be as expensive. If you want to win the Championship you've got to be realistic about how much that's going to cost; I think at the end of the grid that we want to operate on, it's pretty clear to us what we need to do.

"The team has operated on a modest budget over the last four years and been very successful in the sport. We're expecting this year that the budget's going to be between £60–62m. For the first time ever the team is going to be entitled to prize fund money, which covers more than half the budget, and it's a completely different financial prospect given that the team has finished in tenth place or higher in the last two seasons.

"While there's been an investment so far, I guess the idea is not for this to be the black hole of the Fitzpatrick family finances for three years and then I have to downsize my house and so on!

"For me the sport is to know what your budget is and to get the best of that budget. It's not to spend money that you don't have. I feel very strongly that there's a sporting aspect to Formula 1 that I love and there's a business aspect to Formula 1 that earns you a right to be there and there's no point thinking that Formula 1 is [just] any kind of business."

The team also appointed Justin King (ex-Sainsbury's CEO), whose son is a GP2 driver, as new interim chairman in a move that further steadies the ship: "This [Manor] was a unique opportunity for me to get involved in the sport, which I'm thoroughly enthusiastic about. I intend to do something a bit different from what I've done in the past and I'm working on a few things for my day job."

MANOR MARUSSIA IN THEIR OWN WORDS ...

History

The Manor team has helped to develop some of the finest drivers in the world.

Manor has a strong heritage in motor racing—formed in 1990 by John Booth, the team initially competed in Formula Renault. They recorded championship wins in the 1991, 1994-95 and 1997-2000 seasons. When the team switched to Formula 3 in 1999, they won the Championship at the first attempt. Manor Racing was guided by a set of key principles, which involved focusing on performance, doing more with less and providing a launch pad for young drivers—including former World Champion, Kimi Räikkönen, and current Formula One World Champion, Lewis Hamilton.

In 2009, the FIA opened the F1 door to four new contenders, and Manor secured entry to the highest echelon of motor sport for the start of the 2010 season. In Bahrain, with a brand new team and two rookie drivers, they lined up on the Formula One grid for the first time. They set out to demonstrate that with the same passion and determination as some of the more established players, but with a fraction of the spend, the team could succeed in this highly competitive arena.

In February 2015, after a very uncertain time for the team, and after significant support from new backers, Manor emerged from administration and is now participating in the 2015 FIA Formula One World Championship, with John Booth and Graeme Lowdon continuing to lead the team.

Graeme Lowdon—President & Sporting Director

Graeme graduated from the University of Sheffield with an MEng in Mechanical Engineering, and then earned an MBA from the University of Newcastle. Graeme co-founded a business that became a pioneer in the market for delivering rapid data communications to high-speed vehicles and in 2000 joined Manor Motorsport as commercial director, became CEO in 2009, and then President & Sporting Director; a position he continues to hold at the revived team.

John Booth—Team Principal

John's racing career began in 1977, and by 1979 he had won his first race at Oulton Park—racing with the late, great Ayrton Senna. John retired from motorsport racing in 1989 and in 1990 Manor Motorsport was born. After considerable successes in Formula Renault, and later Formula 3, the next obvious step was Formula One and in 2009, when the FIA opened the door to four new contenders, Manor submitted an entry for the 2010 season.

John McQuilliam—Technical Director

John graduated from Salford University with a degree in Mechanical Engineering. In 1986 he joined the Williams team, but then took time out to gain the first-ever Masters Degree in Composite Materials. When he returned it was with Arrows, and then the Jordan team; pioneering the use of composite suspension. He was promoted to Chief Designer—responsible for all of their race-winning cars. In 2008 John took a break from the sport, but was back in 2010—commissioned to design and build Manor's first car.

Will Stevens—Driver

Will is a British racing driver, hailing from Rochford in Essex. He started his racing career in go-karting at the age of twelve. After racing in Formula Renault 2.0 and Formula 3.5, Will made his debut in Formula One at the end of the 2014 season, racing with Caterham, at the Abu Dhabi Grand Prix. Will joined the Manor team in February 2015.

Roberto Merhi—Driver

Roberto is a Spanish racing driver from Castellon in Spain. Roberto competed in Formula Renault, Formula Three and Formula Renault 3.5. His first participation in F1 was with Caterham, in a practice session at the 2014 Italian Grand Prix, and he joined the Manor Racing team in March 2015.

As the team develops throughout the 2015 season more up-to-date information on the team and drivers can be found at:

www.manorf1team.com
www.willstevens.com
www.robertomerhi.es

"This will be a great test of resolve for all of the team. Formula One is an absolutely relentless business and for some reason we keep coming back." (Graeme Lowdon)

DRIVERS

LEWIS HAMILTON

Lewis Hamilton, son of an English mother and a father whose parents originally came from Grenada, was born on 7 January 1985 in Stevenage, England. At the tender age of eight, Lewis sat in a kart for the first time and was immediately bitten by the racing bug. Mercedes-Benz soon recognized the British youngster's enormous talent, becoming one of his sponsors right at the start of his career.

Lewis not only showed incredible pace on the track but also climbed the career ladder at an impressive rate. After he had blown the competition out of the water in every category of karting and won the McLaren Mercedes Champions of the Future series, McLaren Mercedes enrolled him on their young driver programme in 1997. In 2001, he moved up into British Formula Renault, which he won two years later.

Lewis's winning streak continued unabated in the Formula 3 Euro series; he was crowned champion of the competition for juniors while still in his second year and subsequently progressed up into GP2 with ART Grand Prix. The rookie dominated GP2 by posting five victories and was crowned champion at the end of the season, a success that immediately led to his promotion to the very top of motor racing, F1.

He made his Formula 1 debut at McLaren Mercedes on 18 March 2007 as teammate to Fernando Alonso while still only twenty-two years old. Despite his youth, he fought confidently for the world title right up until the season finale in Brazil, taking on top drivers who were considerably more experienced. Numerous records set by Lewis in his rookie season are testimony to his class, and include the most race wins, the most pole positions as well as the most points in a debut season.

What had been an extraordinary career up until then reached its climax in the 2008 season with his World Championship success. At the time of his title victory, Lewis was twenty-three years, nine months and twenty-six days old and so prised from Fernando Alonso the accolade of being the youngest ever Formula 1 World Champion. After six successful years with McLaren Mercedes, Lewis was looking for a new challenge and joined the Silver Arrows MERCEDES AMG PETRONAS works team for the 2013 season.

Lewis secured third place on the podium in only his second race for the team and achieved five pole positions in F1 W04. Having taken his first victory in a works Silver Arrow at the Hungarian Grand Prix, Lewis finished his first season for MERCEDES AMG PETRONAS in fourth place overall on 189 World Championship Points.

In 2014 he clinched his second drivers' World Championship with victory in the season-ending Abu Dhabi Grand Prix, finishing on 384 World Championship points, with eleven wins, to become the fourth Briton to win two titles, and Mercedes' first World Champion since Juan Manuel Fangio's back-to-back titles for the manufacturer in 1954 and 1955.

Nationality: British
Date of birth: 7 January 1985
Place of birth: Stevenage, UK
Lives: Monaco
Height: 1.74 m
Weight: 66 kg
Debut: Australia, 18 March 2007
Best championship: World Champion—2008 & 2014
First win: Canada, 10 June 2007
2014: Formula One: Mercedes AMG Petronas—World Champion
2013: Formula One: Mercedes AMG Petronas—4th
2012: Formula One: McLaren Mercedes—4th
2011: Formula One: McLaren Mercedes—5th
2010: Formula One: McLaren Mercedes—4th
2009: Formula One: McLaren Mercedes—5th
2008: Formula One: McLaren Mercedes—Champion

2007: Formula One: McLaren Mercedes—2nd
2006: GP2 Series: ART Grand Prix—Champion
2005: Formula Three Euro Series: ASM—Champion
2004: Formula Three Euro Series: Manor Motorsport—5th
2003: Formula Renault UK: Manor Motorsport—Champion
2002: Formula Renault UK: Manor Motorsport—3rd

NICO ROSBERG

Born on 27 June 1985 in Wiesbaden, Germany, Nico Rosberg, son of Finnish Formula 1 World Champion Keke Rosberg, had motor racing in his DNA right from birth. He gained his first experience in karting when just six years old. By the time he was sixteen he had fought his way up to the highest echelon of kart racing, the international Formula Super A.

In 2002, his huge talent earned him a Formula BMW cockpit, racing for Team Rosberg, his father's outfit. Nico proved that he had not been given the drive simply by virtue of his name when he won the championship. That same year, he was rewarded for his achievements with his first Formula 1 test for the Williams F1 team. Aged seventeen, he was, at the time, the youngest driver ever to get behind the wheel of a Formula 1 racing car.

In 2014 he achieved his highest position to date in the World Championship, scoring 317 World Championship points to finish second to teammate Lewis Hamilton. Nico claimed victories in Australia, Monaco, Austria, Germany and Brazil.

Nico did not get his driver's licence until one year later but had by then already made quite a name for himself in the Formula 3 Euro Series. In 2006, French team ART Grand Prix signed Nico to drive in GP2, and there was no stopping him. He left the competition trailing far behind him in his first year and was crowned GP2 champion. Again, Frank Williams knocked on his door, this time to offer Nico a Formula 1 contract.

On 12 March 2006, Nico made his Formula 1 debut for Williams as teammate to Mark Webber, and impressed by going on a sensational charge and posting the fastest lap of the race. Aged just twenty years, eight months and thirteen days at the time, he is still the youngest driver ever to have achieved this feat. In addition, he belongs to an elite group of drivers (Giuseppe Farina, Masahiro

Hasemi and Jacques Villeneuve) who all succeeded in posting the fastest lap in their first Grand Prix.

After four years with Williams, Nico was looking for a new challenge and, in 2010, found his way to the Silver Arrows MERCEDES GP PETRONAS works outfit. Nico finished third on the podium in only his third race for the team. He repeated this success in China as well as at his team's home race in Silverstone, proving to all his critics that he could hold his own against teammate and seven-time World Champion Michael Schumacher.

On 14 April 2012, Nico secured his first pole position and then his first Grand Prix win one day later at the Chinese Grand Prix. Nico took a historic victory at the 2013 Monaco Grand Prix, exactly twenty years after his father Keke had won the prestigious race in the Principality. Nico finished his fourth season for MERCEDES AMG PETRONAS in sixth place overall on 171 World Championship Points, having claimed victories in Monaco and in the UK.

Nationality: German
Date of birth: 27 June 1985
Place of birth: Wiesbaden, Germany
Lives: Monaco
Height: 1.78 m
Weight: 67 kg
Debut: Bahrain, 12 March 2006
Best championship: 2nd—2014
First win: China, 15 April 2012
2014: Formula One: MERCEDES AMG PETRONAS—2nd
2013: Formula One: MERCEDES AMG PETRONAS—6th
2012: Formula One: MERCEDES AMG PETRONAS—9th
2011: Formula One: MERCEDES AMG PETRONAS—7th
2010: Formula One: MERCEDES AMG PETRONAS—7th
2009: Formula One: Williams—7th
2008: Formula One: Williams—13th
2007: Formula One: Williams—9th
2006: Formula One: Williams—17th
2005: GP2 Series: ART Grand Prix—Champion
2004: Formula Three Euro Series: Team Rosberg—4th
2003: Formula Three Euro Series: Team Rosberg—8th
2002: German Formula BMW: VIVA Racing—Champion

DANIEL RICCIARDO

Nationality   Australian
Hometown   Perth, Western Australia
Date of birth   1 July 1989

Daniel Ricciardo's first season in the colours of Infiniti Red Bull Racing was an unqualified success. The young Australian broke into the ranks of Grand Prix winners, burnished his reputation as a mighty performer in qualifying and displayed a combination of strategic acumen and decisive overtaking skills that propelled him to a string of impressive results. His three victories in 2014 plus five other podium finishes were enough to secure third place in the Drivers' Championship.

Daniel came through Red Bull's ranks, launched on the path to a successful motorsports career via the well-established Junior Team. After learning his trade in Formula Ford and Asia-Pacific Formula BMW competitions, he took the plunge and followed in the footsteps of the southern hemisphere greats by moving to Europe to pit his skills against the best of his generation.

Daniel initially took residence in Italy, contesting the 2007 season in Formula Renault 2.0. He joined the Red Bull programme in 2008 and won the Formula Renault 2.0 WEC championship. The next year he took the prestigious British Formula 3 title— traditionally a gateway to great things in Formula One. He finished 2009 with a three-day test at Jerez in the RB5, finished top of the timesheets, completed nearly 300 faultless laps and made the Red Bull engineers sit up and take notice with his speed, confidence and precise feedback. They also noticed he seemed to be enjoying it quite a bit, too.

Off the back of his Formula 3 record and his speedy acclimatization to the F1 cockpit, Daniel stepped up for 2010, becoming test and reserve driver for Scuderia Toro Rosso while also competing in the Renault World Series. Daniel narrowly missed out on the FR3.5 title after impressive victories at Hockenheim, the Hüngaroring, the Circuit de Catalunya and the showpiece round in Monaco.

Daniel stayed with Toro Rosso for the 2011 season, advancing his education with regular drives in the first free practice sessions of Grand Prix weekends. He also raced a truncated campaign in FR3.5 and again won the prestigious Monaco Grand Prix support race, also taking pole position and fastest lap. These striking performances in the first half of the year led to Daniel being loaned to HRT for the final eleven rounds of the 2011 F1 season. He made his F1 race debut at Silverstone, and strong performances for the Spanish backmarker led to confirmation of a race seat for Toro Rosso in 2012.

In two years with Toro Rosso, Daniel established himself as a future star of Formula One. Lightning qualifying performances were backed up with Grand Prix drives that blended aggression and intelligence: Daniel pushed the limit—but he also brought the car home.

During his two years at Infiniti Red Bull Racing's sister team, Daniel scored points in thirteen of his thirty-nine Grands Prix and established a reputation as a gritty, confident racer, sure in his decisions and capable of getting the best out of his machinery. In qualifying he had demonstrated real quality, frequently hauling his Toro Rosso to the head of the midfield.

Eight times in 2013 Daniel made it through into the Q3 qualifying shoot-out, pulling out scintillating laps when the pressure was on. When Mark Webber announced his retirement from F1, Daniel was the natural and ready-made replacement.

The immediate question in the minds of many was how Daniel would shape up with four-times World Champion Sebastian Vettel as a teammate. "A lot of people have been asking the question," said Daniel at the time. "I'd love to believe that I'll be up there with him but I don't want to start saying I'm going to beat him only for him to come out on track and kick my butt. I'll let the driving do the talking and hopefully it'll do some positive talking. Obviously it's my biggest challenge yet—but hopefully it's his biggest challenge, too."

To the outside world, Daniel exceeded expectations in 2014 by proving a match for Sebastian but inside Infiniti Red Bull Racing he did the job that was expected of him. Daniel did not come into the team cold: he had tested the Red Bull cars on many occasions;

had frequently been the outfit's driver for marketing work and show car appearances and, perhaps most pertinently, had served his time as their simulator driver, logging long days in Milton Keynes, providing factory support as the race team travelled the globe chasing World Championship glory. To the Red Bull engineers he was a known quantity: fast, intelligent and capable of providing the type of feedback that would prove crucial in a season liable to be dominated by technical development.

In his initial season with a front-running team, Daniel recorded a number of firsts. At his home Grand Prix and on debut for Infiniti Red Bull Racing, he qualified in second place—his first front row Grand Prix start. He also finished the race second and made his first podium appearance—only for that to be struck from the record for a technical infringement. Strong performances and an obvious affinity with the RB10, however, meant it was not long before he was back on the podium. After fourth place finishes in Bahrain and China, he moved up to third in Spain and Monaco—but then things got really interesting.

Daniel took his maiden F1 victory in Canada. A good strategy and overtaking moves right on the ragged edge saw him charge through the field in the closing laps to gain the lead shortly before the chequered flag, becoming the fourth Australian to win a Grand Prix, following in the footsteps of Sir Jack Brabham, Alan Jones and Mark Webber.

Two more victories followed: a finely judged tactical triumph in Hungary and a flat-out blast around the mighty Spa-Francorchamps. He also made the podium at Silverstone, in Singapore and on the Circuit of the Americas. One of his finest performances, however, was saved until last—a fourth place in Abu Dhabi after starting in the pitlane.

The standout for Daniel is naturally his performance at the Circuit Gilles Villeneuve. "My first victory was a lot of fun," he says. "And it definitely wasn't a boring race. Yeah, it was pretty cool."

"Fun" is a word that occupies a lot of Daniel's conversation and, according to race engineer Simon Rennie, for Daniel it's all about the overtaking: "I haven't had a driver who's been so excited about overtaking people for a long time! It's not the qualifying or the race for Daniel—they're simply the journey to get to the overtake."

When the visor's down Daniel radiates focused determination—but out of the car he's a very laid-back Aussie, keen on spending as much time as possible outdoors, always searching for the next adventure and eager to get into pretty much any sport you care to mention—but predominantly anything involving bikes or water.

This is slightly at odds with his alter ego, The Honey Badger. It features prominently in his helmet design and perhaps needs a bit of explaining. In his own words: "I came across the Honey Badger a while back. It's not the biggest animal in the world, in fact it looks a bit like a wombat. It's pretty cute, you wouldn't think much of it—but in reality it's a raging ball of anger that tears things apart. It's a bit like me: don't be fooled by the sunshine exterior, press the right buttons and I can be a very dark individual." Based upon all the available evidence, nobody believes this to be true.

DANIIL KVYAT

Nationality	Russian
Hometown	Ufa, Russia
Date of birth	26 April 1994

Whatever awaits Daniil Kvyat in the coming years, he is unlikely to forget the last few months of 2014 when, in the wake of the announcement that Sebastian Vettel would be leaving the team at the end of the season, the young Russian found out that following his debut season for sister team Toro Rosso, he would be replacing the four-time champion at Infiniti Red Bull Racing.

Admittedly they're big shoes to fill, but ever since he was a youngster the kid from Ufa in Russia has been taking on big challenges and overcoming them.

Daniil first encountered motorsport when, aged nine, he saw a local go-kart track on the way home from school and begged to have a go. Bitten by the racing bug, he was soon competing seriously in Russia, but when the time came to step up to the next level, there were few top-line options in his home country. And so, in 2007, the entire Kvyat family moved to Rome to be nearer to the heart of European karting.

In 2008 he finished third in the KF3 European series and was runner-up in the WSK international series. He eventually made the switch to single-seaters in 2010, with support from the Red Bull Junior Team programme, which found him a berth in Formula BMW in Europe and in the Pacific series. Over the following winter he honed his skills in the Toyota Racing Series in New Zealand and then, in 2011, he returned to Europe to move up to Formula Renault 2.0, in which he raced full seasons in both the Eurocup and North European Cup competitions of the category.

Daniil came third in the Eurocup and second in the NEC and opted to stick with Formula Renault 2.0 for 2012. This time he finished as runner-up in the Eurocup series, but won the category's Alps series title, taking seven wins along the way.

With a title in the bag, Daniil's next challenge was the 2013 GP3 series. The rookie began in fine style finishing just off the podium three times in the first three race weekends. As the season hit its midpoint, Daniil didn't look like a championship contender, but in the second half of the campaign he blossomed, winning three of the final six races to lift the title. The performance, in his debut season, was enough to fast track him to Formula One and soon after his G3 triumph he was named as Toro Rosso's 2014 replacement for the Infiniti Red Bull Racing-bound Daniel Ricciardo.

Kvyat made his F1 debut, aged nineteen, in the 2014 Australian Grand Prix, where he finished ninth, breaking Sebastian Vettel's record as the youngest points-scorer in Formula One. He went on to score points in the Malaysian, Chinese, British and Belgian Grands Prix, finishing fifteenth in the World Championship.

FELIPE MASSA

Nationality: Brazilian
Born: 25 April 1981
Height: 1.66 m
Weight: 60 kg

CAREER HIGHLIGHTS
2014: Race Driver for the Williams F1 Team
2006–13: Race Driver for Scuderia Ferrari
2004–5: Race Driver for the Sauber Petronas F1 Team
2003: Test Driver for Scuderia Ferrari
2002: Race Driver for the Sauber Petronas F1 Team
2001: Euro F3000 Champion. Test with the Sauber Petronas F1
   Team
2000: Italy and European Formula Renault Champion
1999: Formula Chevrolet Champion (Brazil)
1998: Debut in single-seater
1990: Debut in karting

Born on 25 April 1981 in Sao Paulo, Brazil, not far away from
the Interlagos circuit, Felipe Massa was already behind a wheel
at the age of nine. In 1990 he came fourth in the Sao Paulo
Micro-Kart Championship; sixth in 1991; and finished on the
podium in 1992.

After his successes in kart racing he participated in Formula
Chevrolet in 1998, concluding the Brazilian Championship that
year in fifth place and winning the title in 1999. Felipe then moved
to Europe and won the Italian Formula Renault Championship in
2000 and the following year he won the European F3000
Championship. Also in 2001 he signed a contract with Ferrari and
took part in what was his first test behind the wheel of a Formula
One single-seater, with Sauber, at the Mugello circuit.

In 2002 Felipe made his Formula One debut at the Australian
GP with Sauber-Petronas. By the end of the season he had
claimed four points and started work as a test driver for Ferrari,
his full-time occupation in the 2003 season.

Felipe returned to a race seat with Sauber-Petronas in 2004, scoring twelve points that season and eleven points the following year. He then signed as a race driver for Ferrari in 2006. In his first year he claimed his first race win at the Turkish Grand Prix and finished third in the Drivers' Championship. 2008 would be a standout year for Felipe, coming close to winning the Drivers' Championship only to be beaten by Lewis Hamilton by one point.

Felipe suffered a severe injury during qualifying for the 2009 Hungarian Grand Prix that forced him to sit out the remaining seven races of that season. Fully recovered, he returned for Ferrari in 2010 and finished sixth in the Drivers' Championship, taking five podium finishes. In his sixth season with Ferrari and again he finished sixth in the 2011 Drivers' Championship, and in 2012 Felipe finished seventh in the Drivers' standings with 122 points.

2013 was Felipe's last season with Ferrari, moving to the Williams F1 Team for the 2014 and 2015 Formula One World Championships.

VALTTERI BOTTAS

Nationality: Finnish
Born: 28 August 1989
Height: 1.73 m
Weight: 70 kg

CAREER HIGHLIGHTS
2013 FIA Formula One World Championship, Williams F1 Team
2012 Official Reserve Driver, Williams F1 Team
2011 GP3 Series champion
2010 Formula 3 Euroseries, 3rd overall
Masters of Formula 3, series champion—first driver to win the
    title twice
2009 Formula 3 Euroseries, 3rd overall
Masters of Formula 3, series champion
2008 Formula Renault 2.0, series champion: 12 wins, 13 poles
Formula Renault 2.0 Eurocup, series champion: 5 wins, 7 poles
2007 Formula Renault 2.0, 3rd overall: 2 wins, 2 poles

1994–2006 Karting: multiple title winner and seven years in the National Karting squad

Valtteri Bottas began his career racing karts at the age of six. Over the next decade he amassed countless race wins, many championships and a seven-year stint in Finland's National Karting squad.

After winning three prestigious kart titles in 2006, Valtteri moved into single-seaters. He won both the 2008 Formula Renault Eurocup and the 2008 Formula Renault Northern European Cup, winning seventeen races from twenty-eight starts.

Despite not winning a race in 2009, Valtteri claimed two poles and secured third in the Formula 3 Euroseries. In June of the same year he won the Masters of Formula 3 at Zandvoort, a title he won again in 2010 and he remains the only driver ever to have won the title twice.

Valtteri remained with ART's F3 team in 2010, securing two victories en route to third place in the championship. In 2011 he won the GP3 Series at the first attempt, claiming a win at each of the last four race weekends to secure the title in the penultimate race.

He was assigned test driver for Williams in 2010 and 2011, before becoming Official Reserve Driver in 2012 taking part in fifteen practice sessions over the course of the season. After impressing the team, he was duly promoted to Race Driver and competed in his debut FIA Formula One World Championship season with the Williams F1 Team in 2013.

SEBASTIAN VETTEL

When Sebastian Vettel secured his first pole position and first victory at the rain-affected 2008 Italian Grand Prix it heralded the arrival of a unique talent. In a Toro Rosso car that was much more midfield than front-runner he became the youngest driver in history to win a Formula One Grand Prix, aged only twenty-one years and seventy-four days. In a season in which he was named Rookie of the Year, Vettel proved that he was fast, intelligent and technically adept. More importantly he showed that could drive an F1 car quickly and that he had the ability to do this consistently.

In 2009 Vettel graduated to Red Bull Racing where he went on to win the Chinese Grand Prix to become, at the age of twenty-one years and 287 days, the youngest Grand Prix driver in history to win for two different teams. He claimed second place in the title race behind Jenson Button by winning the Abu Dhabi Grand Prix, giving him four victories along with four more podiums for the season.

In what was a landmark season, Red Bull Racing won its first Constructors' Championship in 2010 and Sebastian Vettel became the youngest ever World Drivers' Champion with a dramatic victory at the Abu Dhabi Grand Prix to beat Fernando Alonso to the title by a mere four points.

Following a contract extension Vettel was utterly dominant in 2011 finishing the season with fifteen poles, eleven victories, seventeen podiums from nineteen races and a record total of 392 points. His closest rival Jenson Button could only achieve 270 points.

Seven different drivers won the first seven races of the 2012 season, and in a tense season finale in Abu Dhabi Vettel finished sixth to secure his third consecutive World Drivers' Championship by only three points over Fernando Alonso.

Once again in 2013 Vettel was supreme with a run of nine consecutive victories at the end of the season as he secured the World Drivers' Championship for a record fourth time with three races in hand at the Indian Grand Prix. He became only the third driver in the sixty-four years of Formula One, along with Fangio and Schumacher, to win four consecutive championships.

The change in engine regulations in 2014 saw Vettel struggle to come to terms with the new car and with his new teammate Daniel Ricciardo, who won three races, while the German became the first reigning champion since Jacques Villeneuve to fail to win a race.

Vettel's lifelong dream of driving for Ferrari came true when it was announced that he would take up a three-year contract to drive for the Scuderia from 2015 where he will partner close friend Kimi Räikkönen.

CAREER HIGHLIGHTS
1995–2002 Karting winning various titles
2003 2nd in F.BMW ADAC with Eifelland Racing
2004 F.BMW ADAC Champion with ADAC Berlin-Brandenburg e.V.
2005 5th in F3 Europe with ASL-Mucke Motorsport
2006 2nd in F3 Europe with ASM Formule 3 and tester for BMW Sauber F1 Team
2007 14th in the F1 World Championship with BMW Sauber and Scuderia Toro Rosso (6 points)
2008 8th in the F1 World Championship with Scuderia Toro Rosso (35 points)
2009 2nd in the F1 World Championship with Red Bull Racing (84 points)
2010 F1 World Champion with Red Bull Racing (256 points)
2011 F1 World Champion with Red Bull Racing (392 points)
2012 F1 World Champion with Red Bull Racing (281 points)
2013 F1 World Champion with Red Bull Racing (397 points)
2014 5th in the F1 World Championship with Red Bull Racing (167 points)

Birthdate:	7 March 1987
Birth place:	Heppenheim, Germany
Height:	1.75 m
Weight:	62 kg
Civil Status	Single

Hobbies: Football, sports in general
Races: 139
Wins: 39
Pole positions: 45
Podiums: 66 (1: 39 2: 14 3: 13)
Fastest race laps: 24
Points: 1614

KIMI RÄIKKÖNEN

Kimi Räikkönen was born to race. From his very first seat time in a pedal kart at the tender age of three, he showed a passion and talent for speed. Whether it is a Formula One car, roadcar, motorbike, kart, pushbike, snowmobile or pair of skis, Kimi has only one speed: flat out.

Kimi's sky-rocket rise to Formula One World Champion was dramatic. His graduation to Formula One direct from Formula Renault redefined fast-track career progression; to this day no driver in history has made such a sensational entry. From his very first Grand Prix with Sauber, Kimi made an impression on both drivers and fans, and not long into his rookie season McLaren made a successful bid to sign Kimi on a five-year contract. For Kimi, this was not enough and with fame in his sights he joined Ferrari, the most illustrious name in Formula One.

The 2007 season witnessed one of the most exciting battles in history, and Kimi's fight to win the drivers' title was an international hit. His title defence in 2008 was hampered by bad luck, despite finishing third in the championship and winning the DHL fastest Lap Award. He achieved an historic victory at Spa in 2009 in a year when the established teams struggled due to new regulations.

The Flying Finn had a new challenge in 2010: a move to the World Rally Championship where he competed for a second season in 2011, doing what he does best—breaking the speedometer and making sure his co-driver had a change of underwear.

Kimi returned to Formula One for the 2012 season, signing a two-year contract with the popular Lotus team. His return proved without doubt that he had not forgotten how to drive an F1 car. He was the only driver to finish every race and scored points in all but one; his season including seven podiums, an historic win in Abu Dhabi and third place in the Drivers' Championship. It marked a highly successful and much longed-for return to the F1 cockpit, reigniting the passion of the Iceman's fans across the globe.

Kimi returned to the Prancing Horse for 2014 with high expectations and his loyal fans cheering the laconic Flying Finn to twelfth position. And remember: leave him alone, he knows what he's doing.

CAREER HIGHLIGHTS

1999 Runner-up in the European Formula Super A Karting Championship

2000 Formula Renault UK Champion. Two test sessions with Sauber-Ferrari F1 car.

2001 10th in the F1 World Championship (9 points) with Sauber

2002 6th in the F1 World Championship (24 points) with McLaren

2003 2nd in the F1 World Championship (91 points) with McLaren

2004 7th in the F1 World Championship (45 points) with McLaren

2005 2nd the F1 World Championship (112 points) with McLaren

2006 5th in the F1 World Championship (65 points) with McLaren

2007 Formula One World Champion (110 points) with Ferrari

2008 3rd in the F1 World Championship (75 points) with Ferrari

2009 6th in the F1 World Championship (48 points) with Ferrari

2010 10th in the World Rally Championship (25 points) with Citroen Junior WRC

2011 10th in the World Rally Championship (34 points) with ICE1 RACING Team, 2 races in NASCAR Championship with Kyle Busch Motorsports

2012 3rd in the F1 World Championship (207 points) with Lotus

2013 5th in the F1 World Championship (183 points) with Lotus

2014 12th in the F1 World Championship (55 points) with Scuderia Ferrari

JENSON BUTTON

From an early age, McLaren driver Jenson Button showed the potential to realize his dream of driving in Formula 1 and one day becoming World Drivers' Champion.

Jenson began his racing career on the karting track, as the majority of drivers do, and achieved instant success. Winning the British Super Prix in 1989, aged nine, was just the beginning. In 1991, he won all thirty-four races of the British Cadet Kart Championship before becoming Junior TKM Champion a year later.

Further successes followed until Jenson became the youngest driver ever to win the European Super A Championship, aged just seventeen. The following year he moved to cars and won the British Formula Ford Championship, which helped him win the prestigious McLaren Autosport BRDC Young Driver Award.

He then secured a race seat in the British Formula 3 Championship in 1999 and finished third in his first year. Also that year, as part of his prize for winning the McLaren Autosport BRDC Young Driver Award, he had the opportunity to drive a McLaren F1 car. Little did he know that it would not be the last time Jenson jumped into a McLaren F1 cockpit.

Jenson got his big break in Formula 1 in 2000 with Williams, and became the youngest British driver ever to start an F1 race in Australia, aged just twenty years and fifty-three days. That year, he produced some standout performances including finishing fourth at the chaotic German Grand Prix in Hockenheim, in addition to stunning the Formula 1 paddock by qualifying third for the Belgian Grand Prix.

After being loaned to Benetton for two years (although the team became Renault in 2002), Button joined British American Racing in 2003, where he immediately made his mark by outscoring his championship-winning teammate Jacques Villeneuve by seventeen points to six. Button managed his first F1 podium in Malaysia in 2004, the first of ten that season which drove him into third spot in the Drivers' Championship.

However, Jenson's breakthrough year was 2006. He drove

brilliantly at the Hungarian Grand Prix to take his maiden win after starting down in fourteenth place.

The following two campaigns were more testing seasons for Jenson and it was not sure whether he would be racing in 2009 due to Honda's decision to cease involvement in the sport. The outfit was saved just a few weeks before the start of the season and renamed Brawn GP. After winning six of the first seven races, he continued to secure strong points-scoring finishes to realise his dream and become World Drivers' Champion.

Jenson arrived at McLaren as reigning world champion in 2010 and immediately got off to a great start with wins in Australia and China. The following year he won the Canadian Grand Prix in spectacular fashion, contending with a sodden Circuit Gilles Villeneuve, six Safety Car periods and having to fight through the field three times to take one of the all-time greatest F1 victories.

He won further races in 2011 in Hungary and Japan to finish second in the World Drivers' Championship and achieved similar success in 2012, winning in Australia, Belgium and Brazil.

Held in high esteem by his fellow F1 drivers, Jenson recently became a director of the Grand Prix Drivers' Association. He also dedicates time away from the circuit to help charities. In 2010, he set up the Jenson Button Trust, which supports and fundraises for charitable causes and campaigns close to his heart. He is therefore highly regarded throughout the sporting world not only for his motor racing achievements but also for trying to help those who need it most.

FERNANDO ALONSO

The giant-killing performances were what caught your attention first: hustling an uncompetitive Minardi around Suzuka to finish a barely credible eleventh in 2001; seemingly appearing from nowhere to grab his first pole position (Malaysia 2003); and becoming the then-youngest-ever Grand Prix winner (Hungary, again in 2003) in truly effortless fashion.

With his intent signified, his move to the Renault team gave him the firepower to fulfil his ambition.

Armed with 2005's R25, the greatness that had been glimpsed

in snatches was quickly and thoroughly refined. It was immediately apparent at that year's San Marino Grand Prix, where his incredibly precise defence of the lead kept no less than keening, hungry world champion Michael Schumacher at bay. It was a performance marked by the skill of an old veteran rather than an eager newcomer.

The trickle of victories quickly turned into a torrent: seven wins by the end of the season, but—just as important—a steady stream of podium positions (five runner-up spots and three third places) that cemented his ascent to the title. Underlining the point, his was a world crown won with the seasoned experience of a master, not that of a fresh-faced youngster feeling his way nervously toward his first championship.

He was crowned in Brazil, finishing third behind—presciently—two McLarens. Standing on the podium, his ear was turned by Ron Dennis, who quietly assured the Spaniard that his future surely lay in one of Woking's silver cars.

A deal was quickly signed—but for 2007, leaving him to once again race for Renault in 2006. That season, the old enemy—Schumacher—was back in contention, and both he and his Ferrari team used every weapon in their sizeable armoury to peg back Fernando's progress.

It made for a tense, nervy and paranoid season—but one where Fernando once again triumphed by playing the numbers game whenever he lacked the outright competitiveness to win. For the record, he still scored seven victories, and backed those up with seven further runner-up spots.

His 2006 title made him the sport's then-youngest-ever double world champion. Buoyed by this momentum, he quickly made his mark at McLaren in 2007, winning his second race for the marque and quickly re-establishing the team at the competitive vanguard after a disappointing 2006 season.

More victories followed. He led home an emotional McLaren one-two at Monaco, showcased his controlled aggression to snatch victory at the Nürburgring, and pummelled the opposition into submission at Monza. But his winning progress was matched by his rookie teammate Lewis Hamilton, who also took four victories—and, at season's end, the McLaren challenge wasn't concerted enough to stem the singular charge of Ferrari's Kimi Räikkönen, who took the title by just one point at the final race in Brazil.

If the title near-miss was a blow, it was not the most problematic issue in a season that was overshadowed by competitive rancour both on and off the track. The fallout was intense, both McLaren and Fernando parted company—the Spaniard returning to Renault for two largely uncompetitive seasons before joining Ferrari for 2010.

Fernando's time at the Scuderia was a rollercoaster of highs and lows—he won his very first race in a red car, at the 2010 Bahrain Grand Prix, but went on to lose the title by the narrowest of margins after a strategical error cost him dearly at the final race of the season in Abu Dhabi.

In 2011, he scored a solitary victory at Silverstone, then wrestled a less-than-competitive Ferrari to three magnificent victories in 2012 as he spearheaded the charge to usurp world champion Sebastian Vettel. While Fernando gave his all, his campaign once again came undone at the final race.

While his final two seasons at Ferrari coincided with a dip in the Scuderia's competitive fortunes, his period with the Maranello squad would repeatedly underline his credentials as the greatest, and most respected, driver in the sport. And while the record books will not fully reflect his successes, history will tell us that Fernando Alonso stood a shoulder above his peers in terms of reputation and ability.

For 2015, he takes on an ambitious new challenge with McLaren and Honda, tasked with returning the reunited giants back to the top step of the podium—while at the same time aiming to replicate the past successes of his favourite driver and his childhood idol, Ayrton Senna.

NICO HÜLKENBERG

After a season away, Nico Hülkenberg returned in 2014 to the team where he impressed during the 2012 season.

Born in 1987, Nico made a huge impression in his karting career, winning the German Junior title in 2002 and the senior version the following year. In 2005, aged just seventeen, he graduated to the domestic Formula BMW series, taking eight race wins and beating Sebastian Buemi to the title.

In 2006 he competed in German F3, finishing fifth in the championship. By now under the management of Willi Weber—the man who discovered Michael Schumacher—Nico really made his mark in A1 GP, taking nine wins and earning the series title for Germany.

In 2007 he made a sideways move to Euro F3 with the ASM team. In his first year in the highly competitive series he finished a strong third, behind only teammate Romain Grosjean and Buemi. He also won the F3 Masters event at Zandvoort, and experienced his first F1 mileage when he tested for Williams at Jerez in December.

He remained with the renamed ART team in 2008, dominating the championship with eight wins. He also had a taste of the GP2 Asia Series, winning in Qatar, while he continued to log F1 mileage in testing for Williams.

In 2009 Nico contested the GP2 Series for ART, earning five wins and taking the title in his rookie year—a feat previously matched only by Lewis Hamilton and Nico Rosberg, both of whom also drove for ART. He continued to impress the Williams team with his testing performances and duly earned a full-time seat for 2010, alongside Rubens Barrichello.

On only his third outing in Malaysia Nico qualified fifth and scored his first point with tenth place. He finished in the points seven times in total, with a best of sixth place in Hungary. The undoubted highlight, however, was a stunning pole position on a drying track at the penultimate race of the season in Brazil.

Unfortunately Williams could not offer Nico a race seat for 2011, and instead he switched to Sahara Force India to take up the role of reserve driver, sampling the car for the first time in Valencia in February. He subsequently took part in fourteen Friday practice sessions on race weekends.

Nico impressed the team with his speed, feedback and approach, and in December 2011 he was confirmed as a race driver for 2012, alongside Paul Di Resta. The season confirmed Nico's talent as a driver, with eleven point finishes and a career-best fourth place in Belgium. During the last race of the season, he led the race under treacherous conditions for thirty laps before settling for fifth at the chequered flag.

A move to Sauber for 2013 saw Nico match his best-ever result of fourth in Korea, where he held off Hamilton and Fernando Alonso in the closing stages of the race. Following the season, Nico announced his return to Sahara Force India for the 2014 season and beyond.

Date of birth: 19 August 1987
Car Number: 27
Born: Emmerich, Germany
Lives in: Switzerland
Weight: 74 kg
Height: 1.84 m
Marital status: Single

2014	F1 Force India, 9th place
2013	F1, Sauber F1, 10th place
2012	F1, Sahara Force India Formula One Team, 11th Place
2011	F1, Sahara Force India Formula One Team reserve driver
2010	F1, Williams F1, 14th place
2009	GP2, champion; F1, test driver, Williams F1
2008	F3 Euroseries, Champion; 2nd, Masters in Zolder
2006–7	A1GP, Champion, nine wins; Winner, Masters in Zolder (2007)
2006	5th, German F3, one win, three poles
2001–4	Vice European Champion Kart Cadets (2001); Italian Junior Champion (2001/2002); German Junior Champion (2002); 8th, European Championship (2002); 5th,

Italian Championship (2003); German Champion (2003); German Vice Champion (2004)

SERGIO PÉREZ

One of the most promising drivers in Formula One, Sergio Pérez—known as "Checo"—arrived at Sahara Force India in 2014 determined to help the team's progress towards the front of the grid.

Born in 1990 in Mexico, Sergio comes from a motorsport family: having raced cars himself, his father Antonio was active in driver management. In this environment, it was natural for young Sergio to start his career in karting at the age of six, winning junior categories and quickly progressing to shifter karts.

A move to single seaters at the age of fourteen and the start of a long-standing partnership with Escuderia Telmex saw him take part in the Skip Barber National Championship in the United States; the following year, in 2005, Sergio moved to Europe to compete in Formula BMW.

After two years in this category, including a two-race stint in A1GP for Team Mexico, Sergio graduated to British Formula Three, dominating the National Class in 2007 and claiming four wins on his way to fourth in the International Class in 2008. A first appearance in the GP2 Asia Series saw him complete a lights-to-flag win in Bahrain and a call-up to GP2.

In his second year in the Formula One feeder series, Sergio won races in prestigious venues such as Monaco, Silverstone, Hockenheim, Spa-Francorchamps and Abu Dhabi to mount a title challenge and finish runner-up to Pastor Maldonado. His performance earned him promotion to Formula One with Sauber.

In his first season in the pinnacle of motorsport, in 2011, five finishes in the points helped cement his position in Formula One. Confirmed at Sauber for 2012, he claimed three podiums including two second places, on his way to tenth in the Drivers' Championship (with sixty-six points). Sergio demonstrated an incredible ability to extract the best out of the car in changing weather conditions, pushing eventual winner Fernando Alonso closely in Malaysia and performing incredible comebacks in Canada and Italy.

A move to McLaren for 2013, replacing Lewis Hamilton alongside former World Champion Jenson Button, gave Sergio vital experience of the workings of a top team: eleven points finishes, including four consecutive ones in the final four races in the season, set him up as a consistent driver, earning him a place in the Sahara Force India Formula One team line-up for 2014 and beyond.

Date of birth: 26 January 1990
Car Number: 11
Born; Guadalajara, Jalisco, Mexico
Lives in: Switzerland
Weight: 64 kg
Height: 1.73m
Marital status: Single

2014	F1 Force India, 10th place, one podium
2013	F1, McLaren, 11th place
2012	F1, Sauber F1, three podiums, 10th place
2011	F1, Sauber F1, 16th place
2010	2nd, GP2 Series, five wins, one pole, seven podiums
2009	12th, GP2 Series, two podiums; (GP2 Asia Series 2009-2010: four races)
2008	4th, British Formula Three, four wins, seven podiums
2007	1st, British Formula Three—National Class, 14 wins, 14 poles, 19 podiums
2006–7	A1GP, Team Mexico, two races
2005–6	Formula BMW ADAC, three podiums
2004	11th, Skip Barber National Championship, 77 points

MAX VERSTAPPEN

At the age of seventeen years and 166 days Max Verstappen had the opportunity to become the youngest driver to start a Formula One race at the 2015 Australian Grand Prix, breaking the record held by Jaime Alguersuari who made his debut at nineteen years and 125 days.

Verstappen, son of former Formula One driver Jos, is considered to be one of the most skilled young drivers of the new generation with the necessary maturity and mental strength to take on the challenge of Formula One.

He is a product of the Red Bull Junior Programme and became the youngest driver to participate at a Grand Prix weekend during the first free practice of the 2014 Japanese Grand Prix.

Verstappen began karting at the age of four-and-a-half and by the age of fifteen, after securing many titles in lower categories, won the 2013 World KZ Championship at Varennes-sur Allier in France, in KZ1, the highest karting category.

He made his single-seater debut in the inaugural Florida Winter Series before going on to join Van Amersfoot Racing to take part in the FIA Formula 3 Championship.

At the age of sixteen, Verstappen won his sixth race before going on to win ten races for the season including a run of six consecutive victories. He finished third in the overall rankings to Esteban Ocon and Ton Blomqvist.

In August of 2014 it was confirmed that Max would join Scuderia Toro Rosso in 2015 partnering Carlos Sainz Jr following Daniil Kvyat's promotion to Red Bull Racing making them the youngest pairing on the F1 grid with a combined age of just thirty-seven years.

Verstappen enters Formula One with a distinct lack of experience in the junior categories having not driven in Formula Renault or either GP2 or GP3, but is highly rated nonetheless.

Born: 30 September 1997
Hometown: Hasselt, Belgium

Nationality: Dutch
Car Number: 33

CAREER HIGHLIGHTS
2014 Masters of Formula 3 First
2014 Formula 3 Europe Third
2013 World Karting Championship KZ First
2013 World Karting Championship KF1 Third
2012 WSK Karting Master Series KF2 First
2012 World Karting Cup KF2 Second
2011 WSK Karting Euro Series KF3 First
2010 Bridgestone Karting Cup Europe KF3 First
2009 Belgium Karting Champioship KF5 First
2008 Belgium Karting Championship-Cadet First
2007 Rotax Max Challenge Belgium National Minimax First
2006 Rotax Max Challenge Belgium Minimax First
2005 Limburgs Kart Championship-Mini-Junior Second

CARLOS SAINZ JR

At seven years old Carlos junior began to drive in Karts at his father's Indoor Karting in Las Rozas, Madrid, but he started to compete seriously in the summer of 2005 when he was ten.

In 2006 racing for the teams Alemany, MGM and Benikart, he won the Cadete Category of the Madrid Championship, finishing third in the Industry Trophy (Parma-Italy) and second in the Champions Race at the end of the year.

During the three following years, he contested the Junior Category. In his first year with the Genikart Team, he won the City of Alcañiz Automobile International Trophy and finished third at Macau in the Asian-Pacific Championship.

In 2008 in the same category, he won the Asian-Pacific Championship at Macau and finished second in the Spanish Championship, being the best Spaniard classified in the Europe Championship, the German Championship, the Trophy Andrea Margutti (Italy) and the Monaco Kart Cup.

In 2009 he was part of the Tony Kart Junior Team and contested some races at Genikart Team, winning the famous

Monaco Kart Cup as well as the Classification for the West
European area (England, France, Switzerland, Portugal, Andorra
and Spain). He also finished second in the European
Championship's Final and second in the Spanish Championship,
adding a final third at WSK International Series.

In 2010, and after several tests, Red Bull chose him to be a
member of their Junior Team and after four days of testing he
won the BMW Scholarship following the qualifying competition.
At just fifteen years old he contested the Formula BMW European
Championship and took part in three races of the Formula BMW
Asian-Pacific Championship with the Antonio Ferrari's
Eurointernational team clinching the Formula BMW European
Championship. During the same season, he achieved victory in
the Formula BMW Asian-Pacific Championship held in Macau.
John Smith continued to support him in his career as they had
done in the previous season.

In the 2011 season his evolution as a driver continued as the
young Spaniard contested two of the most prestigious championships at European level: the Northern European Cup (NEC) and
the Eurocup Formula Renault 2.0

In 2012, the Red Bull Junior Team driver competed with
Carlin Motorsport in the British Formula 3 Championship and
the European FIA F3 earning in his debut year a brilliant victory
at the legendary Spa-Francorchamps circuit under a heavy
downpour. The wins did not stop there, as Carlos achieved four
victories in the British Formula 3 Series to finish in sixth place,
while in the continental competition he finished fifth. Carlos
ended his season at the prestigious Macau Grand Prix Formula
3, where despite qualifying in fourth a bad start left him with few
options to fight for the win as he ended the race in seventh. At the
end of 2012 the oil company Cepsa came on board as a sponsor
for all competitions.

In 2013, Carlos combined GP3 Series racing with several
meetings of the World Series. In an amazing debut at the Monaco
Grand Prix the Spanish driver fought for the top positions in his
first outing at the wheel of the powerful Formula Renault 3.5.
However his GP3 season was marked by misfortune and bad
luck, but the young Spaniard repeatedly demonstrated his talents

achieving two podiums. The key moment arrived in midsummer when Carlos Sainz sat behind the wheel of a Formula 1 car for the first time during the young driver test held at the Silverstone Circuit. The Spanish driver enjoyed a day with the Scuderia Toro Rosso car and also at the wheel of the Red Bull RB9, the premier car in the Championship.

In 2014, Carlos made history becoming the first Spanish driver and member of the Red Bull Junior Team to win the World Series by Renault. The Spaniard won the title for French team, DAMS Racing, to become the youngest driver to win the title, and in record-breaking fashion, winning seven races.

After driving in the Abu Dhabi test for Infiniti Red Bull Racing it was announced in November of 2014 that Carlos would join the Scuderia Toro Rosso completing the driver line-up with Max Verstappen.

International participations in single-seaters

Races	165
Wins	29
Podiums	61
Poles	30
Fastest Laps	36

2014

	Races	Wins	Podiums	Poles	Fastest Laps
World Series	17	7	7	7	6
TOTAL 2014	17	7	7	7	6

SUMMARY BY YEARS
2013

	Races	Wins	Podiums	Poles	Fastest Laps
GP3 Series	16	0	2	1	2
World Series	9	0	0	0	1
F3 Cup	2	2	2	2	2
Macau Grand Prix	2	0	0	0	0
TOTAL 2013	29	2	4	3	5

2012

	Races	Wins	Podiums	Poles	Fastest Laps
British Formula 3	25	5	9	2	2
European FIA F3	24	1	5	2	1
GP Masters F3	1	1	0	0	0
Macau Grand Prix	1	0	0	0	0
TOTAL 2012	51	6	14	4	3

2011

	Races	Wins	Podiums	Poles	Fastest Laps
Formula Renault 2.0	34	10	24	11	16
Formula 3	3	0	0	0	0
TOTAL 2011	37	10	24	11	16

2010

	Races	Wins	Podiums	Poles	Fastest Laps
Formula BMW	25	6	10	5	4
Formula 3	4	0	1	0	4
Formula Renault 2.0	6	0	1	0	2
TOTAL 2010	35	4	12	5	6

ROMAIN GROSJEAN

Romain made a return to Formula 1 with Lotus F1 Team in 2012 having previously contested the final seven Grands Prix of 2009 for the Enstone outfit (at the time known as Renault F1 Team). Prior to that, the Frenchman had secured the 2007 Formula Three Euroseries championship and won the inaugural GP2 Asia Series championship in 2008.

After his 2009 Formula 1 foray, Romain contested the inaugural FIA GT1 World Championship in 2010, winning the opening race of the season while also competing in his first 24 Hours of Le Mans. It wasn't long before he returned to single seaters, with partial campaigns in the Auto GP and GP2 Series championships that same year before taking both the GP2 Asia Series and GP2 Series championship titles in 2011.

Date of birth: 17 April 1986
Place of birth: Geneva, Switzerland
Nationality: French
Official Website: http://www.romaingrosjean.com/
Official Twitter Feed: https://twitter.com/RGrosjean
Official Facebook Fan Page: https://www.facebook.com/grosjeanromain
Grands Prix Contested: 45
Podiums: 9
Fastest Laps: 1

KEY DATES
2013: Coming of Age
Last year saw Romain establish himself as one of the strongest all-round drivers in Formula 1. His pace was harnessed and strengthened by an acute awareness of the tactical acumen required to get the best out of car and tyres in all conditions. A maiden victory looked to be very close more than once in a season where his performances just got better and better.

2012: BACK IN THE BIG TIME

Having impressed during his two test sessions the previous year, Romain was handed a return to the Formula 1 grid by Lotus F1 Team for the 2012 season. The team's faith in the Frenchman was soon justified, with a second row start at the season opening Australian Grand Prix followed by a string of impressive qualifying performances and three podium finishes throughout the season.

2010–11: REGROUP AND REFOCUS

Romain contested the inaugural FIA GT1 World Championship in 2010, winning the first Championship Race. In that same year, he also made a return to single seaters, scoring a first and second in his debut event in the Auto GP series and taking the title within three further races. In July he returned to the GP2 Series, which proved a precursor to a full-time return the following season. Late in the year, Romain also became Pirelli's test driver.

The following year, 2011, proved to be highly successful, with the GP2 Asia Series and the GP2 Series titles won and a return to the Formula 1 cockpit with Friday runs in Abu Dhabi and Brazil for Lotus Renault GP.

2009: GP2 SERIES, RENAULT F1 TEAM TEST AND RACE DRIVER

2009 was to prove to be a year of opportunity for Romain. He started the year as test driver for Renault F1 Team as well as contesting the GP2 Series. By the European Grand Prix he was a full-time race driver for the team, starting seven Grands Prix as teammate to Fernando Alonso.

2008: GP2 SERIES AND RENAULT F1 TEAM TEST DRIVER

In 2008, Romain combined his responsibilities as test driver for Renault F1 Team with campaigns in the GP2 Asia Series and GP2 Series. He won the Asia series, taking four wins along the way. Romain was also a frontrunner in the main Series, winning two races.

2006–07: F3 EUROSERIES
Romain finished thirteenth in the F3 Euroseries, which included two wins during the British F3 Championship rounds. For 2007, he took the title in impressive style in a closely fought series with a total of six wins, six podiums and four pole positions.

2004–05: FORMULA RENAULT 2.0
Romain completed partial seasons in both the French and European Formula Renault championships, finishing as second best rookie in the 2004 French championship and taking one win and three podiums along the way. He won the title with ten wins the following year.

2000–03: KARTING AND FORMULA RENAULT
Romain raced karts from junior categories through to ICA in addition to completing Formula A races in 2002. He combined Formula ICA in 2003 with the start of his car racing career, where ten wins from ten races saw the young Frenchman crowned Swiss Formula Renault champion.

PASTOR MALDONADO

Maldonado brings with him a wealth of experience and plenty of podium achievements following his success with his former Formula 1 team in 2012. Despite a tricky year in 2013, the career and accolades of the Venezuelan driver speak for themselves and the future looks extremely positive.

Date of birth: 9 March 1985
Age: 28
Place of birth: Maracay, Venezuela
Nationality: Venezuelan
Official Website: http://www.pastormaldonado.com/
Official Twitter Feed: https://twitter.com/pastormaldo
Official Facebook Fan Page: https://www.facebook.com/Pastor MaldonadoOficial
Grands Prix contested: 58

Podiums: 1
Wins: 1

KEY DATES
2011–13: RACING WITH WILLIAMS F1 TEAM
Fresh from his GP2 Series title success, Pastor made his full Formula 1 debut for the Williams team at the start of the 2011 season; it was a tough debut year and the Venezuelan took just a single World Championship point at the Belgian Grand Prix. The following season proved significantly more fruitful, with six top-ten finishes—including pole position and victory at the Spanish Grand Prix—establishing him as a strong points contender. After a frustrating 2013 campaign in largely uncompetitive machinery, Pastor made the switch to Enstone to partner Romain Grosjean at Lotus F1 Team.

2007–10: GP2 SERIES
Following a successful post-season test during the winter of 2006, Pastor began to attract interest from the GP2 Series paddock; opting for a switch to the category for the 2007 season alongside selected Euroseries 3000 races. The young Venezuelan quickly got to grips with his new machinery, taking pole position and victory in only his fourth race at Monaco, before a training injury curtailed his campaign with four rounds remaining. A string of race wins, podiums and impressive drives saw Pastor finish fifth and sixth respectively in 2008 and 2009, before finally clinching the GP2 Series crown in 2010 with a run of six consecutive feature race triumphs—setting a new record of ten series victories in the process.

2005–06: WORLD SERIES BY RENAULT
After a half-season in the category during 2005, Pastor secured a full-time WSR seat for the 2006 season in what would prove to be an impressive year for the young driver. Three wins, six podiums and five pole positions saw him finish third in the overall standings, and he was deprived of the title only at the final round having been stripped of a race victory on technical grounds.

2003–04: FORMULA RENAULT 2.0

Pastor's first foray into single-seater racing came at the wheel of a Formula Renault 2.0, competing in the Italian, German and European Championships. Highlights included a series victory in the 2003 Italian Winter Series, matched the following year in the full series. Alongside his regular racing commitments during the latter season, the young Venezuelan also acted as test driver to Minardi F1 Team (now Scuderia Toro Rosso).

MARCUS ERICSSON

RACING HIGHLIGHTS
2011 2nd GP2 Final—Abu Dhabi
2009 1st Formula 3 Japan
2007 1st Formula BMW UK

KARTING HIGHLIGHTS
2005 1st Swedish Championship—ICA Junior
2005 3rd Italian Open Masters ICA Junior
2005 1st Nordic Championship ICA Junior
2005 2nd Torneo Industrie ICA
2004 3rd Tom Trana Trophy—ICA Junior
2003 1st MKR series Sweden Formula Mini
2001 3rd MKR series Sweden Formula Micro

It all started when nine-year-old Marcus Ericsson turned up at the local rental go-kart track. First time in a go-kart he almost broke the lap record. Owner of the track was Fredrik Ekblom, himself an accomplished Swedish racing driver with experience from touring cars, Champ Car and Le Mans. Fredrik took notice and convinced Marcus's father, Tomas, to buy his son a kart and so the karting career begun.

What made this impressive is that Marcus does not come from a family with a motorsport background. But then again maybe it was not such a surprise after all as he hails from Kumla in Örebro, a motorsport region in the middle of Sweden from which many Swedish racing drivers have emerged. Maybe the most famous is Ronnie Peterson, the Swedish F1 legend from the seventies.

Such was Marcus's talent that Fredrik Ekblom recommended him to Fortec Motorsport team boss Richard Dutton. A Swedish Indy500 and IndyCar champion Kenny Bräck had also taken notice of the young talent at a go-kart race in Gothenburg. Both Ekblom and Bräck had been driving for Fortec and Kenny Bräck now helped to put a deal together for Marcus Ericsson to take the step up to cars with Fortec Motorsport in Formula BMW UK—just seven years after he first showed up at the local go-kart track.

At sixteen, Marcus made a sensational start to his motor racing career in Formula BMW, winning the championship in his rookie season with seven victories and eleven pole positions.

After that initial success Marcus took another big step and moved up with Fortec to Formula 3 the following year. He finished fifth overall in British F3 in 2008 and showed great speed during the season, gaining five podium finishes. For 2009, and at just eighteen years old, he took the bold decision to move to Japan for the Japanese F3 Championship with TOM'S (Tachi Oiwa Motorsport).

With five victories during the season he won the Japanese F3 Championship and claimed his second motorsport title in just three years. During the year he also made several guest appearances at his former British F3 hunting grounds and won at Rockingham and Hockenheim and also finished fourth overall at the Macau Grand Prix.

This set up a big move to GP2 where he towards the end of 2009 in only his third year of car racing raced in the GP2 Asia series. With Super Nova he moved to the GP2 series for 2010 and at Valencia he took his first GP2 victory. A switch to iSport for 2011 and 2012 resulted in another victory at Spa in 2012 during two seasons that showed bags of promise but that were plagued by bad luck.

In 2013 Ericsson joined reigning champion team DAMS (Driot-Arnoux Motorsport) for his fourth season in GP2. Despite being in the lead, and with several front row starts, bad luck and unfortunate circumstances led to a disastrous start to the season. But in the second half of the campaign he was the ace in the pack and made a brilliant comeback to finish sixth overall after victory at Nürburgring as well as another five podiums and two pole positions.

At this point, at twenty-three years old and only fourteen years after he first showed up at Ekblom's go-kart track, Marcus joined Caterham to make his Formula 1 debut and to become Sweden's first F1 driver since 1991. He made sixteen starts, finishing nineteenth overall.

At the US Grand Prix in 2014, the Sauber F1 Team announced it had signed Ericsson for the 2015 campaign.

FELIPE NASR

Nationality: Brazilian
Born: 21 August 1992
Job description: Test and Reserve Driver
Height: 1.75 m
Weight: 67 kg

CAREER HIGHLIGHTS
2014—Test and Reserve Driver, Williams
2013—GP2 Series with Carlin Motorsport. 4th in Championship standings
2012—GP2 Series with DAMS. 10th in Championship standings with 4 podiums. 3rd in Rolex 24 Hours at Daytona
2011—British Formula 3 Championship Winner with 7 wins and 17 podiums. Winner of the Sunoco Daytona Challenge
2010—British Formula 3 Championship, 5th in standings with 1 win and 4 podiums
2009—Formula BMW Europe Champion with 7 wins and 14 podiums
2008—Formula BMW Americas, competed in 2 races with 1 podium finish
2000–2007—Karting, 6-time Brazilian Champion.

Felipe began karting in Brazil at the age of seven and won several national titles before making his open-wheel racing debut in 2008 in the final round of the Formula BMW Americas season. For 2009 he moved to Europe and competed in the Formula BMW Championship, claiming the title in his first season with seven wins in sixteen starts. In 2010 and 2011 he competed in the British Formula Three Championship, finishing in fifth position in his first year and claiming the title in 2011 after securing seven wins.

Felipe moved to the GP2 Championship with DAMS for 2012 where he finished the championship as the second-highest-placed rookie and in tenth place overall. That year he also competed in the Daytona 24 Hour, finishing third overall with Michael Shanks

Racing. For the following season Felipe joined Carlin Motorsport, finishing in fourth place in the GP2 Championship.

For 2014 Felipe reached Formula One as the official Test and Reserve Driver for Williams. He drove the Williams Mercedes FW36 in three testing and five FP1 sessions during the campaign and conducted simulator testing at the Williams factory in Grove, Oxfordshire.

At the end of 2014 it was announced that Nasr had signed for the Sauber F1 team.

CIRCUITS

AUSTRALIA

THE AUSTRALIAN GRAND PRIX
13–15 March 2015

The Formula One season traditionally begins with Melbourne hosting the Australian Grand Prix, the first round of the FIA Formula One World Championship. The 2015 race marks the twentieth anniversary of the event being held at the street circuit in Albert Park and the sixty-sixth season of the Championship. The race is a firm favourite with teams and drivers as well as the fans who welcomed the news that the contract to host the race in the Victorian capital was extended in 2013 until 2020.

The temporary track around the Albert Park Lake provides a stern challenge for drivers and cars alike. As is traditional with street circuits the surface lacks the grip of a permanent racing facility. This leads to high track evolution as the circuit "rubbers-in" over the weekend making set-up a moving target. The circuit also has various bumps and undulations that have been known to catch drivers out, particularly coming into braking zones, with gravel traps and unyielding walls waiting for the unwary or the unlucky.

Being the first race of the season the teams are still on a learning curve with their new equipment. They acknowledge that the first Grand Prix of the season is a slightly unpredictable voyage into the unknown with many honing the capabilities of their race-specification cars during the practice sessions.

The race also represents the first opportunity for tyre supplier

Australian GP
ROUND 01

RACE DATE .... 15 March 2015
CIRCUIT NAME .... Albert Park Circuit
NO. OF LAPS 58
START TIME .... 17.00 local/6.00 GMT
CIRCUIT LENGTH 5.303km
RACE DISTANCE 307.574k
LAP RECORD 1:24.125
                     Michael Schumacher
                     (2004)

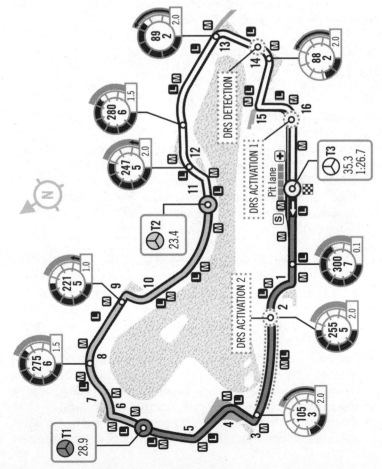

Legend

Speed kmh / Braking / Lateral G-force / Gear
- 150
- 3
- -1.0

Timing sector / Sector time / Lap time
- T3
- 35.3
- 1:26.7

Sector 1
Sector 2
Sector 3

- (S) Safety car
- (+) Medical car
- (M) Marshals

- Circuit
- Start
- Finish
- (L) Light panels
- Run-off areas
- Gravel traps

DRS DETECTION
DRS ACTIVATION 1
DRS ACTIVATION 2
Pit lane

T1 — 28.9
T2 — 23.4
T3 — 35.3 / 1:26.7

Turn data:
- 89 / 2 / 2.0
- 88 / 2 / 2.0
- 280 / 6 / 1.5
- 247 / 5 / 2.0
- 221 / 5 / 1.0
- 275 / 6 / 1.5
- 105 / 3 / 2.0
- 255 / 5 / 2.0
- 300 / 7 / 0.1

Pirelli to showcase their new range of tyres in a competitive situation. The new compounds and constructions that were developed last season and into winter testing can be put to the test with the key as always being to cut down on degradation while maintaining the same level of performance.

The race also marks the return to Formula One of Honda who renew their successful relationship with McLaren as an engine supplier following on from their partnership from 1988 to 1992, a period that garnered four world titles.

2015 would see a Hamilton/Rosberg Mercedes one-two with Vettel 34 seconds behind the winner. Sauber would pick up an impressive fourteen points, with Felipe Massa in 5th.

CIRCUIT DATA—ALBERT PARK CIRCUIT
Length of lap: 5.303 km
Lap record: 1:24.125 (Michael Schumacher, Ferrari, 2004)
Start line/finish line offset: 0.000 km
Total number of race laps: 58
Total race distance: 307.574 km
Pitlane speed limits: 60 km/h in practice, qualifying, and the race

CIRCUIT NOTES
- The kerb on the exit of Turn Twelve has been levelled and renewed (with the same design of kerb).

DRS ZONE
DRS sectors are the same as those used in 2014. Activation One is 762 m before Turn One, Activation Two is 510 m before Turn Three. They share a single detection point, located 13 m before Turn Fourteen.

AUSTRALIAN GRAND PRIX FAST FACTS
- Albert Park first hosted an F1 World Championship Grand Prix in 1996. It twice held the (Formula Libre) Australian Grand Prix in the 1950s—the 1956 edition being won by Stirling Moss driving a Maserati 250F.
- Prior to the Melbourne event the race was held in Adelaide, also on a street circuit, from 1985 to 1995.

- Michael Schumacher has the most wins, with four, followed by Jenson Button with three while McLaren is the most successful constructor with twelve wins to date.
- Of the twenty F1 races held at Albert Park, thirteen have been won from the front row. David Coulthard's victory from eleventh in 2003 is the win from furthest back; next is Kimi Räikkönen's surprise victory in 2013 from seventh on the grid.
- Before 2015 (Rosberg won for Mercedes in 2014), the last back-to-back constructor victories were by Renault in 2005 and 2006 with each of the other nine races being won by a different team.

MALAYSIA

2015 MALAYSIAN GRAND PRIX
27–29 March 2015

Equatorial Sepang provides a very different environment to the previous round in Albert Park, Melbourne, and is always liable to produce a very different race as machinery and drivers are pushed to their limits.

The Malaysian Grand Prix has the potential to be the hottest race of the year and, with around two-thirds of the lap conducted at full throttle, cooling performance is expected to play a big part.

Added to that are the traditional challenges of racing at Sepang. While visually dominated by the long pit and back straights, Sepang has great variety, with a twisting middle sector and several high-speed corners requiring a relatively high level of downforce. It also demands excellent traction, with speed onto the straights being a key factor in setting a good lap time.

It is tyres, however, that have a tendency to dictate performance at the Malaysian Grand Prix. Pirelli will usually favour its two hardest compounds but wear rates are still expected to be high. Of course in Malaysia the ever-present threat of rain means the Intermediate and Wet tyres could also see some use. The forecast is normally for thunderstorms across all three days at some stage although the rain has historically bracketed the sessions rather than fallen during them.

CIRCUIT DATA—SEPANG CIRCUIT
Length of lap: 5.543 km
Lap record: 1:34.223 (Juan Pablo Montoya, BMW-Williams, 2004)
Start line/finish line offset: 0.000 km
Total number of race laps: 56
Total race distance: 310.408 km
Pitlane speed limits: 80 km/h in practice, qualifying and the race.

Malaysian GP
ROUND 02

RACE DATE ....29 March 2015
CIRCUIT NAME ......Sepang Circuit
NO. OF LAPS.........56
START TIME..........16.00 local/08.00 GMT
CIRCUIT LENGTH....5.543km
RACE DISTANCE 310.408km
LAP RECORD.........1.34.223
                    Juan Pablo Montoya
                    (2004)

Timing sector
Sector time
Lap time

T3
38.6
1.35.1

Speed kmh
Braking
Lateral G-force
150
3
-1.0
Gear

Circuit
Start
Finish
Light panels

Sector 1
Sector 2
Sector 3

Safety car [S]
Medical car [+]
Marshals [M]

Run-off areas
Gravel traps

N

96
2
2.0

260
6
0.1

200
4
-3.0

DRS ACTIVATION 2

285
6
-0.1

170
4
-2.2

110
3
-1.0

15

7

8

6

9

10

T2
31.7

11

5

Pit lane

DRS DETECTION 2

4

T3
38.6
1.35.1

DRS ACTIVATION 1

12

255
5
2.0

13

DRS DETECTION 1

3

14

270
6
1.5

80
2
-2.0

1

2

300
7
0.1

200
4
-3.0

CIRCUIT NOTES

- The kerb on the exit of Turn Five has been extended further towards the apex of Turn Six.
- Artificial grass has been removed from areas around the outside of Turns One, Five, Seven, Eight and Twelve.
- A new kerb has been installed on the Turn Fifteen entry.
- Attention has been paid to levelling the grass verges as bumps were noted during December's inspection.

DRS ZONE

The DRS sectors will be between Turns Fourteen and Fifteen and Turns Fifteen and One. The first zone has detection 54 m after Turn Twelve and activation 104 m after Turn Fourteen. Zone two has detection 16 m after the Turn Fifteen apex followed by activation 28 m after Turn Fifteen.

MALAYSIAN GRAND PRIX FAST FACTS

- Michael Schumacher (2000, '01, '04), Fernando Alonso (2005, '07, '12) and Sebastian Vettel (2010, '11, '13), each have three victories at Sepang. From fifteen Malaysian Grands Prix the other winners are Eddie Irvine (1999), Ralf Schumacher (2002), Kimi Räikkönen (2003, '08), Giancarlo Fisichella (2006), Jenson Button (2009) and Lewis Hamilton (2014). Alonso has the distinction of taking his three victories with three different teams.
- Button's win in 2009 was the last Formula One race in which half-points were awarded. Due to torrential rain, the race was red flagged after thirty-one laps and a restart was not possible. Button was awarded the victory and the five points that went with it.
- Eight of the fifteen grands prix at Sepang have been won from pole. The race has been won from as far down as eighth (Alonso 2012), while podium finishers have come from as far down as tenth (Nick Heidfeld 2005, '09).

CHINA

2015 CHINESE GRAND PRIX
10–12 April 2015

The Shanghai International Circuit, located in the Jiading district of one of China's most vibrant cities, presents a markedly different set of challenges to those offered up in Bahrain. Gone is the high heat and the point and squirt nature of the BIC's layout. Instead, Shanghai offers up cooler climes, two long straights and a tough mix of slow, medium and high-speed corners, all of which test a team's ability to find a good balance for their cars.

It is a layout that in the past has given tyres a good workout—particularly rear tyres. That issue was exacerbated in 2014 by the torque available from the new power units and the reduced aerodynamic grip available under the new regulations. It is a circuit where race strategy can play a major role and in particular making the best possible combined use of the available tyre compounds.

CIRCUIT DATA—SHANGHAI INTERNATIONAL CIRCUIT
Length of lap: 5.451 km
Lap record: 1:32.238 (Michael Schumacher, Ferrari, 2004)
Start line/finish line offset: 0.190 km
Total number of race laps: 56
Total race distance: 305.066 km
Pitlane speed limits: 80 km/h in practice, qualifying and the race.

CIRCUIT NOTES
- Other than routine maintenance no changes have been made to the circuit since 2014.

DRS ZONES
The DRS sectors at the Shanghai International Circuit will be as last year. The detection point of the first zone is at Turn Twelve and the activation point is 752 m before Turn Fourteen. The

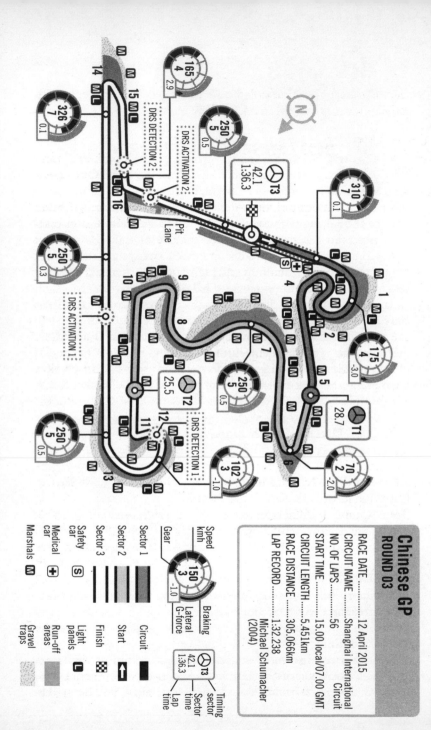

Chinese GP
ROUND 03

RACE DATE .... 12 April 2015
CIRCUIT NAME .... Shanghai International Circuit
NO. OF LAPS 56
START TIME .... 15.00 local/07.00 GMT
CIRCUIT LENGTH 5.451km
RACE DISTANCE 305.066km
LAP RECORD 1:32.238
Michael Schumacher (2004)

second zone's detection point is 35 m before Turn Sixteen, with activation occurring 98 m after Turn Sixteen.

CHINESE GRAND PRIX FAST FACTS
- This will be the eleventh running of the Chinese Grand Prix, the race having been added to the Formula One calendar in 2004.
- The first five editions of the race took place towards the end of the campaign, with the 2005 race being the season-ending event. At that race, Renault wrapped up its first Constructors' Championship title as a manufacturer thanks to Fernando Alonso's win and Giancarlo Fisichella's fourth place. The race moved to the front end of the season in 2009 in an April slot that has become now traditional for the event.
- The winner of the inaugural race was Rubens Barrichello, who took his ninth career victory in Shanghai for Ferrari. It would be the Brazilian's last win for five years. Barrichello's tenth win came when driving for Brawn GP at the 2009 European Grand Prix in Valencia, some 1,793 days after his Chinese Grand Prix win.
- Lewis Hamilton and Fernando Alonso are the only multiple winners. Alonso won the 2005 race for Renault and the 2013 event for Ferrari. Both of Hamilton's first two wins were for McLaren in 2008 and 2011, which he added to with last year's victory for Mercedes.
- The other winners here are: Michael Schumacher in 2006; Kimi Räikkönen (2007); Sebastian Vettel (2009); Jenson Button (2010) and Nico Rosberg (2012). Vettel's win here was his first for Red Bull Racing, while Rosberg's was, of course, his maiden Grand Prix victory. Rosberg's 2012 victory was Mercedes' first Grand Prix win as a manufacturer since the 1955 Italian event when Juan Manuel Fangio took victory ahead of Piero Taruffi, also driving for the three-pointed star.
- Ferrari are the most successful constructors at this race, with four victories (Barrichello 2004, Schumacher '06, Räikkönen '07 and Alonso '13). McLaren are the Italian

squad's closest rivals, with three successes (Hamilton 2008 and '11 and Button '10). Mercedes have two wins while Renault and Red Bull Racing and have one win each.

- Six of the eleven events held so far have been won from pole position—in 2004, '05, '08, '09, '12 and '14. Michael Schumacher's 2006 victory was from the furthest back on the grid so far; the Ferrari driver started sixth. The man in pole position has only failed to finish on the podium twice in the ten runnings to date. Sebastian Vettel finished sixth in 2010, and in 2007 Lewis Hamilton famously crashed out at the pit lane entrance.

BAHRAIN

2015 BAHRAIN GRAND PRIX
17–19 April 2015

In 2014 the Bahrain International Circuit joined Singapore and Abu Dhabi in staging a race fully or partially at night and the 6 p.m. start changed how the race was approached. For instance, the cooler temperatures of the evening led to changes in balance and grip levels. The bonus, too, is that the lower temperatures place less stress on the cooling capabilities of the power units.

Elsewhere, the Bahrain track features a mix of slow-speed corners at the end of straights, which means that the BIC is one of the most severe on brakes all season. However, this also means that tyres take some punishment, especially in terms of longitudinal energy going into the tyres.

The event in 2014, the 900th Grand Prix since the start of the World Championship, produced a stunning race. The spectacular wheel-to-wheel battle between Lewis Hamilton and Nico Rosberg was hailed as a classic that well and truly put to rest any criticism of the new regulations.

CIRCUIT DATA—BAHRAIN INTERNATIONAL
CIRCUIT
Length of lap: 5.412 km
Lap record: 1:31.447 (Pedro De la Rosa, McLaren, 2005)
Start line/finish line offset: 0.246 km
Total number of race laps: 57
Total race distance: 308.238 km
Pitlane speed limits: 80 km/h in practice, qualifying and the race.

DRS ZONES
The DRS sectors at the Bahrain International Circuit will be as last year. The detection point of the first zone is 10 m before Turn Nine and the activation point is 50 m after Turn Ten. The second zone's detection point is 108 m before Turn Fourteen, with activation occurring 270 m after Turn Fifteen.

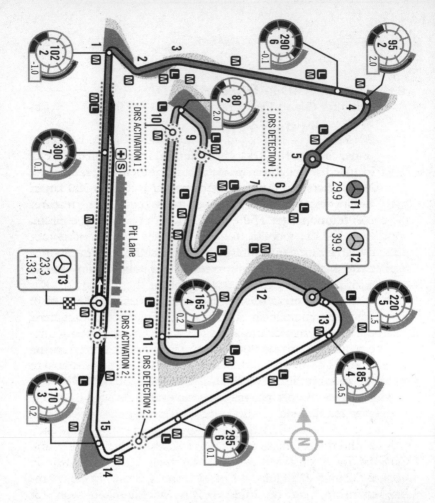

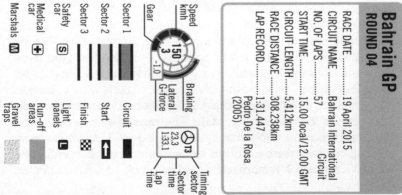

Bahrain GP
ROUND 04

RACE DATE ....19 April 2015
CIRCUIT NAME ....Bahrain International
Circuit
NO. OF LAPS 57
START TIME............15.00 local/12.00 GMT
CIRCUIT LENGTH......5.412km
RACE DISTANCE.......308.238km
LAP RECORD...........1:31.447
Pedro De la Rosa
(2005)

Speed
kmh

Gear

Sector 1
Sector 2
Sector 3
Finish
Start
Circuit

Safety
car
Medical
car

Marshals M

Braking
Lateral
G-force

Light
panels
Run-off
areas

Gravel
traps

Pit Lane

DRS ACTIVATION 1
DRS DETECTION 1
DRS ACTIVATION 2
DRS DETECTION 2

Timing
sector
Sector
time
Lap
time

BAHRAIN GRAND PRIX FAST FACTS
- The first Grand Prix took place on 4 April 2004 and was won by Michael Schumacher.
- 2015's race is the second Bahrain Grand Prix to be staged under lights. In 2014 the Bahrain International Circuit installed 495 lighting poles around the circuit, ranging in height from ten to forty-five metres. More than 500 km of cabling was installed to power the system and it features 5,000 luminaries. The entire lighting project at the circuit took just six months to complete. This is F1's third race under lights. Singapore has hosted a full night race since 2008 and Abu Dhabi's race begins in twilight and ends in darkness.
- The first corner at the BIC was in 2014 named in honour of Michael Schumacher. As well as winning the first race here, the seven-times champion advised the Grand Prix organizers during construction phase of the circuit. After retiring from F1 at the end of 2006, Schumacher made his return to the sport at the 2010 Bahrain GP.
- Fernando Alonso has the most victories here, with three. He won for Renault in 2005 and 2006 and then for Ferrari in 2010. Sebastian Vettel and Felipe Massa have two wins each at the track. Massa won back-to-back events in 2007 and '08 for Ferrari, while Red Bull Racing driver Vettel won in 2012 and 2013. Jenson Button won for Brawn in his championship year of 2009 and Lewis Hamilton won for Mercedes last year.
- Vettel and Schumacher as well as Rosberg are the only multiple pole position winners here. Schumacher was on pole at the first race and again in 2006, Vettel started from the front in 2010 and 2011 and Rosberg in 2013 and 2014. The other pole position men are: Massa ('07), Robert Kubica ('08) Jarno Trulli ('09). Kubica's pole was the only one of his F1 career to date. The race has never been won from further back than fourth on the grid. Alonso won in 2006 from the back of the second row, as did Button in '09.
- Victories for Schumacher, Massa (two) and Alonso mean

that Ferrari is the most successful constructor here with four wins. Renault (2005 and 2006) and Red Bull Racing (2012, 2013) are the next closest challengers.

SPAIN

2015 SPANISH GRAND PRIX
8–10 May 2015

Barcelona is famed as a bellwether circuit: cars that race well here are expected to race well in any and all conditions. The logic behind this is that the Circuit de Catalunya provides a well-rounded examination of every aspect of car design, requiring maximum downforce, strong power delivery, excellent traction and handling.

Particularly interesting is how tyre compounds cope with the conditions. The Circuit de Catalunya has a combination of high-energy corners, an abrasive surface and only one long straight on which tyres can cool. In warm weather degradation is expected to be high, with the left side of the car particularly vulnerable.

Barcelona is the race venue at which teams traditionally reveal their first major upgrade of the season and most teams circle this race for something substantial.

CIRCUIT DATA—CIRCUIT DE CATALUNYA
Length of lap: 4.655 km
Lap record: 1:21.670 (Kimi Räikkönen, Ferrari, 2008)
Start line/finish line offset: 0.126 km
Total number of race laps: 66
Total race distance: 307.104 km
Pitlane speed limits: 80 km/h in practice, qualifying and the race.

CIRCUIT NOTES
- The kerb on the approach to Turn One has been renewed and the verge behind it is now laid with asphalt.
- The kerbs on the exit of Turns Nine and Twelve have been renewed and the artificial grass behind them extended.
- A seven-metre-wide strip of gravel around the outside of Turn Eleven has been replaced with asphalt.

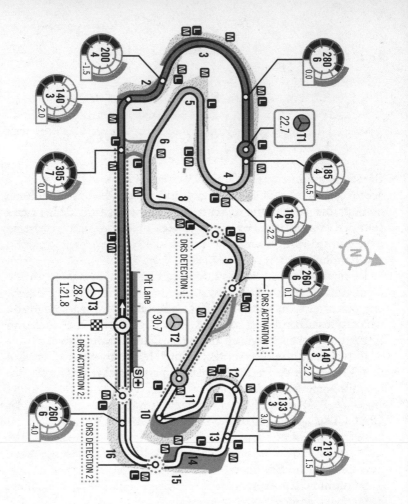

Spanish GP
ROUND 05

RACE DATE .... 10 May 2015
CIRCUIT NAME .... Circuit de Catalunya
NO. OF LAPS 66
START TIME .... 14.00 local/12.00 GMT
CIRCUIT LENGTH 4.655km
RACE DISTANCE 307.104km
LAP RECORD 1:21.670
Kimi Räikkönen
(2008)

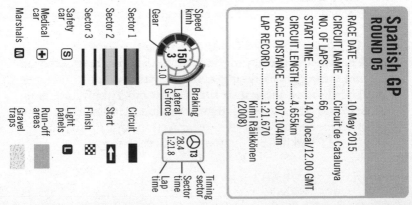

Speed
kmh

Gear

Sector 1
Sector 2
Sector 3

Safety
car   S

Medical
car   +

Marshals M

Braking

Lateral
G-force

Circuit

Start

Finish

Light
panels   L

Run-off
areas

Gravel
traps

Timing
sector

Sector
time

Lap
time

DRS ZONES

Two DRS zones exist at the Circuit de Catalunya. The first has a detection point 86 m before Turn Nine and an activation point 40 m after. DRS detection point two is at the Safety Car line, with activation 157 m after Turn Sixteen.

SPANISH GRAND PRIX FAST FACTS

- The F1 World Championship Spanish Grand Prix has been held at the Circuit de Catalunya every year since the circuit first opened in 1991. The race has also been held at Jerez (1986–90), Jarama (1968, '70, '72, '74, '76-79, '81), Montjuïc (1969, '71, '73, '75) and the Pedralbes Street Circuit (1951, '54).

- Barcelona is the most familiar circuit for F1 teams and, after a break in 2014, they arrive for the Spanish Grand Prix with full testing data from the Circuit de Catalunya with 2014 being the only time it has not featured as a winter testing venue.

- Jenson Button holds the record for the number of test kilometres completed at this circuit with a mighty 34,706 km. The record for most days testing completed, however, belongs to David Coulthard with 118.

- Pole position is a priority at this circuit. On seventeen of the twenty-three occasions the Spanish Grand Prix has been held at the Circuit de Catalunya, the driver starting in pole position has won the race. Damon Hill, Mika Häkkinen, Nigel Mansell, and Sebastian Vettel have won from P2, Michael Schumacher from P3 and Fernando Alonso in 2013 from P5.

- Alonso is the only Spanish winner of the F1 World Championship Spanish Grand Prix. He raced to victory in 2013 for Ferrari, using a four-stop strategy to advance from his third-row grid slot. Alonso, driving for Renault, also won the race in 2006, on that occasion starting from pole position.

MONACO

2015 MONACO GRAND PRIX
21–24 May 2015

Glamorous, thrilling and unlike any other race on the calendar, the Monaco Grand Prix is truly a race apart and always one of the most fascinating of the season.

With the circuit featuring the lowest average speed of the season (158 kph/98 mph) the Circuit de Monaco levels the power playing field and, with limited overtaking opportunities, the importance of qualifying is paramount.

The ability of teams to be effective in Monaco rests to a significant degree on their ability to meet the circuit's need for high downforce. Teams always bring one-off developments to this race in a bid to obtain maximum aerodynamic grip. In terms of mechanical grip, Monaco's bumpy streets make car set-up tricky but a softer suspension helps in ensuring a good contact patch from tyres and thus improved grip.

Elsewhere, cooling could be another point of interest. Heat has been an issue at several previous races with cooling opportunities hard to come by on the tight streets of the Principality.

These are just some of the variables at play on the race weekend. How teams respond to the low-speed nature of the track in their set-ups, the closeness of the barriers, the risk of costly driver error and the high probability of a Safety Car appearance make Monaco an exciting mix of possibilities.

CIRCUIT DATA—CIRCUIT DE MONACO
Length of lap: 3.340 km
Lap record: 1:14.439 (Michael Schumacher, Ferrari, 2004)
Start line/finish line offset: 0.000 km
Total number of race laps: 78
Total race distance: 260.520 km
Pitlane speed limits: 60 km/h in practice, qualifying and the race.

Monaco GP
ROUND 06

RACE DATE ....24 May 2015
CIRCUIT NAME ....Circuit de Monaco
NO. OF LAPS.........78
START TIME.........14.00 local/12.00 GMT
CIRCUIT LENGTH.....3.340km
RACE DISTANCE 260.520km
LAP RECORD 1:14.439
Michael Schumacher
(2004)

Timing sector
Sector time
Lap time

T3
37.4
1:15.7

Speed kmh
Braking
Lateral G-force
150
3
-1.0
Gear

Sector 1
Sector 2
Sector 3

Safety car — S
Medical car — ✛
Marshals — M

Circuit
Start
Finish
Light panels — L
Run-off areas
Gravel traps

DRS DETECTION

DRS ACTIVATION

T1
19.8

T2
18.5

T3
37.4
1:15.7

CIRCUIT NOTES

- The track has been resurfaced from the exit of the Casino until the start of the tunnel. Additionally, small sections before the Nouvelle Chicane and Tabac (Turn Twelve) have been resurfaced.
- The entire pit wall and debris fence has been renewed.
- The TecPro barrier at Turn Twelve has been more efficiently constrained.

DRS ZONES

There is a single DRS zone in Monaco, with the detection point located 80 m after Turn Sixteen and the activation point located 18 m after Turn Nineteen.

MONACO GRAND PRIX FAST FACTS

- This years's event will be the sixty-second Grand Prix of the Formula One era. The race dates back to 1929, however, when the first event was won by William Grover-Williams.
- The driver with the most wins here is Ayrton Senna, with six. His first Monaco victory came in 1987 at the wheel of a Lotus 99T. Two years later he began a streak of five straight wins in the Principality, all at the wheel of McLaren machinery.
- The next most successful drivers at the circuit are Graham Hill and Michael Schumacher, both of whom scored five wins. Hill scored a hat-trick of wins between 1963 and 1965 and then landed back-to-back victories in 1968 and 1969. Schumacher began with a double in 1994 and 1995 and then won in 1997, 1999 and 2001.
- The record for pole positions in Monaco is also held by Senna. The Brazilian started from the front of the grid five times: in 1985, '88, '89, '90 and '91. The next most successful in qualifying are Juan Manuel Fangio, Jim Clark, Jackie Stewart and Alain Prost with four poles each.
- On the current grid there are six Monaco winners: Kimi Räikkönen (2005), Fernando Alonso (2006 and 2007), Lewis Hamilton (2008), Jenson Button (2009), Sebastian Vettel (2011) and Nico Rosberg (2013 and 2014). Alonso

scored his wins with two different teams, Renault in 2006 and McLaren in 2007.

- With few overtaking opportunities, grid position is all-important here. On the evidence of the past ten Grands Prix that would appear to be the case, with the race being won from further back than pole just once—Lewis Hamilton claiming victory from third on the grid in 2008. However, look back at the ten races prior to that period and the race was won from pole just twice—Michael Schumacher in 1994 and Mika Häkkinen in 1998.

- Olivier Panis holds the record for victory from the lowest starting position. In 1996, he won the rain-lashed race from fourteenth on the grid. You have to go back to 1970 to find the next lowest-starting winner, Jochen Rindt, who won from eighth. Prior to that, the 1955 race was won by Maurice Trintignant from ninth.

- 2015 marks the thirty-first anniversary of McLaren's first win in Monaco. Alain Prost claimed victory ahead of Ayrton Senna and Stefan Bellof (whose was later disqualified) when the race was red-flagged after thirty-one laps because of heavy rain. Since then the team has won on fourteen other occasions making it the most successful here, with Ferrari in second with eight wins.

CANADA

2015 CANADIAN GRAND PRIX
5–7 June 2015

Following Monaco, Montreal's Circuit Gilles Villeneuve is another temporary track with unforgiving walls millimetres off the racing line—but that's where the similarities end. From the slowest race of the year, F1 moves to one of its fastest and teams unleash medium-low downforce packages in an attempt to stay competitive on the long straights of the Île Notre-Dame.

In essence, the long, thin circuit is a series of high-speed straights linked by slow corners. The start-stop nature of the lap, in which cars may hit more than 300km/h on four separate occasions before braking down to first or second gear, has long been recognised as exceptionally harsh on brakes and engines, not to mention the added demands of the MGU-K, which will have to deal with heavy braking loads, and the MGU-H which will be kept busy with unrelenting demand from the turbocharger.

Cars are set up for high top speeds, but the demands of the three chicanes and the hairpin prevent use of ultra-low downforce packages. There is much fine-tuning to be done as teams seek to find the right balance between low downforce and good stability in the all-important braking zones. With the added requirements of riding the kerbs well and getting good traction from low speed, Circuit Gilles Villeneuve has plenty to keep engineers occupied.

CIRCUIT DATA—CIRCUIT GILLES-VILLENEUVE
Length of lap: 4.361 km
Lap record: 1:13.622 (Rubens Barrichello, Ferrari, 2004)
Start line/finish line offset: 0.000 km
Total number of race laps: 70
Total race distance: 305.270 km
Pitlane speed limits: 80 km/h in practice, qualifying and the race.

Canadian GP
ROUND 07

RACE DATE ....7 June 2015
CIRCUIT NAME ....Circuit Gilles Villeneuve
NO. OF LAPS 70
START TIME ....14.00 local/18.00 GMT
CIRCUIT LENGTH 4.361km
RACE DISTANCE 305.270km
LAP RECORD...........1.13.622
Rubens Barrichello
(2004)

Speed kmh — Braking — Timing sector
Gear — Lateral G-force — Sector time — Lap time

Sector 1 — Circuit
Sector 2 — Start
Sector 3 — Finish

S Safety car — Light panels
✚ Medical car — Run-off areas
M Marshals — Gravel traps

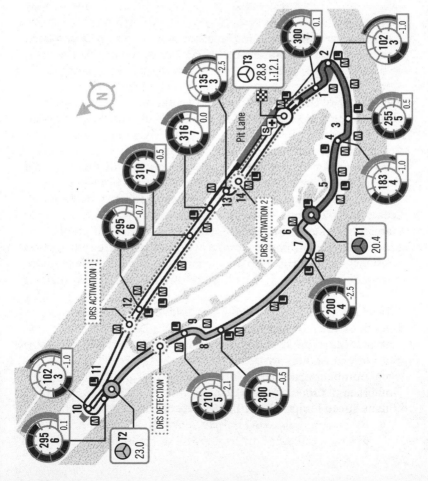

CIRCUIT NOTES
- Asphalt has replaced gravel around the outside of Turns Ten and Thirteen.
- The barrier on the outside of Turn Thirteen (right of the pit entry) has been moved further back and a new debris fence has been installed.
- New guardrail posts have been added in order to ensure that no spacing is greater than two metres.
- The speed bumps used in 2013 at the exit of Turn Nine and the final chicane will be in place again this year.

DRS ZONES
There are two DRS zones in Canada. They share a single detection point, located 110 m after Turn Nine. The first activation point is 55 m before Turn Twelve, the second 70 m after Turn Fourteen.

CANADIAN GRAND PRIX FAST FACTS
- McLaren lead the way with thirteen F1 World Championship victories at the Canadian Grand Prix. Ferrari are second with eleven.
- Michael Schumacher is the most successful driver in the history of the race, winning at the Circuit Gilles Villeneuve on seven occasions. Of the current field, Lewis Hamilton has the best record with three victories.
- Hamilton triumphed in 2007, 2010 and 2012. The first of those was his debut victory in F1. Of the eight other grand prix winners in the current field, he is one of five to have taken a maiden victory on the same weekend as a maiden pole position—the others being Felipe Massa (Turkey 2006), Sebastian Vettel (Italy 2008), Nico Rosberg (China 2010) and Pastor Maldonado (Spain 2010).
- Mosport Park was the original venue in 1967 and also staged the race in 1969, 1971–74 and 1976–77. The Mont-Tremblant circuit held the race in 1968 and 1970. The first Grand Prix at the track now known as the Circuit Gilles Villneuve was held in 1978.
- In total there have been forty-four Canadian Grands Prix—the race did not run in 1975, 1987 nor 2009. A dispute

between sponsors led to the race being cancelled in 1987 but the circuit made use of the hiatus, moving the pitlane to its present home from an original location after the hairpin. Other significant changes that have brought the circuit to its present state include the gradual straightening of the Casino Straight and a tightening of the chicane to lower speeds (1991) plus a relocation of the pitlane exit to prevent cars merging directly onto the racing line (2002).

- Jenson Button won the race in 2011 in a time of 4 hours 4 minutes 39.537seconds. This stands as the record for the longest (duration) race in the history of the Formula One World Championship and was caused by a two-hour suspension brought about by torrential rain. That race also holds the records for the most appearances of the Safety Car (six) and—technically—the lowest average race-winning speed (74.864 kph).

- The duration of that race led to an amendment to the F1 Sporting Regulations. A limit of four hours duration has been placed on a race interrupted by red flag suspensions. At the four-hour mark, the chequered flag will be waved the next time the leader crosses the finishing line.

AUSTRIA

2015 AUSTRIAN GRAND PRIX
19–21 June 2015

Prior to the 2014 race, the previous incarnation of the Austrian Grand Prix was held from 1997 to 2003 at the circuit known as the A1 Ring though the facility has seen major trackside updates since.

The track, however, remains much as it was in 2003. One of the shortest on the calendar at 4.326 km, the "Red Bull Ring" features just nine corners. It is a circuit of fast straights and slow-to medium-speed corners with, historically, just the first three taken in low gear. As such, lap times are low—with sub-1 minute 10 second laps the norm. Average speeds are high, with the 2003 event seeing Michael Schumacher win with an average speed of 213 km/h.

The race in 2014 was won by Nico Rosberg of Mercedes in 1:27:54.976 from third on the grid with an average speed of 215 km/h, while Sergio Pérez set the fastest lap time of 1:12.142.

CIRCUIT DATA—RED BULL RING
Length of lap: 4.326 km
Lap record: 1:08.337 (Michael Schumacher, Ferrari, 2003)
Start line/finish line offset: 0.126 km
Total number of race laps: 71
Total race distance: 307.020 km
Pitlane speed limits: 80 km/h in practice, qualifying and the race.

CIRCUIT NOTES
- This is Formula One's second visit to the circuit since 2003 and there were many upgrades made in the intervening years with the circuit being fully refurbished for the 2014 race.

DRS ZONES
There are two DRS zones in Austria. The detection point of the first zone is 360 m before Turn Two with the activation point 85

Austrian GP
ROUND 08

RACE DATE ....21 June 2015
CIRCUIT NAME ....Red Bull Ring
NO. OF LAPS 71
START TIME ....14.00 local/12.00 GMT
CIRCUIT LENGTH......4.326k
RACE DISTANCE 307.020k
LAP RECORD...........1:08.337
Michael Schumacher
(2003)

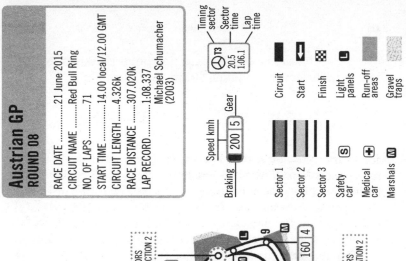

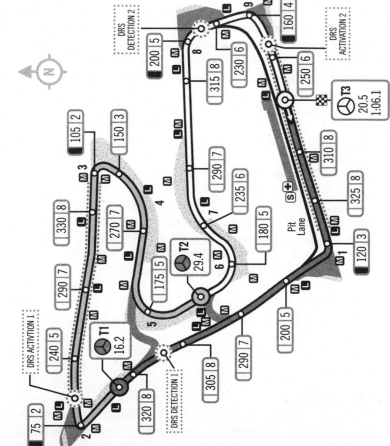

m after Turn Two. The second detection point is 10 m after Turn Eight with the activation point 110 m after Turn Nine.

AUSTRIAN GRAND PRIX FAST FACTS

- Prior to 2014 the last Austrian Grand Prix was twelve years ago. There are four drivers on the grid who have F1 racing experience at this circuit in its A1 Ring guise. Jenson Button has made four appearances here—from 2000 to 2003. His best result came at the most recent race in 2003 in which he qualified seventh and finished fourth for BAR/Honda.
- Kimi Räikkönen has three Austrian Grands Prix on his CV. In 2001 he started in ninth place for Sauber and finished fourth to claim the second of four points finishes in his debut season. In 2002, having moved to McLaren, his race was stopped after seven laps with engine failure but the following year, again with McLaren, he qualified and finished in second place.
- Fernando Alonso has raced here twice, with Minardi in 2001 and with Renault in 2003. On both occasions he failed to finish. His first attempt was ended by gearbox trouble after thirty-eight laps and an engine problem saw him exit the 2003 race after forty-four laps.
- Felipe Massa, meanwhile, raced at the A1 Ring for Sauber in his debut season, 2002. He qualified seventh but retired from the contest with a suspension problem after seven laps
- McLaren is the most successful team at the Austrian GP with six wins. The victories came in 1984 with Niki Lauda, in 1985 and 1986 with Alain Prost, in 1998 and 2000 with Mika Häkkinen and in 2001 courtesy of David Coulthard. The next most successful are Ferrari with five wins (1964, '70, '99, 2002 and '03) and Lotus with four (1972, '73, '78 and '82).
- Alain Prost has the most victories of any driver at the Austrian Grand Prix. The Frenchman took the chequered flag in 1983 with Renault and then claimed back-to-back wins for McLaren in 1985 and 1986.
- The current circuit configuration largely mirrors that of the A1 Ring, where the race was won from pole position three

times in seven events. Jacques Villeneuve won from the front of the grid in 1997, as did Häkkinen in 2000 and Michael Schumacher in 2003. The A1 Ring race has only been won from further back than third on the grid once, in 2001, when David Coulthard started in seventh position.

GREAT BRITAIN

2015 BRITISH GRAND PRIX
3–5 July 2015

Silverstone sets a different challenge to the preceding host circuits; its defining characteristic is the need for high-speed changes of direction that will severely test 2015's reduced downforce specification.

There is more, however, to the modern Silverstone than just its fast corners. The Arena layout, introduced in 2010, has subtly changed the nature of the circuit. Drivers were slow to appreciate the changes but today the infield section has acquired many fans, with the variety of lines through Turns Three and Four creating more overtaking opportunities leading into the first DRS zone in the Wellington Straight.

The succession of medium- and high-speed corners puts huge amounts of energy through the tyres. Something else to factor in when considering those corners is the fact that Silverstone is not especially demanding on the brakes. With drivers braking for only around 8 per cent of the long lap, recovering the maximum allowed amount of energy through the MGU-K every lap proves difficult.

CIRCUIT DATA—SILVERSTONE CIRCUIT
Length of lap: 5.891 km
Lap record: 1:33.401 (Mark Webber, Red Bull Racing, 2013)
Start line/finish line offset: 0.134 km
Total number of race laps: 52
Total race distance: 306.198 km
Pitlane speed limits: 80 km/h in practice, qualifying and the race.

CIRCUIT NOTES
- Artificial grass has been removed from the exits of Turns Five, Eight and Nine.
- The wall to the driver's left before Turn Six has been extended.

British GP
ROUND 09

RACE DATE .... 5 July 2015
CIRCUIT NAME ...... Silverstone
NO. OF LAPS 52
START TIME .... 13.00 local/12.00 GMT
CIRCUIT LENGTH 5.891km
RACE DISTANCE 306.198km
LAP RECORD 1:33.401
                      Mark Webber
                      (2013)

Timing sector
Sector time
Lap time

⊘ T3
24.9
1:30.4

Speed kmh
Gear

150
3

Braking
Lateral G-force
-1.0

Circuit
Start
Finish
Light panels
Run-off areas
Gravel traps

Sector 1
Sector 2
Sector 3

Safety car S
Medical car +
Marshals M

N

DRS ACTIVATION 1
DRS DETECTION 2
DRS DETECTION 1
DRS ACTIVATION 2

Pit Lane

⊘ T1
28.6

⊘ T2
36.9

⊘ T3
24.9
1:30.4

280 6 -0.1
295 6 0.1
275 6 -4.5
140 3 -1.3
305 7 -4.5
220 5 1.0
295 6 0.0
105 3 2.0
300 7 0.1
215 5 -3.0
240 5 -3.0

8 9 10 11 12 13 14 5 6 7 3 4 1 2 15 16 17 18

- Drainage has been improved in a number of places around the circuit.

DRS ZONES
There are two DRS zones at Silverstone. The detection point of the first is 25 m before Turn Three with the activation point 30 m after Turn Five. The second detection point is at Turn Eleven with the activation point 55 m after Turn Fourteen.

BRITISH GRAND PRIX FAST FACTS
- The British Grand Prix is one of two ever-present races on the Formula One World Championship calendar. The other race to feature every year since 1950 is the Italian Grand Prix.
- Three venues have hosted the British Grand Prix during the World Championship era. Silverstone shared the early races with Aintree, which held races in 1955, '57, '59 and 1961–62. Aintree was replaced by Brands Hatch, which staged the British Grand Prix in even years between 1964 and 1986. Silverstone has hosted all of the other races.
- 2015 marks the forty-ninth running of the Formula One World Championship British Grand Prix at Silverstone. The circuit, however, is celebrating its fifty-first, having hosted pre-World Championship Grands Prix in 1948 and 1949. Both of those races were won by Maserati, courtesy of drivers Luigi Villoresi and Baron Emmanuel "Toulo" de Graffenried respectively.
- Giuseppe "Nino" Farina won the inaugural World Championship race in 1950. Before Silverstone, a British Grand Prix was held at the Brooklands circuit in 1926 and 1927.
- Silverstone is situated in an area known as "Motorsport Valley". Six of the nine F1 teams are clustered within 125 km of the track. In order of distance they are McLaren (125 km), Williams (65 km), Lotus (40 km), Red Bull (33 km) and Mercedes (13 km) with Force India based just a few hundred metres from the front gates of the circuit.

Additionally, Mercedes High Performance Powertrains' manufacturing facility is based 33 km from the circuit, while Toro Rosso's wind tunnel is 23 km away.

- Jim Clark (1962, '63, '64, '65, '67) and Alain Prost (1983, '85, '89, '90, '93) share top billing at the British Grand Prix with five victories each. One victory adrift is Nigel Mansell who won in 1986, '87, '91 and '92. Mansell did, however, claim five victories in all on home soil, winning the 1985 European Grand Prix at Brands Hatch a year before winning his first British Grand Prix at the same circuit. Mansell is one of only two drivers to have won differently titled Grands Prix at the same circuit (Nelson Piquet won the 1980 Italian and 1981 San Marino Grands Prix at Imola.)

HUNGARY

2015 HUNGARIAN GRAND PRIX
24–26 July 2015

The twisting Hüngaroring is similar in characteristic to a street circuit—lacking the walls but retaining the tight radius corners, bumpy surface and low grip. It has something of a mixed reputation among drivers; common consensus suggests it is a wonderful track for a qualifying lap but a difficult place to race, given the paucity of overtaking opportunities. In close battles, good strategy has frequently been the decisive factor, more so than at other permanent circuits.

Teams run their maximum downforce packages in Hungary to cope with the many slow corners. The issue that will occupy the minds of engineers during the practice sessions is the need to maximize traction to get the best return from the many low-gear acceleration points.

CIRCUIT DATA—HÜNGARORING
Length of lap: 4.381 km
Lap record: 1:19.071 (Michael Schumacher, Ferrari, 2004)
Start line/finish line offset: 0.040 km
Total number of race laps: 70
Total race distance: 306.630 km
Pitlane speed limits: 80 km/h in practice, qualifying and the race.

CIRCUIT NOTES
- The guardrail to the left of the run-off area at Turn Three has been re-aligned to better protect the recovery vehicle and to allow space for a car that has been recovered.
- Speed bumps 50 mm high have been installed two metres from the track edge in the run-off area at Turns Six and Seven.
- New debris fencing has been installed close to the guardrail on the left between Turns Eleven and Twelve and around the outside of Turn Fourteen.

Hungarian GP
ROUND 11

RACE DATE ....26 July 2015
CIRCUIT NAME ....Hungaroring
NO. OF LAPS 70
START TIME ....14.00 local/12.00 GMT
CIRCUIT LENGTH 4.381km
RACE DISTANCE 306.630km
LAP RECORD..........1:19.071
Michael Schumacher (2004)

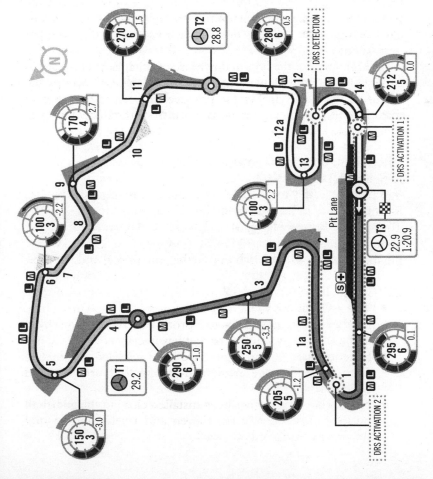

DRS ZONES

There are two DRS zones sharing a detection point 5 m before Turn Fourteen. Activation points are 130 m after the apex of Turn Fourteen and 6 m after the apex of Turn One.

HUNGARIAN GRAND PRIX FAST FACTS

- The Hungarian Grand Prix made its Formula One World Championship debut in 1986 at the newly-constructed Hüngaroring. It has been held at this venue every year since. Monza and Monte Carlo are the only circuits with a longer run of consecutive races.
- The race has been staged twenty-eight times. Michael Schumacher and Lewis Hamilton are the most successful drivers in the history of the event with four wins each. McLaren are the most successful team with eleven victories on this circuit, including six of the last nine Hungarian Grands Prix.
- In the battle for dominance between engine suppliers, Mercedes lead the way with nine victories; Renault have eight, Honda six, Ferrari five and Ford (Cosworth) one. Honda and Ferrari, however, share the distinction of having a victory in each decade of the race's operation.
- Amid its last ten outings, the Hungarian Grand Prix has provided debut victories for Fernando Alonso (2003), Jenson Button (2006) and Heikki Kovalainen (2008).
- Sebastian Vettel has a blind spot when it comes to the Hungarian Grand Prix, never having won there. Prior to 2015, during his first six seasons as a Red Bull Racing driver, he took at least one victory in every other country to host a Grand Prix.
- The 2011 Hungarian Grand Prix holds the distinction of being the race with the most pit stops—there were eighty-eight in total.
- Both Nigel Mansell, in 1992, and Michael Schumacher (2001) clinched the Drivers' World Championship at the Hungarian Grand Prix. In Mansell's case it was the eleventh race of a sixteen-race season, for Schumacher it was the thirteenth of seventeen. Schumacher holds the record for

the earliest conclusion to the Championship, taking the title in 2002 at the French Grand Prix with six races remaining.

- Williams secured the 1996 Constructors' World Championship in Hungary with a one-two formation finish, Jacques Villeneuve leading Damon Hill over the line. Ferrari repeated both the one-two finish and securing the Championship in 2001, 2002 and 2004.
- The 1992 Grand Prix was memorable for more than Mansell claiming the Drivers' crown. It was the last F1 Grand Prix to feature pre-qualifying and also the final race for the Brabham. Damon Hill qualified twenty-fifth and finished eleventh (last).
- Hamilton made a small piece of history at the 2009 Hungarian Grand Prix by becoming the first driver to win a Grand Prix in a hybrid car.

BELGIUM

2015 BELGIAN GRAND PRIX
21–23 August 2015

The Belgian Grand Prix at Spa-Francorchamps is one of the calendar's true classics and one of its toughest tests. The 7.004-km circuit features every kind of challenge. From the run down through Eau Rouge and up the steep incline towards the blind Raidillon corner, to the flat-out blast of the Kemmel Straight. Through Les Combes and the technically difficult stretch down through Rivage, Pouhon and Fagnes and on to the fearsomely fast Blanchimont left-hander, Spa-Francorchamps is a circuit that, despite myriad alterations over the years, still pushes man and machine to the limit.

Spa is one of the season's fastest tracks, with average speeds of 230 km/h, and the stretch from the exit of La Source to Les Combes sees the throttle wide open for twenty-three seconds—the longest single period on the calendar.

Set-up is tricky too, with the key to success being the right balance between low downforce for the high-speed first and third sectors and good grip for the twistier middle sector.

And then there's the weather. The Ardennes defines the phrase "four seasons in one day" and while one end of the circuit can be bathed in sunshine, the opposite side can be drenched with rain. The changeable conditions can present a real headache for teams, especially with regard to tyre choice.

CIRCUIT DATA—CIRCUIT SPA-FRANCORCHAMPS
Length of lap: 7.004 km
Lap record: 1:47.263 (Sebastian Vettel, Red Bull Racing, 2009)
Start line/finish line offset: 0.124 km
Total number of race laps: 44
Total race distance: 308.052 km
Pitlane speed limits: 80 km/h in practice, qualifying and the race.

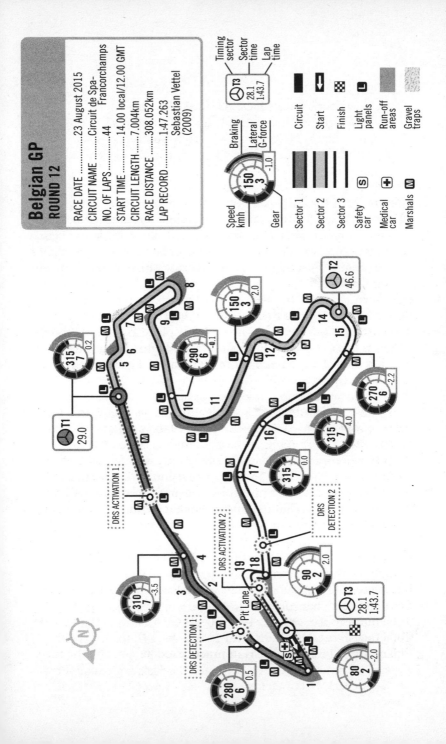

Belgian GP
ROUND 12

RACE DATE ....23 August 2015
CIRCUIT NAME ....Circuit de Spa-Francorchamps
NO. OF LAPS 44
START TIME ....14.00 local/12.00 GMT
CIRCUIT LENGTH......7.004km
RACE DISTANCE 308.052km
LAP RECORD ....1:47.263 Sebastian Vettel (2009)

Timing sector
Sector time
Lap time

T3
28.1
1:43.7

Braking
Lateral G-force

Speed kmh
150
3
-1.0
Gear

Circuit
Start
Finish
Light panels
Run-off areas
Gravel traps

Sector 1
Sector 2
Sector 3

S Safety car
✚ Medical car
M Marshals

CIRCUIT NOTES
- New debris fences have been installed at Turn One.
- Drainage has been installed or improved at Turns Two, Four, Eight, Eleven, Sixteen and Seventeen.
- The wall on the driver's left after Turn Eleven has been renewed.

DRS ZONES
There will be two DRS zones in Belgium. The detection point for the first zone will be 240 m before Turn Two, with the activation point 310 m after Turn Four. The second detection point will be 160 m before Turn Eighteen, with the activation point 30 m after Turn Nineteen.

BELGIAN GRAND PRIX FAST FACTS
- 2015 is the sixtieth Formula One Belgian Grand Prix, the first having been held in 1950 when it formed the fifth round of the inaugural F1 championship. That race was won by Juan Manuel Fangio.
- That first edition of the race was held at Spa-Francorchamps, on what was then a 14.1-km circuit. That track, with minor modifications, was used from 1950–56, 1958 and from 1960–70, after which the track was deemed too dangerous for Formula One. The Nivelles circuit was used in 1972 and 1974, with Zolder hosting the race in 1973 and then from 1975 until 1982. Spa hosted the 1983 race on a shortened 6.949-km circuit and Zolder hosted its final Belgian Grand Prix in 1984.
- With just two exceptions (2003 and 2006, when the Belgian Grand Prix was not held) the race has been staged at Spa-Francorchamps every year since 1985. The circuit has been subject to frequent modifications since then, the largest being the construction of new pit and paddock facilities prior to the 2007 race. The resultant 7.004-km layout is the one currently in use. This will be the forthy-eighth F1 Grand Prix held at Spa-Francorchamps.
- The most successful driver at the Belgian Grand Prix is Michael Schumacher, with six wins. The first of the

German's ninety-one career wins was scored here in 1992 with Benetton. He won with the same team in 1995 before recording a quartet of victories for Ferrari in 1996, 1997, 2001 and 2002. Ayrton Senna comes next, with five wins, in 1985, '88, '89, '90 and '91.

- No driver has won the Belgian Grand Prix at more than one venue. Emerson Fittipaldi, however, is the only man to have won at Nivelles. The unloved 3.7-km circuit near Brussels hosted just two Formula One Grands Prix and the Brazilian won both, first with Lotus in 1972 and then in 1974 with McLaren.

- Ferrari is the most successful constructor at the Belgian Grand Prix. The Italian squad has sixteen wins followed by fourteen for McLaren and eight for Lotus. Williams have four wins while Red Bull Racing have three. Mercedes also have a win to their credit too, in 1955, courtesy of Fangio.

ITALY

2015 ITALIAN GRAND PRIX
4–6 September 2015

Monza is F1's premier speed circuit, the last of its kind, a flat-out blast through parkland that sees cars, configured for low downforce, reach their highest velocities of the year. Its reputation as Formula One's fastest track is likely to be enhanced this year. While 2014's technical regulations produced cars with less downforce than those of recent years, the corresponding decrease in drag sees the cars hitting 360 km/h on the long straights.

But Monza isn't simply about top speeds. Recent races have seen winners emerge from among the slowest through the speed traps, preferring a set-up that possesses sufficient downforce to carry speed through the circuit's few corners and onto the long straights. Other requirements include a car that is stable under braking, rides kerbs well and has good traction out of the chicanes.

The fast Parabolica corner places high lateral energy demands on tyres, while the stop-go nature of the chicanes means Monza also makes high longitudinal demands on the rubber.

CIRCUIT DATA—AUTODROMO NAZIONALE MONZA
Length of lap: 5.793 km
Lap record: 1:21.046 (Rubens Barrichello, Ferrari, 2004)
Start line/finish line offset: 0.309 km
Total number of race laps: 53
Total race distance: 306.720 km
Pitlane speed limits: 80 km/h in practice, qualifying and the race.

CIRCUIT NOTES
- To enhance safety, the inner half of the gravel trap at Parabolica has been replaced with an asphalt run-off.

DRS ZONES
There are two DRS zones in Italy. The detection point for the first zone is 95 m before Turn Seven, with the activation point

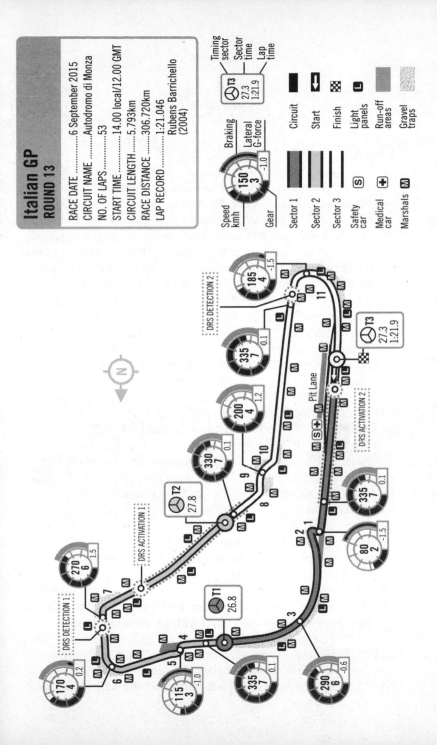

Italian GP
ROUND 13

RACE DATE .... 6 September 2015
CIRCUIT NAME .... Autodromo di Monza
NO. OF LAPS 53
START TIME .... 14.00 local/12.00 GMT
CIRCUIT LENGTH 5.793km
RACE DISTANCE 306.720km
LAP RECORD 1:21.046
Rubens Barrichello (2004)

Timing sector
Sector time
Lap time

T3
27.3
1:21.9

Braking
Lateral G-force
Speed kmh
150
3
−1.0
Gear

Circuit
Start
Finish
Light panels
Run-off areas
Gravel traps

Sector 1
Sector 2
Sector 3

S Safety car
+ Medical car
M Marshals

N

DRS DETECTION 2

185
4
−1.5

335
7
0.1

200
4
1.2

330
7
0.1

T2
27.8

270
6
1.5

DRS ACTIVATION 1

DRS DETECTION 1

170
4
0.2

115
3
−1.0

335
7
0.1

290
6
−0.6

80
2
−1.5

335
7
0.1

DRS ACTIVATION 2

T3
27.3
1:21.9

Pit Lane

T1
26.8

11

10

9

8

7

6

5

4

3

2 1

210 m after Turn Seven. The second detection point is 20 m before Turn Eleven, with the activation point 115 m after the finish line.

ITALIAN GRAND PRIX FAST FACTS

- The Italian Grand Prix is one of only two ever-present races on the Formula One World Championship calendar. The other is the British Grand Prix.
- 2015 is the sixty-sixth running of the Italian Grand Prix as part of the F1 World Championship. Sixty-four of the previous sixty-five were held at Monza, the exception being 1980, when the race was held at Imola and won by Nelson Piquet. Piquet also has three Italian Grand Prix victories at Monza (1983, '86, '87), placing him second on the all-time list. Michael Schumacher holds the record with five wins (1996, '98, 2000, '03, '06), all for Ferrari.
- Ferrari holds the record as a winning constructor, having taken victory eighteen times. Alongside Schumacher there have been Ferrari wins at Monza for Alberto Ascari (1951, '52), Phil Hill (1960, '61), John Surtees (1964), Ludovico Scarfiotti (1966), Clay Regazzoni (1970, '75), Jody Scheckter (1979), Gerhard Berger (1988), Rubens Barrichello (2002, '04) and Fernando Alonso (2010). Alonso also won the race in 2007 while driving for McLaren. Lewis Hamilton (2012, '14) and Sebastian Vettel (2008, '11, '13) are the only other Italian Grand Prix winners on the current grid.
- Vettel's win for Toro Rosso in 2008 makes him the youngest winner in Championship history. He was twenty-one years and seventy-four days old. A day earlier he became the youngest driver to secure pole position.
- The Italian Grand Prix at Monza has been won from pole position only twenty-one times. Interestingly, more than half of those victories have come since the turn of the century. Only in 2002 (Rubens Barrichello from fourth), '06 (Michael Schumacher from second) and '09 (Barrichello from fifth) has the sequence been interrupted.
- Monza's speed records are many and varied, particularly

from the latter years of the V10 era. Michael Schumacher holds the record for the highest average race speed, winning the 2003 Italian Grand Prix with an average speed of 247.585 km/h.

- Unsurprisingly, this race is also timed as the shortest duration Grand Prix (of those going the full distance) with Schumacher finishing in a time of 1 hour 14 minutes 19.838 seconds. Rubens Barrichello set F1's highest average lap speed in qualifying, taking pole position for the 2004 race at an average of 260.395 km/h, though Juan Pablo Montoya went faster that same weekend, taking the record for the fastest average lap speed overall, with a lap averaging 262.242 km/h, set during a practice session.
- Montoya also holds the record for the highest top speed achieved during a Formula One race, hitting 372.6 km/ during the 2005 Italian Grand Prix.

SINGAPORE

2015 SINGAPORE GRAND PRIX
18–20 September 2015

One of only two full night races on the calendar—the other being Bahrain—Singapore's race winds through twenty-three corners over 5 km and is the season's longest in terms of duration, regularly nudging the two-hour mark. Run in temperatures topping 30°C, it's a gruelling test of man and machine. Singapore's twisting streets require a high downforce configuration. The circuit has historically not been a tough race on powerplants, with the previous generation of cars running at just 46 per cent full throttle over a lap, but that changed considerably last year due to F1's new hybrid power units. As one of the least fuel efficient races of the year, due to its stop-start nature, it is also an interesting challenge in terms of how teams manage fuel restrictions.

Marina Bay is also tough on brakes. About a quarter of the lap is spent braking and while wear is not a major difficulty, problems are caused by the temperatures the brakes reach due to a lack of opportunities for cooling in the tight confines. After the power and speed of Monza and Spa, Singapore requires different attributes which benefits the teams who produce a car nimble enough to claw back competitiveness around the vastly twistier track.

CIRCUIT DATA—MARINA BAY STREET CIRCUIT
Length of lap: 5.065 km
Lap record: 1:48.574 (Sebastian Vettel, Red Bull Racing, 2013)
Start line/finish line offset: 0.137 km
Total number of race laps: 61
Total race distance: 308.828 km
Pitlane speed limits: 60 km/h in practice, qualifying and the race.

CIRCUIT NOTES
- Resurfacing has been carried out in various sections of the circuit. The Track Jet sweeper will be used in order to ensure

Singapore GP
ROUND 14

RACE DATE .... 20 September 2015
CIRCUIT NAME .... Marina Bay Street Circuit
NO. OF LAPS 61
START TIME .... 20.00 local/12.00 GMT
CIRCUIT LENGTH 5.073km
RACE DISTANCE 309.316km
LAP RECORD .... 1:48.574 Sebastian Vettel (2013)

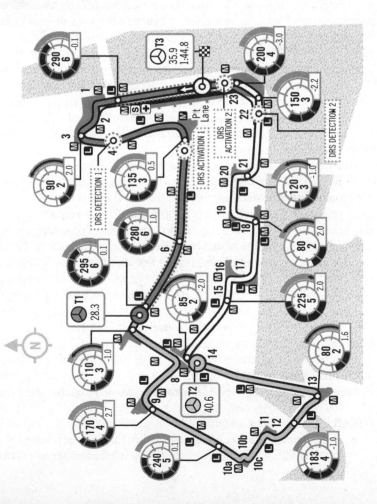

that grip on the new surfaces will match the older asphalt. The fast lane in the pit lane has been resurfaced.

- The wall on the right of the run-off area at Turn Seven has been re-aligned to allow one lane of traffic to circulate behind.

DRS ZONES

There are two DRS zones in Singapore. The first detection zone will be 230 m before Turn Five and the first activation point will be 50 m after the same corner. The second detection point will be 80 m before the apex of Turn Twenty-two and the activation point will be 45 m after the apex of Turn Twenty-three.

SINGAPORE GRAND PRIX FAST FACTS

- 2015 is the seventh running of the Singapore Grand Prix, the race having joined the F1 calendar in 2008. The circuit has been raced in three incarnations, with the first configuration being subject to minor changes for 2009. However, 2013's reprofiling of the Turn Ten "Singapore Sling" chicane as a more flowing corner led to faster lap times and a new lap record for Sebastian Vettel.
- Only three drivers have won this race. Fernando Alonso won the inaugural event in 2008 with Renault and then repeated the feat in 2010 for Ferrari. Lewis Hamilton won the 2009 race with McLaren and the 2014 race with Mercedes. Vettel, however, is the most successful driver at Marina Bay having won three Grands Prix here. Thus the Singapore Grand Prix has only ever been won by Formula One World Champions. Vettel's three wins make Red Bull Racing the most successful constructor here and the only team to score multiple victories.
- While Vettel has three wins and a second place to his credit he is not the driver with the most podium finishes here. That honour goes to Alonso who has finished in the top three in all but two Singapore Grands Prix. In 2011 he finished fourth having started from fifth place for Ferrari. Aside from his two wins, Alonso finished third in 2009 and 2012, second in 2013 and fourth last year.
- With Singapore's streets being tough to overtake on, pole

position is important. The race has been won from pole four out of the six times the grand prix has been held. Fernando Alonso won from as far back as fifteenth on the grid in 2008, albeit it in highly controversial circumstances, and Sebastian Vettel won from a starting place of third in 2012.

- The Singapore Grand Prix is always one of the season's longest races, regularly running close to the two-hour limit. The quickest race here so far was the 2009 edition, won by Lewis Hamilton in 1 hour 56 minutes 6.337 seconds. The longest was the 2012 race, won by Sebastian Vettel in a time of 2 hours 00 minutes 26.144 seconds. Having reached maximum duration, the race was ended two laps early at the end of lap fifty-nine.

JAPAN

2015 JAPANESE GRAND PRIX
25–27 September 2015

Few circuits can match Suzuka for elegance or excitement. The John Hugenholtz-designed track in Mie Prefecture defies easy categorization. It is a power circuit but also contains intricate low- and medium-speed corners. It features wide sweeping turns but a racing line that threads the eye of a needle. It is a well-known and well-loved venue but presents new set-up challenges every year. Suzuka is a difficult place to master.

The primary challenge is finding the right compromise on wing settings. The long flat-out back and pit straights require teams to eschew a maximum downforce set-up but lap time is frequently to be found in the twisty sections, both through the high speed of the Esses and Dunlop and around the low-speed hairpin and Casio Triangle Chicane. There is no right way to do it but there are plenty of opportunities to get it wrong. It is also very tough on tyres particularly due to the amount of lateral energy going through the rubber.

CIRCUIT DATA—SUZUKA CIRCUIT
Length of lap: 5.807 km
Lap record: 1:31.540 (Kimi Räikkönen, McLaren, 2005)
Start/finish line offset: 0.300 km
Total number of race laps: 53
Total race distance: 307.471 km
Pitlane speed limits: 80 km/h throughout the weekend.

CHANGES TO THE CIRCUIT SINCE 2014:
- TecPro barriers have been added on the driver's left after Turn Fifteen.
- A section of the track has been re-surfaced between Turns Fifteen and Sixteen.
- The tall lampposts that were very close to the debris fences around the outside of Turns Thirteen and Fourteen have been moved further back from the guardrail.

Japanese GP
ROUND 15

RACE DATE ....27 September 2015
CIRCUIT NAME ....Suzuka Circuit
NO. OF LAPS 53
START TIME ....15.00 local/06.00 GMT
CIRCUIT LENGTH......5.807km
RACE DISTANCE 307.471km
LAP RECORD...........1:31.540
Kimi Räikkönen
(2005)

Timing sector
Sector time
Lap time

T3
18.8
1:29.8

Speed kmh
Braking
Lateral G-force

150
3
-1.0

Gear

Sector 1
Sector 2
Sector 3

Safety car S
Medical car +
Marshals M

Circuit
Start
Finish
Light panels
Run-off areas
Gravel traps

N

T3
18.8
1:29.8

DRS ACTIVATION
DRS DETECTION

Pit Lane

T1
30.4

T2
40.4

300 7 0.1
160 4 -2.2
240 5 2.0
210 5 1.0
210 5 3.6
260 6 -4.5
140 3 -2.5
260 6 -1.4
95 2 -2.1
70 2 -2.0
285 6 -0.3
305 7 0.1
295 6 0.1
230 5 2.0
185 4 2.1

1 2 3 4 5 6 7 8 9 10 11 12 13 14 15 16 17 18

DRS ZONES

The single DRS zone at Suzuka is on the main straight. The detection point is 50 m before Turn Sixteen and the activation point is 100 m before the control line.

JAPANESE GRAND PRIX FAST FACTS

- There have been thirty F1 Japanese Grands Prix. The Fuji Speedway played host in 1976 and 1977 before a decade-long hiatus. The event returned in 1987 to Suzuka and stayed until 2006. The races went back to Fuji for 2007 and 2008 before reverting back to Suzuka in 2009.
- Of the twenty-six Japanese Grands Prix held at Suzuka, twelve have been won from pole position and a further ten from P2 on the grid. Alessandro Nannini (1989) and Nelson Picquet (1990) both won from sixth on the grid, the former after the disqualification of Ayrton Senna. Fernando Alonso won from P5 in 2006 but the standout performance is surely that of Kimi Räikkönen who triumphed from a starting position of seventeenth in a thrilling 2005 race.
- Only two Japanese drivers have ever stood on the podium. Aguri Suzuki secured third place for Lola in 1990 and Kamui Kobayashi did the same for Sauber in 2012. It was Suzuki's best finish in F1 and remains Kobayashi's best result to date.
- The most thrilling corners at Suzuka have been Degner 1 and 2, which were modified in 1983 from one continuous turn into two distinct but linked corners. The two fast turns have a very narrow line and frequently see cars running wide to beach in the gravel or hit the wall at the exit of Degner 2.

RUSSIA

2015 RUSSIAN GRAND PRIX
9–11 October 2015

While a Russian Grand Prix has been held before, in St Petersburg, 2015's race marks Grand Prix racing's return to the country for only the second time in a century. F1's second visit to the Russian Federation will be hosted by the all-new Sochi Autodrom. Designed by Hermann Tilke, the circuit is built around the Olympic Park site used for last year's 2014 Winter Olympics and runs around the venues used for ice hockey and skating as well as the stadium in which the Games' opening and closing ceremonies took place.

So far, the eighteen-corner Sochi layout has been likened to a mix between the Valencia Street Circuit used by Formula One between 2008 and 2012 and the Korean International Circuit used between 2010 and 2013. Blending medium- and low-speed corners with two long, fast sections, the signature corner has turned out to be the long, fast, left-hand arc around the Medal Arena.

Given the circuit's nature, it seems that establishing a set-up suited to dealing with the higher downforce requirements of the middle and end sections of the circuit and the lower downforce needs of the more flowing, faster sections of the track create quite a few race-engineering challenges.

CIRCUIT DATA—SOCHI AUTODROM
Length of lap: 5.848 km
Lap record: 1:40.896 (Valtteri Bottas, Williams, 2014)
Start line/finish line offset: 0.212 km
Total number of race laps: 53
Total race distance: 309.732 km
Pitlane speed limits: 80 km/h in practice, qualifying and the race.

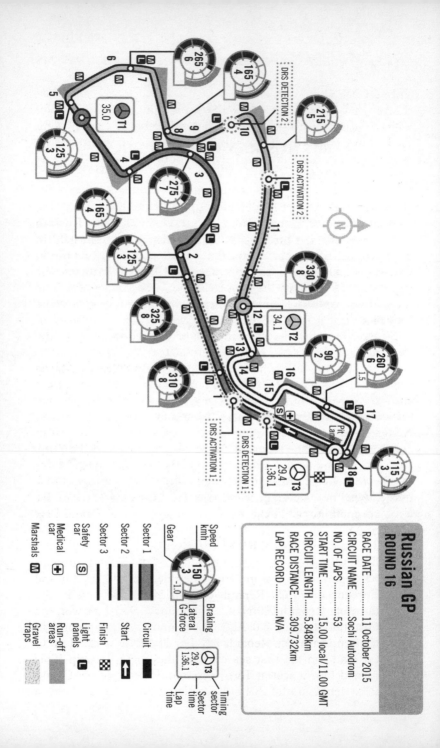

Russian GP
ROUND 16

RACE DATE ....11 October 2015
CIRCUIT NAME ....Sochi Autodrom
NO. OF LAPS...........53
START TIME............15.00 local/11.00 GMT
CIRCUIT LENGTH........5.848km
RACE DISTANCE 309.732km
LAP RECORD ....N/A

Gear
Speed kmh

150
3
-1.0

Braking
Lateral
G-force

Sector 1
Sector 2
Sector 3

Start
Finish
Circuit

Light
panels

Safety
car — S

Medical
car — +

Marshals — M

Run-off
areas

Gravel
traps

T3
29.4
1.36.1

Timing
sector
Sector
time
Lap
time

DRS ZONES

There will be two DRS zones in Sochi. The first detection point will be just after the start-finish line with the activation point located just after Turn One. The second detection point will be before Turn Nine, with the detection point just after the same corner.

RUSSIAN GRAND PRIX FAST FACTS

- 2015's race is only the second Formula One Russian Grand Prix. Two pre-F1 Grands Prix have been held in Russia, both prior to the outbreak of World War One in 1914. The first race took place in 1913 and was won by Russia's Georgy Suvorin driving a Benz 29/60 PS. The 1914 event was won by German driver Willy Scholl at the wheel of a Benz 55/150 HP. It was the last time such an event was held as the outbreak of WWI put a stop to the race.
- So far, Russia has had just two F1 drivers. On 14 March 2010 Vitaly Petrov from Vyborg became the first driver from the country to start a Grand Prix when he lined up for Renault at the season-opening Bahrain Grand Prix. To date, Petrov has scored sixty-four Championship Points from fifty-seven race starts. His best result is third at the 2011 Australian Grand Prix, the only F1 podium finish of his career so far. After thirty-eight races for Renault, Petrov moved to Caterham in 2012. He has not raced in F1 competitiion since the season-ending Brazilian Grand Prix of 2012.
- The other Russian F1 driver is current Red Bull racer Daniil Kvyat. Hailing from the city of Ufa, Kvyat's route to F1 began in earnest in 2011 when he finished third in the Eurocup Formula Renault 2.0 series and second in the Formula Renault 2.0 NEC. The following year he won the Formula Renault 2.0 Alps series and finished second in Eurocup Formula Renault 2.0. For 2013, he stepped up to GP3 and in his first season won the championship. This earned him a seat at Toro Rosso and he made his Grand Prix debut at the 2014 Australian Grand Prix, where a

ninth-place finish made him the sport's youngest-ever points scorer. At last year's Japanese Grand Prix it was announced that as a result of four-time champion Sebastian Vettel's decision to move on from Red Bull Racing at the end of that season, Kvyat would race for the team in 2015.

- On 31 August it was announced that the T4 grandstand at Sochi Autodrom has been renamed the Kvyat grandstand for the 2015 race.

USA

2015 UNITED STATES GRAND PRIX
23–25 October

The Circuit of the Americas (COTA) has become a firm favourite for F1 since opening in 2012. The track, to the south-east of Austin, is considered by many to be a modern masterpiece of circuit design. Much of the layout takes inspiration from established classics of the genre but the signature Turn One, with its steep ascent and blind apex, is unique.

Uppermost in the minds of engineers and drivers will be finding the right trade-off between speed on the straights and cornering grip, with the high-speed run and fast changes of direction from Turn Two down to Turn Ten being particularly sensitive to downforce levels.

Traditionally, the challenges of set-up lessen as teams become more familiar with a venue—but COTA does not make it easy. Staged late in the year, a characteristic of the first two Grands Prix has seen swings of up to 20°C between morning and afternoon track temperatures, altering fundamentally the balance of the cars.

CIRCUIT DATA—CIRCUIT OF THE AMERICAS (COTA)
Length of lap: 5.513 km
Lap record: 1:39.347 (Sebastian Vettel, Red Bull Racing, 2012)
Start line/finish line offset: 0.323 km
Total number of race laps: 56
Total race distance: 308.405 km
Pitlane speed limits: 80 km/h throughout the weekend.

CIRCUIT CHANGES
- Some of the asphalt run-off area around the outside of Turn Ten has been replaced by gravel (at the request of FIM).
- Some light panels will be mounted closer to the ground for improved visibility.

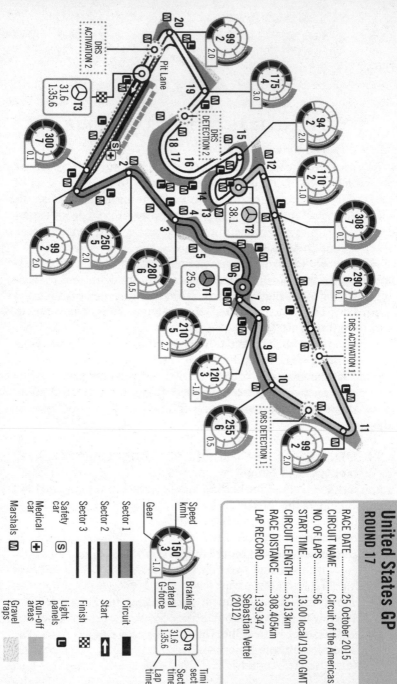

United States GP
ROUND 17

RACE DATE .... 25 October 2015
CIRCUIT NAME .... Circuit of the Americas
NO. OF LAPS 56
START TIME .... 13.00 local/19.00 GMT
CIRCUIT LENGTH 5.513km
RACE DISTANCE 308.405km
LAP RECORD 1:39.347
Sebastian Vettel
(2012)

Speed
km/h

Gear

Sector 1

Sector 2

Sector 3

Safety car — S

Medical car — +

Marshals — M

Braking

Lateral
G-force

Circuit

Start

Finish

Light panels — L

Timing
sector

Sector
time

Lap
time

Circuit

Run-off areas

Gravel traps

DRS ZONES

There will be two DRS zones at COTA. The detection point of the first will be 150 metres after Turn Ten, with the activation point 320 m after Turn Eleven. The second zone's detection point will be 65 m after Turn Eighteen, with the activation point 80 m after Turn Twenty, on the start/finish straight.

UNITED STATES GRAND PRIX FAST FACTS

- 2015 is the fourth running of the US Grand Prix at the Circuit of the Americas. Sebastian Vettel took pole positions in 2013 and 2012, winning in 2013 and finishing second in 2012. Lewis Hamilton won from second on the grid in 2012 and 2014, ensuring that the race has never been won at Austin from beyond the front row.
- Hamilton's victory in 2012 was a back-to-back US Grand Prix win—albeit with five years separating his achievements. His previous US success came in 2007, the last of eight occasions for the race to be run on the road course at the Indianapolis Motor Speedway.
- As a round of the F1 World Championship, the United States Grand Prix has been held at four other venues before IMS and COTA: Sebring (1959), Riverside (1960), Watkins Glen (1961–1980) and Phoenix (1989–91).
- Additional F1 World Championship races in the United States include: the US Grand Prix West, held at Long Beach between 1976 and 1983; the Detroit Grand Prix (1982–88), the Dallas Grand Prix (1984), and the Las Vegas Grand Prix (1982–83). The Indianapolis 500 was also included in the World Championship between 1950 and 1960.
- An American driver has never won the US GP—Mario Andretti, however, won the US Grand Prix West in 1977. Andretti managed two pole positions at the US Grand Prix but his best result was second in 1977. Other American drivers on the podium at their home race include Dan Gurney, second in 1961 and 1965, Ritchie Ginther, second in 1963 and Eddie Cheever, third in 1989.
- The most recent American driver to grace any F1 podium was Michael Andretti. McLaren driver Andretti had a best

F1 result of third place at the 1993 Italian Grand Prix. It was his final race in F1.

- At the 2013 US Grand Prix, the top three finishing cars were all powered by Renault engines, Vettel winning for Red Bull Racing, ahead of Lotus's Romain Grosjean and the second Red Bull of Mark Webber.

BRAZIL

2015 BRAZILIAN GRAND PRIX
13–15 November 2015

The penultimate round of the 2015 Formula One season takes teams and drivers to one of F1's shortest but most demanding tracks, the Autodromo Jose Carlos Pace in the Interlagos district of Sao Paulo, home of the Brazilian Grand Prix.

Although the circuit is just 4.309 km long, making it the second shortest on the calendar after Monaco, it packs plenty of challenges into its fifteen corners. The narrow track runs anti-clockwise around a natural amphitheatre, the contours of which provide a tough combination of technically demanding medium- and slow-speed turns linked by highspeed straights and bends. The topography means many of these are off-camber, thus increasing the challenge.

One characteristic of the track is its notorious bumpiness; this should be tempered somewhat this year as the track has been completely resurfaced. This should help with car set-up, though by how much is still open to question.

The other major set-up consideration is the altitude. At 800 m Interlagos is the highest altitude circuit of the year and the thinner air has in the past led to engines producing 7–8 per cent less power here than would be the case at sea level. How teams will work the new hybrid power units to mitigate against this will be interesting. The altitude affects aerodynamic performance too and as such teams will run high downforce packages to cope with the inefficiency. The lack of aerodynamic grip also means that the importance of good mechanical grip is emphasized. The final variable is, of course, the weather and conditions at Interlagos at this time of year are notoriously hard to read.

CIRCUIT DATA—AUTODROMO JOSE CARLOS PACE (INTERLAGOS)
Length of lap: 4.309 km
Lap record: 1:11.473 (Juan Pablo Montoya, BMW-Williams, 2004)

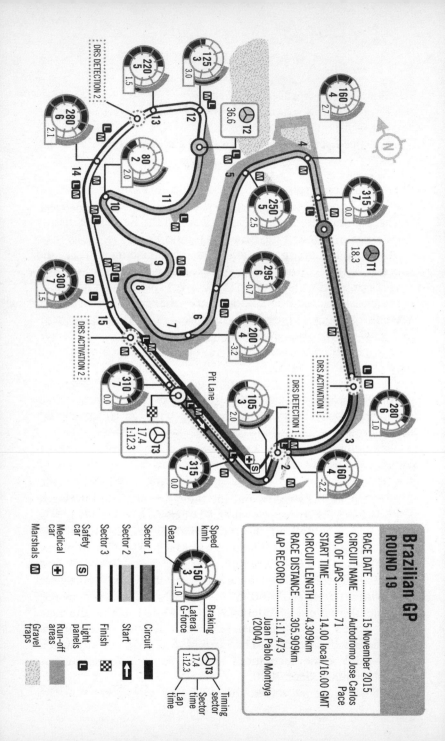

Brazilian GP
ROUND 19

RACE DATE ....15 November 2015
CIRCUIT NAME ....Autodromo Jose Carlos
START TIME ....14.00 local/16.00 GMT   Pace
NO. OF LAPS 71
CIRCUIT LENGTH 4.309km
RACE DISTANCE 305.909km
LAP RECORD 1:11.473
Juan Pablo Montoya
(2004)

Legend

Gear
Speed km/h — 150 / 3 / -1.0 — Braking / Lateral G-force

Sector 1
Sector 2
Sector 3

Safety car — S
Medical car — +
Marshals — M

Circuit
Start
Finish
Light panels — L
Run-off areas
Gravel traps

Timing sector
T3 / 17.4 / 1:12.3 — Sector time / Lap time

DRS DETECTION 2
DRS ACTIVATION 2
DRS DETECTION 1
DRS ACTIVATION 1

Pit Lane

T1 18.3
T12 36.6
T3 17.4 1:12.3

Start line/finish line offset: 0.030 km
Total number of race laps: 71
Total race distance: 305 909 km
Pitlane speed limits: 80 km/h in practice, qualifying and the
   race.

CIRCUIT NOTES
   • The entire circuit, including the pit lane, was resurfaced last
     year.
   • The pit lane exit has been realigned further to the left to
     provide a run-off area around the outside of Turn Two.
   • The pit entry has been realigned to the left in order to
     allow the pit wall start to be moved further away from the
     track.
   • A low kerb has been placed on the apex of Turn Fifteen in
     order to prevent cars from cutting the corner.

DRS ZONES
There are two DRS zones at Interlagos. The first has a detection
point at the apex of Turn Two with activation 20 m after Turn
Three. The second zone has its detection point 30 m after Turn
Thirteen with an activation point 60 m after Turn Fifteen.

BRAZILIAN GRAND PRIX FAST FACTS
   • The 2015 event marks the forty-third Formula One
     Brazilian Grand Prix and the thirty-third at the Interlagos
     Circuit. The first race was held at Interlagos in 1973 on a
     7.96-km version of the track and was won by local hero
     Emerson Fittipaldi. The event moved to Rio's Jacarepagua
     circuit for 1978 before returning to Interlagos in 1979 and
     1980. It then moved back to Jacarepagua for a nine-year
     stint between 1981 and 1989. Following redevelopment and
     the shortening of Interlagos to a 4.3-km layout the race
     returned to a much-changed circuit in 1990 where it has
     remained ever since. The current 4.309-km layout, which
     features minor alterations to the 1990 configuration, has
     been in use since 2000.
   • Alain Prost has more wins that any other driver at the

Brazilian Grand Prix with six. However, just one of the Frenchman's victories was secured at Interlagos, in 1990, driving for Ferrari. His other wins, in 1982, '84, '85, '87 and '88 were all scored at Jacarepagua. The first was for Renault, while the rest in Rio were achieved while racing for McLaren.

- Carlos Reutemann and Michael Schumacher are the next most successful drivers, each having taken four victories. Reutemann won twice in Rio and twice in Sao Paulo while all of Schumacher's wins (1994, 1995, 2000, 2002) came at Interlagos.

- McLaren are the most successful constructor at this race with twelve wins—four at Jacarepagua and eight at Interlagos. Ferrari have also gained eight wins in Sao Paulo but just two in Rio.

- Four current drivers are past Brazilian Grand Prix winners—Kimi Räikkönen, Felipe Massa. Sebastian Vettel and Jenson Button. Only Vettel and Massa are repeat winners with Massa winning in 2006 and 2008 and Vettel doing so in 2010 and 2013.

- Massa's 2006 victory made him just the fifth Brazilian driver to win his home event. The others are Ayrton Senna in 1993 and 1991, Nelson Piquet in 1986 and 1983, Carlos Pace in 1975 and Fittipaldi in 1973 and 1974. Picquet is the only one to win at Jacarepagua.

- Brazil has fielded thirty Formula One race drivers. Rubens Barrichello has contested more Brazilian GPs than any of his countrymen to date. He raced nineteen times at Interlagos with a best finish of third in 2004 for Ferrari.

- Statistically pole position is not important at Interlagos. Since the current circuit layout was introduced in 2000 the race has been won from the front of the grid just three times. Both Massa's wins were scored from pole, as was Vettel's in 2013.

- Glancarlo Fisichella holds the record for winning from furthest back on the grid. The Italian's victory in the rain-lashed 2003 contest was scored from eighth place at the start. A chaotic, red-flagged end to the race initially saw

Räikkönen awarded victory only for post-race analysis to later hand the win to Jordan driver Fisichella; he received his trophy two weeks later at the following round in San Marino.

ABU DHABI

2015 ABU DHABI GRAND PRIX
27–29 November 2015

F1's only day/night race is staged at a circuit consisting of three distinct sectors. The first features fast flowing corners ending in a hairpin that gives way to a second sector consisting of two long straights that end in heavy-braking corners where overtaking is a possibility. The winding final stretch features a complex series of low-speed corners that place a premium on aerodynamic grip and balance. With three such different sections, finding the right set-up to meet its many demands is more of a challenge than at most circuits. The scene of last year's title-deciding climax, it presents a unique challenge to race engineers and stategists.

CIRCUIT DATA—YAS MARINA CIRCUIT
Length of lap: 5.554 km
Lap record: 1:40.279 (Sebastian Vettel, Red Bull Racing, 2009)
Start line/finish line offset: 0.115 km
Total number of race laps: 55
Total race distance: 305.355 km
Pitlane speed limits: 80 km/h in practice, qualifying and the race.

CIRCUIT NOTES
- Other than routine maintenance, no significant work has been carried out.

DRS ZONES
There will be two DRS zones for the Abu Dhabi Grand Prix. The first detection point will be 40 m before Turn Seven, with the activation point 390 m after Turn Seven. The second detection point will be 50 m after Turn Nine with the activation point at Turn Ten.

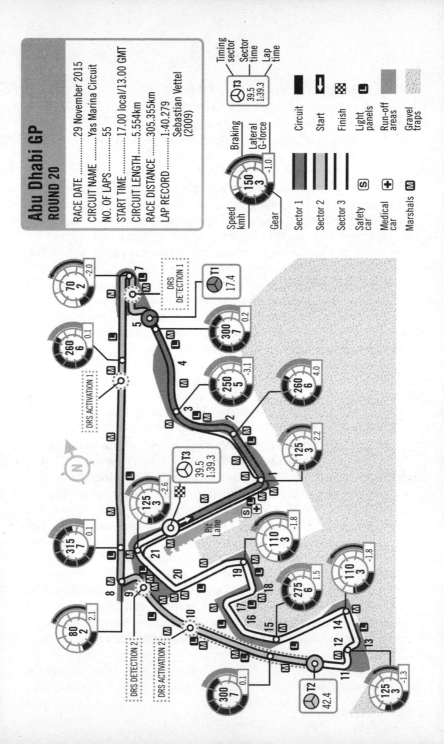

Abu Dhabi GP
ROUND 20

RACE DATE ....29 November 2015
CIRCUIT NAME ......Yas Marina Circuit
NO. OF LAPS 55
START TIME ....17.00 local/13.00 GMT
CIRCUIT LENGTH 5.554km
RACE DISTANCE 305.355km
LAP RECORD 1.40.279
Sebastian Vettel
(2009)

Timing
sector
Sector
time
Lap
time

T3
39.5
1.39.3

Braking
Lateral
G-force
-1.0

Speed
kmh
150
3
Gear

Circuit
Start
Finish
Light
panels
Run-off
areas
Gravel
traps

Sector 1
Sector 2
Sector 3

Safety
car
Medical
car
Marshals

ABU DHABI GRAND PRIX FAST FACTS

- 2015 is the seventh Abu Dhabi Grand Prix. The Yas Marina Circuit made its Formula One calendar debut on 1 November 2009. Abu Dhabi has hosted the season finale three times—in 2009, 2010 and 2014. It has been the venue for two championship deciders. The first came in 2010 when Sebastian Vettel came out on top in a four-way battle for the championship featuring Red Bull Racing teammate Mark Webber, Ferrari's Fernando Alonso and Lewis Hamilton, then driving for McLaren. Vettel took victory to win the title by four points from Alonso, who finished the race in seventh.

- In 2014, Lewis Hamilton carried a seventeen-point lead over Mercedes teammate Nico Rosberg into the final event, which had double points on offer. Hamilton made sure of the title by coming home first, whereas Rosberg was plagued by electrical problems and finished outside the points.

- The 2015 Abu Dhabi Grand Prix will mark Vettel's final race for Red Bull Racing before he heads for Ferrari in 2016. The German driver won the inaugural event here in his first season with Red Bull Racing; he also won here in 2013.

- The other two winners here are also F1 champions. Lewis Hamilton, winner of the 2008 title, won here in 2011 for Mclaren, while 2007 world champion Kimi Räikkönen won for Lotus in 2012. Räikkönen's 2012 victory was his first since Belgium 2009, shortly before he took a two-year hiatus from the sport. He has won just once since, in Australia in 2013, again with Lotus.

- While Vettel is the only multiple winner in Abu Dhabi he is joined on the list of repeat pole position winners by Hamilton. Vettel started from the front in 2010 and 2011, while Hamilton was on pole in 2009 and 2012. The only other men to take pole here are Mark Webber, in 2013, for Red Bull Racing and Nico Rosberg last year for Mercedes.

- The race has been won only once from pole—by Vettel in 2010. In 2009, '11, '13 and '14 it was won from second on the grid and in 2012 from fourth.

- Felipe Massa is the only current driver to score points in every one of his Yas Marina outings. The Brazilian missed the inaugural race but since then has finished tenth, fifth, seventh, eighth and second.

TECHNICAL

AERODYNAMICS OF FORMULA ONE

In strange structures far from the racetrack, seemingly endless hours and excessive amounts of budget are spent in the pursuit of marginal gains. Aerodynamicists search for minute ways to maximize the amount of downforce created while minimizing the amount of drag produced to obtain the maximum results from the aerodynamics of the car. From inside Formula One, Willem Toet, Head of Aerodynamic at the Sauber F1 Team explains how the work done in wind tunnels and in CFD (Computational Fluid Dynamics) can improve track lap times.

Why is so much money and energy spent looking at the shapes of Formula 1 (F1) cars? Fundamentally because it pays dividends. If you can reduce the drag of the car you will go faster on the straights. If you can use the shape of the car to generate some downward pressure (usually called downforce) onto the tyres, then the car will go faster around the corners. Research into aerodynamics has allowed cornering speeds in "high speed" corners to be much higher than that which is possible without the use of aerodynamic aids, although it has reduced ultimate top speeds. Track lap times have improved significantly.

The aerodynamics of racecars is intensely researched and 5–10 per cent downforce increases have been possible if rules do not change too much between seasons. Due to the nature of the vehicles, the aerodynamics of F1 cars are quite different to that of road cars—with drag coefficients of between 0.7 and 1.0 (it used to be even higher but rules restrict how much area can be used for aerodynamic devices)—this is between about two and four times as much as a good modern road car. This is partly due to the rules (running exposed wheels is part of the definition of an open-wheeled racing car) and partly because downforce is usually much more important than drag.

Aerodynamic research in F1 has been an area of high investment in the past thirty years. Assuming no regulatory limitations, this trend would continue while the bodywork rules are changed continually or while changes to the shape of the cars continue to provide significant improvements in lap time around the F1 circuits of the world. However, agreements between teams started to limit investment and now rules have been introduced officially to limit how much research is done into aerodynamics in wind tunnels and in CFD (Computational Fluid Dynamics). Naturally teams will optimize their resources to still obtain the maximum they can from the aerodynamics of the cars.

To investigate the aerodynamics of an F1 car the teams use various methods of research. Tests are conducted using scale models of the cars in a wind tunnel fitted with a "rolling road". Computers are used to mathematically simulate the flow of air around and through the cars and to model vehicle behaviour on the track. In the past, the real cars used to be tested as well in wind tunnels and on special straight-line test facilities, but this is now banned. Cars are tested on the real tracks of course, but that too is limited to fewer test days than was possible in the past. Each of the top ten F1 teams has somewhere between fifty and 150 people working solely on aerodynamic research. It is quite difficult to be sure about how many people work on aerodynamics as generally the teams do not talk about it. Despite the limits on track testing and on aerodynamic research that have been imposed by the rule makers (the FIA) and by agreements between the teams, the amount of research done is "significant". Even a "small" team of fifty people can do a lot of work!

Teams use computer simulation more and more to predict performance and to analyze many "what if?" scenarios. These simulations are then constantly improved by comparing them to the realities of racing and testing. From these tools we know that

Technical

the drag of the cars does slow them down quite a lot but typically, on an average racetrack, it slows them by about 3–5 per cent in lap time. In other words, if drag were to be reduced to zero, the gain in lap time at a typical track would normally be a bit less than 5 per cent. However, if we remove the present levels of downforce then lap times get slower by about 25 per cent or so.

In the present F1 environment, other performance factors that are normally very important have been limited more severely, for example the tyres are all provided by a single supplier and are carefully randomly selected for the teams by the rule makers and tyre supplier together so there is no chance of one team getting an advantage. The tyre supplier selects two of the four dry options they make for the season, plus one intermediate and one extreme wet tyre. So you cannot develop your own tyre to gain an advantage.

In the five years to 2014 engine specifications were, effectively, frozen. For 2014 there was a completely new powertrain formula but the idea is that these powertrains will also be virtually frozen once a reasonable level of parity is established. Cars must race above a certain minimum weight (to protect against dangerous construction as low weight helps lap time). Suspension kinematics is relatively free so this is an area where the teams can make a difference, but suspension must be passive. However, because aerodynamics is so dominant, even this is compromised to ensure aerodynamics benefits are maximized.

The wind tunnel models used for testing are either 50 per cent or 60 per cent (60 per cent upper limit in F1) of full scale and are tested up to sixty hours per week. In the past the tunnels were used twenty-four hours per day, seven days per week and some teams had more than one wind tunnel working, but rules now limit that. The floor of the wind tunnel is replaced with a "rolling road" (a fancy name for conveyor belt) and a boundary layer removal system (that removes the slow-moving air that builds up on a stationary surface) to simulate the fact that the car rolls over the ground.

These rolling roads and boundary layer control systems are essential for racing car work and are a science in themselves. Most teams test their models at 50 m/s (=180 km/h), which is the highest speed the teams are permitted to use by the regulations (more was possible in a few facilities which is why the limit was imposed. The Sauber wind tunnel is one of the best in F1, capable of testing full-scale cars on a rolling road at speeds up to 300 km/h.). To approach a realistic representation of the aerodynamics on a real (full-scale) car, it is best to test a large model as fast as possible, allowing for all the difficulties involved.

All simulation methods, whether computer or wind-tunnel testing, have their strengths and their limitations. One way that teams can make a difference to performance is through the understanding of these factors and the methods employed to take advantage of the strengths and to cover the weaknesses of each type of simulation. The other way is through the quality and/or volume of testing the teams do. Real-car track testing is a vital part of this process but is too slow, imprecise and expensive a method to be the only sort of testing a team should undertake. The variations in temperature, wind, track condition, tyres, driver input variations, and so forth is why track testing is not as precise as testing in a controlled environment. It is, though, the most important reality we try to simulate. Track testing is also limited by regulation so the teams experiment with new parts mainly on Friday practice sessions at a Grand Prix weekend.

The highest proportion of aerodynamic research money and energy is spent by most teams on the wind-tunnel testing of scale models of the car. The models are not usually constructed in the same way or using the same materials as a real car. They are designed to simulate both the internal and external shape of the cars while enabling the teams to change the design of the model shape more simply than would be possible on a miniature replica of the real car. An average example might be a 50 per cent scale model of the car using a wind speed of 50 m/s (=180 km/h). The model is usually suspended from the roof of the tunnel and is packed full of motors, load cells, pressure-measuring equipment, computers and other electronics. Sometimes wheels are not attached directly to the model but are held in place via mounts from outside the model. This has been found to give better overall repeatability of force measurement.

However, wheels-on-model is more accurate and is more

commonly used. The teams routinely conduct tests over a range of ride heights and pitches (differences in front and rear height to the ground) while assessing model (as well as wheel) forces and scanning pressures. The effects of exhaust gas, roll, yaw and steer are conducted regularly as well. Graphical or tabulated results are displayed on monitors during the tests and the final result is seen shortly after the last measurements are taken. Every team is different so this is only a guide. High-speed dynamic movement of the models has been tried by some ambitious teams but the mechanical forces involved are so high that this has not proven to be a reliable development method. However, there has been a strong trend toward continuous motion of the model in the tunnel (a sort of slow-motion Grand Prix simulation).

Mathematical modelling and computational fluid dynamics (CFD) are the areas of the most rapid growth in effort in racing car aerodynamics at the moment. Computers are used to back up real-car and rig testing of things like water and oil cooling, assessing what level of drag and downforce will give the best lap time at a particular track in any given conditions, etc. In-house circuit simulation and lap-time prediction programs are used to assess the effect of aerodynamic gains (as well as engine power, gear ratios, gear-change time, weight, centre of gravity height, cooling, mechanical set-up, etc.) on lap time. Driver simulators take this world of simulation yet another step closer to the reality of a car on track.

CFD is coming into its own as far as racing car aerodynamics are concerned. Modern super computers allow the use of mathematical models that mean complete and reasonably realistic full-vehicle aerodynamic simulations are now possible, if a little slow. The teams have now mainly settled on a CFD method called Navier Stokes which copes well with the realities of racing cars. Some teams are combining the use of commercially available packages with in-house computer programs/enhancements to maximise the gains that can be made using the computers. It will be some time before it is possible to dispense with wind tunnel testing because wind tunnels allow us to very quickly test hundreds of combinations of conditions and vehicle attitudes.

Both wind-tunnel testing and CFD are now limited by the FIA

in its Sporting Regulations. In the wind tunnel, teams are limited by wind-on time (about fifteen hours per week) plus a maximum of eighty runs per week and sixty hours per week tunnel occupancy. CFD simulations are limited to a certain number of teraflops of solver time. Together, wind-on time and teraflops are limited to thirty units in total so, if we use fifteen hours of "wind on" time per week, we can use fifteen teraflops of computer power to solve CFD cases.

All the methods of improving car aerodynamics have their limitations. Testing everything you want to try on a real car is very expensive (engines, tyres, travel to the test tracks, personnel, etc) and has limited precision—plus this sort of activity is strictly limited. Atmospheric, track, tyre and driving changes, for example, mean that small (aerodynamic) steps cannot be reliably assessed. Wind tunnel model testing works reasonably well in a straight line but realistic tyre shape changes at the contact patch are difficult to match to reality and important to aerodynamics. Of course more aerodynamic downforce is only really needed when the driver is not able to drive at full throttle, such as when accelerating at low speed, cornering or braking. To simulate cornering in a wind tunnel is simply not practical as a racing car on the limit of adhesion is sliding all the time and the angle that the air approaches the front and the rear of the car is not the same. It is possible to steer the wheels of the model and to yaw the model.

Of course any limitation irritates engineers, so improvements are constantly sought and we get ever closer to being able to simulate real cornering. The more realistic the simulation of the tyre deformation, the more likely that the model tyres will wear out and this can restrict what a team can achieve because model tyres (also supplied by Pirelli) are restricted to twelve sets per year. While real-cornering simulation is not quite there yet for wind tunnel testing, these sorts of cases are at least theoretically possible using CFD. However, assessing sensitivity of forces to ride height, pitch, roll, yaw, steer and sliding through a corner are significantly slower than in a wind tunnel.

Generally, the approach taken today is to evolve from a baseline using wind tunnel testing and CFD, with most (but not all)

CFD results checked in the wind tunnel, and then to make an update for the real car. This update is then tested on the real car to ensure that there are actually benefits. By conducting thousands of tests in the wind tunnel and using CFD for every real-car update, the aerodynamic step is usually large enough that the improvement is obvious to the drivers and quickly seen as an improvement in lap time. Should an update fail to impress, it is back to the drawing board for the Aerodynamicists to try to understand why. It is for reasons of updates to the car not performing particularly well and research into why this happened that some teams test in the wind tunnel in a yawed and steered condition (for example). Furthermore, it will continue to be the main driving force behind future improvements in the simulation techniques that the teams use.

Much of what the teams do is limited by the (bodywork) rules that govern the sport. It is because of the rules that there are flat-stepped floors on race cars with no car-to-ground bridging devices, the teams run exposed rather that covered wheels, have an open cockpit of certain minimum dimensions and virtually none of the wings or aerodynamic devices is movable or even flexible. Driver-controlled rear wing movement (upper element only, DRS) is allowed with extremely strict limitations and limited driver-initiated deployment under control of the rule makers. No material (even solid steel) is infinitely rigid and most teams have experimented with the limits of flexibility of the so-called rigid aerodynamic devices. As a result, the rule makers regularly adapt and refine specific flexibility limitations where they add a load to certain parts of the car and measure the deflection, which must be less than a rule-dictated limit. Suspension (mainly) parts cannot be wing-shaped and have to be of neutral section, i.e. the same top and bottom shapes, and have to be fitted "horizontally" with a tolerance of ± 5 degrees (more than enough to be able to play with). Rules also govern overhangs, heights, widths, etc. Despite the limitations of the rules, teams are able to work in many areas and continually increase the downforce without increasing the drag. Much of the work done now involves the understanding of airflow in three dimensions and it is mainly as understanding improves that aerodynamic efficiency of the cars is increased. It really is not possible to work on one part of the car in isolation (at least not for long) as everything interacts.

Many of the technical regulations change from time to time for reasons of safety. For example, there are tests for front impact, rear impact, side impact, top crush, floor (fuel cell) puncture— none of which may damage the chassis of the car. The cockpit must be a certain minimum dimension and shape, with energy-absorbing foam around the driver's head, shoulders and legs, and a driver's seat that can be removed from the car with the driver still seated in it. These safety regulations obviously influence what the teams can do aerodynamically speaking with the shape of the car.

Most of the ideas received by various teams from untrained enthusiasts are either illegal or more relevant to supersonic flow (which works more in line with untrained ideas of flow geometry) than to the subsonic flow on racing cars. It is, for example, not intuitive to people without aerodynamic training that there is more likely to be separation around a wing (a stall) at low speed (70 km/h, say) than at high speed (250 km/h). The other most common idea presented to reduce drag is the use of a dimpled or rough surface, via a "golf-ball" or shark skin effect. This and many other surface-finish treatments have been tried by most teams and found to be of no or, at best, little benefit. This is partly because skin friction only changes the drag of a Formula 1 car a very small amount (it is a much bigger deal for aircraft, for example). Research into surface treatments and micro-vortex generators continues.

Racing car wings are installed "upside down" compared to aircraft wings. This means that, on a racing car, the more curved or convex surface faces downward and rearward, while the flatter or sometimes concave surface faces upwards or forwards. On racing cars, it is the underside of the wing which is relatively more important than the top surface. Wings work because of air speed differences (caused by the shape of the wing) which in turn cause pressure differences—the faster the local flow caused by the shape of the wing, the lower the pressure and vice-versa. Wings working in "ground effect", that is wings reasonably close to the ground, are in general more effective (produce more downforce) than those a long way from the ground. Increasing the angle of the wing slows down the air on top of the wing and thus increases its static (surface) pressure, while speeding up the air under the wing leading edge decreases the static pressure there. The mid part of the front wing is a fixed neutral section dictated by regulation. There is no limit to the angle of attack of the rest of the front wings as such other than aerodynamic stall but the rules limit wing position (the wing has to fit into a number of boxes or zones). The optimum front wing design is mainly dictated by the influence the front wing then has on the flow to the underbody of the car and the rear wings. Despite the regulations, the front wing still contributes a high proportion of the downforce of the car. For performance reasons all the front wing shapes are three-dimensional in that the shape is different inboard to outboard.

The aerodynamic set-up of a modern F1 racing car is unlikely to be the same at any two races in a year. Aerodynamic settings (such as the front and rear wings) and hence the drag of the car are optimized to suit individual circuits. In addition, different brake ducts (for circuits where brakes are used more or less than an "average" circuit) and engine cooling exit ducts are fitted. Then, as a result of aerodynamic research, regular aerodynamic updates are made that change some feature of the car. These might be wings or body parts with a totally new shape. There are of course other settings and updates (for example suspension settings and suspension parts, electronics, engine, etc.), that are not directly aerodynamic, that ensure that the car is certainly never raced twice in the same configuration.

The downforce produced by the cars aerodynamically makes a big difference to grip. In a high-downforce configuration the cars produce their own weight (including driver) in aerodynamic download by around 36 m/s (=129 km/h or 80 mph). This means that if, at that speed, you were upside down on the roof of a theoretical "tunnel" you would stay there. In truth, in order to maintain control over the vehicle you would need enough grip to steer the car and to apply enough power to overcome the very high drag of the car. So you would really have to go at about 45 m/s (=162 km/h or 100 mph) to drive one of our cars upside down in a straight line.

BRAKES

IT'S ALL ABOUT HOW FAST YOU CAN SLOW DOWN

Formula One brake discs glow red hot at temperatures up to 1,200°C and drivers experience g-forces in excess of 5.5 when decelerating into corners. The regulation changes of 2014 had a significant impact on F1 braking systems, which are explained below by Brembo, the F1 brakes specialists as well as the most important recent developments in brake technology. Following this are the Brembo circuit-by-circuit guides, which visually detail the essential braking parameters of stopping distance, braking time, deceleration, pedal load and braking power.

Based on new FIA regulations adopted for the 2014 season, several important technical developments (smaller, more fuel-efficient engine, increase in the minimum weight to 691 kg, and the reduction of downforce) had a significant impact on the F1 braking systems, including the new calipers and carbon friction materials.

In 2014 a greater proportion of braking force was transferred to the front axle with the maximum brake torque ideally decreasing due to the reduction of downforce and speed of the cars. Stopping distance, on the contrary, was greater and, consequently, time spent under braking increased. For these same reasons, rear brake discs were smaller in diameter compared to the previous season, with a resulting advantage in terms of weight and speed of response to pressure. Their thickness was also thinner (25 mm), due to the reduction of the energy to dissipate. In 2014 as it will in 2015, the new Energy Recovery System (ERS) created increased drag on the car under braking and therefore required electronic control of the rear brake pressure to insure chassis stability. To accomplish this, an innovative system was incorporated: Brake By Wire (BBW). As the driver brakes, the BBW electronic controller provides the high-pressure hydraulic system with the proper braking force to the rear axle. In case of emergency, the master cylinder remains connected to the brake pedal and the hydraulic connection directly to the calipers for the driver to operate, but this functions only in case of a BBW failure.

Brembo was active in the design and simulation of BBW braking systems, as well as the individual components of Brake By Wire. For some teams, the Italian company developed only the actuator, which acts as an interface between the hydraulics of the car and the rear calipers. For other teams, Brembo developed a more substantial part of the BBW system, involving the valves that influence the BBW switch from normal use to emergency condition.

GREATER VENTILATION FOR BRAKE DISCS

In recent seasons, Brembo engineers have completely changed the brake cooling, which in F1 can reach a maximum temperature of 1,200°C, by redesigning the cooling system to include up to 1,000 ventilation holes. The ventilation of Brembo carbon brake discs went through a development process resulting in a considerable increase in overall performance of the braking system. The increase of airflow was also achieved through CFD calculations (Computational Fluid Dynamics), a synergic study, developed by each race team, of the airflow between the intake and the brake disc. This resulted in an optimal design of ventilation holes, which increased in number but reduced in diameter, thereby increasing exponentially the carbon surface open to airflow and therefore thermal discharge. This structural evolution required a much more complex mechanical processing, along with a growing effort in terms of in-depth analysis of fluid dynamics. The support of each team has been crucial to the design of air intakes for new cars.

LOWER WEAR OF BRAKE SYSTEM THANKS TO CER AND CUSTOMIZATION OF SYSTEMS

Materials are considerably changed as well. For 2014 "CER" was introduced, which represents an evolution of the previous "CCR" material, that considerably reduces wear guaranteeing more effective thermal conductivity. Compared to previous material, CER offers excellent warm-up time; that is, maximum rapidity in reaching more efficient operating temperatures; a wide application range in terms of both pressure and temperature, and very smooth friction performance. All these features provide the driver with a perfect modulation of the braking system. The incredibly low wear results in more reliable performance from the start to the end of race. Disc material is the same for all teams supplied by Brembo, who continue to research and develop composite materials that are more manageable.

SIGNIFICANT NUMBERS

In a full F1 season, Brembo supplies each team with the following material for its two cars: ten sets of calipers (i.e. 4 x 10 components) from 140 to 240 discs and from 280 to 480 pads.

Rolex Australian Grand Prix March 2015
Albert Park (Melbourne)

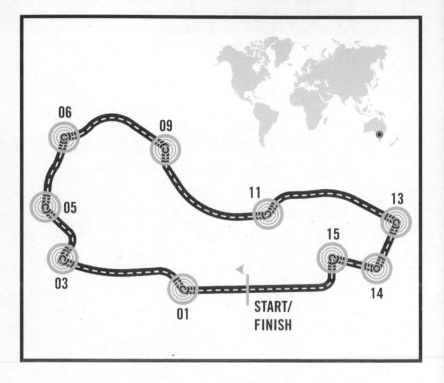

Circuit Braking Data

ALBERT PARK (Melbourne)

Melbourne is an urban track that winds its way through the Albert Park streets. It is a fast track and quite demanding for the brakes. The nine braking zones on the track are all medium-to-high level difficulty for the braking systems and are characterized by variable decelerations. Because it is a non-permanent track, during the race weekend it is gradually rubberized, which causes an increase in deceleration and brake stress in terms of wear and temperature.

- **Length:** 5,303 m
- **Number of laps:** 58
- **Type of circuit:** Hard
- **Number of brakings:** 9
- **Time spent under braking per lap:** 15%

01 *

Initial speed..........................304 (km/h)
Final speed............................135 (km/h)
Stopping distance.................92 (m)
Braking time..........................1.75 (sec)
Max deceleration...................5.78 (g)
Max pedal load.....................148 (kg)
Braking power ····················3004 (kw)

03

Initial speed..........................295 (km/h)
Final speed............................85 (km/h)
Stopping distance.................101 (m)
Braking time..........................2.34 (sec)
Max deceleration...................5.58 (g)
Max pedal load.....................143 (kg)
Braking power ····················2883 (kw)

05

Initial speed..........................156 (km/h)
Final speed............................141 (km/h)
Stopping distance.................21 (m)
Braking time..........................0.51 (sec)
Max deceleration...................1.07 (g)
Max pedal load.......................15 (kg)
Braking power.......................166 (kw)

06

Initial speed..........................275 (km/h)
Final speed............................126 (km/h)
Stopping distance.................80 (m)
Braking time..........................1.63 (sec)
Max deceleration...................5.09 (g)
Max pedal load.....................131 (kg)
Braking power.......................2462 (kw)

09

Initial speed..........................281 (km/h)
Final speed............................107 (km/h)
Stopping distance.................88 (m)
Braking time..........................1.92 (sec)
Max deceleration...................5.23 (g)
Max pedal load.....................135 (kg)
Braking power.......................2620 (kw)

11

Initial speed..........................288 (km/h)
Final speed............................201 (km/h)
Stopping distance.................55 (m)
Braking time..........................0.85 (sec)
Max deceleration...................5.31 (g)
Max pedal load.....................135 (kg)
Braking power.......................2714 (kw)

13

Initial speed..........................289 (km/h)
Final speed............................120 (km/h)
Stopping distance.................91 (m)
Braking time..........................1.88 (sec)
Max deceleration...................5.43 (g)
Max pedal load.....................161 (kg)
Braking power.......................2780 (kw)

14

Initial speed..........................228 (km/h)
Final speed............................182 (km/h)
Stopping distance.................45 (m)
Braking time..........................0.81 (sec)
Max deceleration...................2.53 (g)
Max pedal load.......................55 (kg)
Braking power.......................1258 (kw)

15

Initial speed..........................229 (km/h)
Final speed............................78 (km/h)
Stopping distance.................75 (m)
Braking time..........................2.03 (sec)
Max deceleration...................3.90 (g)
Max pedal load.....................100 (kg)
Braking power.......................1566 (kw)

* Turn 01 is considered the most demanding for the braking system.

Petronas Malaysian Grand Prix March 2015
Sepang International Circuit (Kuala Lumpur)

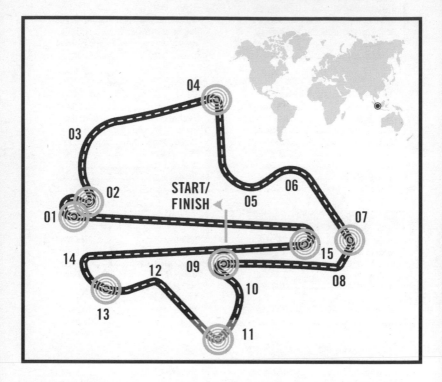

Circuit Braking Data

SEPANG INTERNATIONAL CIRCUIT (Kuala Lumpur)

This is a medium difficulty circuit for brakes except for the first and last braking sections. These sections, although characterized by deceleration close to 5 Gs, are preceded by very long straight stretches during which the friction material has plenty of time to cool efficiently. The major areas of concern relate to the correct sizing of the air intakes, which must allow optimum management of braking system operating temperatures on all the track sections.

- **Length:** 5,543 m
- **Number of laps:** 56
- **Type of circuit:** Medium
- **Number of brakings:** 8
- **Time spent under braking per lap:** 15%

01 *

Initial speed ....314 (km/h)
Final speed ....81 (km/h)
Stopping distance ....112 (m)
Braking time ....1.92 (sec)
Max deceleration ....4.9 (g)
Max pedal load ....115 (kg)
Braking power ···················2290 (kw)

02

Initial speed ....119 (km/h)
Final speed ....77 (km/h)
Stopping distance ....34 (m)
Braking time ....0.19 (sec)
Max deceleration ....1.6 (g)
Max pedal load ....33 (kg)
Braking power ···················264 (kw)

04

Initial speed ....293 (km/h)
Final speed ....109 (km/h)
Stopping distance ....104 (m)
Braking time ....1.98 (sec)
Max deceleration ....4.4 (g)
Max pedal load ....104 (kg)
Braking power ....1911 (kw)

07

Initial speed ....277 (km/h)
Final speed ....182 (km/h)
Stopping distance ....55 (m)
Braking time ....0.87 (sec)
Max deceleration ....4.1 (g)
Max pedal load ....96 (kg)
Braking power ....1665 (kw)

09

Initial speed ....284 (km/h)
Final speed ....71 (km/h)
Stopping distance ....115 (m)
Braking time ....2.49 (sec)
Max deceleration ....4.2 (g)
Max pedal load ....98 (kg)
Braking power ....1759 (kw)

11

Initial speed ....230 (km/h)
Final speed ....160 (km/h)
Stopping distance ....42 (m)
Braking time ....0.78 (sec)
Max deceleration ....3.1 (g)
Max pedal load ....72 (kg)
Braking power ....1048 (kw)

13

Initial speed ....267 (km/h)
Final speed ....127 (km/h)
Stopping distance ....87 (m)
Braking time ....1.24 (sec)
Max deceleration ....3.9 (g)
Max pedal load ....84 (kg)
Braking power ....1525 (kw)

15

Initial speed ....311 (km/h)
Final speed ....111 (km/h)
Stopping distance ....116 (m)
Braking time ....1.74 (sec)
Max deceleration ....4.9 (g)
Max pedal load ....112 (kg)
Braking power ....2237 (kw)

* Turn 01 is considered the most demanding for the braking system.

UBS Chinese Grand Prix April 2015
Shanghai International Circuit (Shanghai)

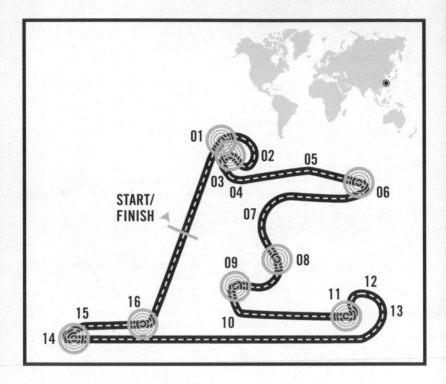

Circuit Braking Data

SHANGHAI INTERNATIONAL CIRCUIT (Shanghai)

It is Turn 14 at the end of the longest back straight of any circuit on the calendar that presents the biggest challenge. Drivers are subjected to nearly 5 Gs as they violently decelerate from 320 Km/h to a mere 58 Km/h for a corner that stands out in a circuit which is relatively easy on the braking systems. Aerodynamic resistance contributes to the deceleration of the single-seaters, helping the braking action. However, the remaining braking sections are relatively light and free of any particular difficulties for braking systems.

- **Length:** 5,451 m
- **Number of laps:** 56
- **Type of circuit:** Light
- **Number of brakings:** 8
- **Time spent under braking per lap:** 15%

01

Initial speed .... 308 (km/h)
Final speed .... 184 (km/h)
Stopping distance .... 56 (m)
Braking time .... 0.78 (sec)
Max deceleration .... 4.5 (g)
Max pedal load .... 113 (kg)
Braking power ·················· 1703 (kw)

03

Initial speed .... 125 (km/h)
Final speed .... 81 (km/h)
Stopping distance .... 8 (m)
Braking time .... 0.26 (sec)
Max deceleration .... 1.64 (g)
Max pedal load .... 40 (kg)
Braking power ·················· 248 (kw)

06

Initial speed .... 282 (km/h)
Final speed .... 70 (km/h)
Stopping distance .... 109 (m)
Braking time .... 2.21 (sec)
Max deceleration .... 3.96 (g)
Max pedal load .... 103 (kg)
Braking power .... 1376 (kw)

08

Initial speed .... 244 (km/h)
Final speed .... 179 (km/h)
Stopping distance .... 40 (m)
Braking time .... 0.68 (sec)
Max deceleration .... 3.24 (g)
Max pedal load .... 83 (kg)
Braking power .... 973 (kw)

09

Initial speed .... 187 (km/h)
Final speed .... 107 (km/h)
Stopping distance .... 43 (m)
Braking time .... 1.01 (sec)
Max deceleration .... 2.34 (g)
Max pedal load .... 60 (kg)
Braking power .... 526 (kw)

11

Initial speed .... 276 (km/h)
Final speed .... 79 (km/h)
Stopping distance .... 109 (m)
Braking time .... 2.29 (sec)
Max deceleration .... 3.83 (g)
Max pedal load .... 100 (kg)
Braking power .... 1304 (kw)

14 *

Initial speed .... 320 (km/h)
Final speed .... 58 (km/h)
Stopping distance .... 139 (m)
Braking time .... 2.84 (sec)
Max deceleration .... 4.77 (g)
Max pedal load .... 139 (kg)
Braking power .... 1852 (kw)

16

Initial speed .... 248 (km/h)
Final speed .... 148 (km/h)
Stopping distance .... 62 (m)
Braking time .... 1.15 (sec)
Max deceleration .... 3.31 (g)
Max pedal load .... 84 (kg)
Braking power .... 1004 (kw)

* Turn 14 is considered the most demanding for the braking system.

Gulf Air Bahrain Grand Prix April 2015
Bahrain International Circuit (Sakhir)

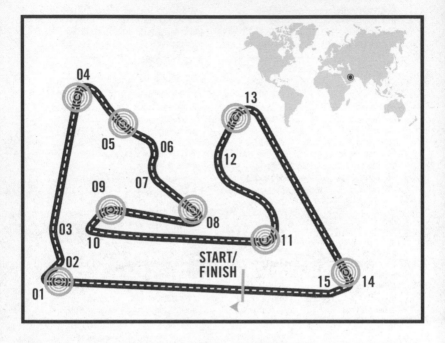

Circuit Braking Data

BAHRAIN INTERNATIONAL CIRCUIT (Sakhir)

This is definitely one of the most demanding circuits for brakes. The races on the Sakhir track, surrounded by the desert, are characterized by high temperatures that increase mechanical grip and make it difficult to dissipate the heat generated during braking. Combined with the presence of numerous high-energy braking sections that are quite close together, this makes Sakhir a hard test bench for all the braking system components, which are continuously stressed by the high-energy forces and the hellishly hot temperatures. The major obstacle is the high wear of the friction material — good management is needed to finish the race.

- **Length:** 5,412 m
- **Number of laps:** 57
- **Type of circuit:** Hard
- **Number of brakings:** 8
- **Time spent under braking per lap:** 17%

01 *

Initial speed..........................319 (km/h)
Final speed...........................60 (km/h)
Stopping distance.................137 (m)
Braking time.........................2.89 (sec)
Max deceleration....................5.0 (g)
Max pedal load.....................119 (kg)
Braking power ····················1940 (kw)

04

Initial speed..........................294 (km/h)
Final speed...........................112 (km/h)
Stopping distance.................99 (m)
Braking time.........................1.82 (sec)
Max deceleration....................4.4 (g)
Max pedal load.....................106 (kg)
Braking power ····················1604 (kw)

05

Initial speed..........................249 (km/h)
Final speed...........................195 (km/h)
Stopping distance.................32 (m)
Braking time.........................0.51 (sec)
Max deceleration....................3.5 (g)
Max pedal load.....................83 (kg)
Braking power.......................1065 (kw)

08

Initial speed..........................240 (km/h)
Final speed...........................71 (km/h)
Stopping distance.................85 (m)
Braking time.........................1.91 (sec)
Max deceleration....................3.3 (g)
Max pedal load.....................80 (kg)
Braking power.......................980 (kw)

10

Initial speed..........................231 (km/h)
Final speed...........................65 (km/h)
Stopping distance.................92 (m)
Braking time.........................2.37 (sec)
Max deceleration....................3.2 (g)
Max pedal load.....................74 (kg)
Braking power.......................901 (kw)

11

Initial speed..........................302 (km/h)
Final speed...........................146 (km/h)
Stopping distance.................83 (m)
Braking time.........................1.37 (sec)
Max deceleration....................4.6 (g)
Max pedal load.....................110 (kg)
Braking power.......................1704 (kw)

13

Initial speed..........................266 (km/h)
Final speed...........................140 (km/h)
Stopping distance.................75 (m)
Braking time.........................1.38 (sec)
Max deceleration....................3.84 (g)
Max pedal load.....................93 (kg)
Braking power.......................1259 (kw)

14

Initial speed..........................299 (km/h)
Final speed...........................120 (km/h)
Stopping distance.................97 (m)
Braking time.........................1.72 (sec)
Max deceleration....................4.52 (g)
Max pedal load.....................109 (kg)
Braking power.......................1661 (kw)

* Turn 01 is considered the most demanding for the braking system.

Spanish Grand Prix May 2015
Circuit de Barcelona-Catalunya (Catalunya)

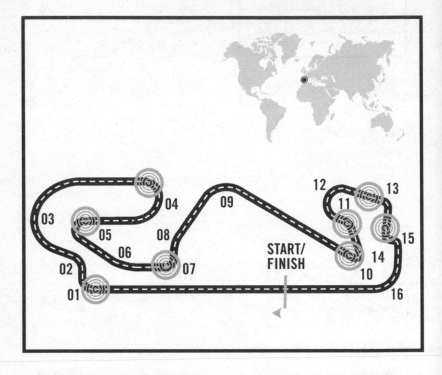

Circuit Braking Data

CIRCUIT DE BARCELONA-CATALUNYA (CATALUNYA)

The is the circuit which is used most for winter testing so the teams and drivers know it quite well. The track's level of grip is always very high and, with about 13 per cent of every lap spent on the brakes, the demand for the brakes on this track can be put in the medium category. On the other hand, the straight stretches allow efficient heat dissipation between one braking section and the next. The most demanding braking section is made up of the ELF turn, which is also on a downhill and has deceleration greater than five Gs.

- **Length:** 4,655 m
- **Number of laps:** 66
- **Type of circuit:** Medium
- **Number of brakings:** 8
- **Time spent under braking per lap:** 13%

01 *

Initial speed ....321 (km/h)
Final speed ....136 (km/h)
Stopping distance ....78 (m)
Braking time ....1.17 (sec)
Max deceleration...................5.36 (g)
Max pedal load ....129 (kg)
Braking power ·······················2107 (kw)

04

Initial speed ....288 (km/h)
Final speed ....153 (km/h)
Stopping distance ....64 (m)
Braking time ....1.03 (sec)
Max deceleration...................4.53 (g)
Max pedal load ....111 (kg)
Braking power ·······················1634 (kw)

05

Initial speed ....250 (km/h)
Final speed ....105 (km/h)
Stopping distance ....73 (m)
Braking time ....1.49 (sec)
Max deceleration...................3.72 (g)
Max pedal load ....91 (kg)
Braking power.......................1176 (kw)

07

Initial speed ....260 (km/h)
Final speed ....149 (km/h)
Stopping distance ....62 (m)
Braking time ....1.12 (sec)
Max deceleration...................3.91 (g)
Max pedal load ....96 (kg)
Braking power.......................1278 (kw)

10

Initial speed ....301 (km/h)
Final speed ....68 (km/h)
Stopping distance ....110 (m)
Braking time ....2.20 (sec)
Max deceleration...................4.84 (g)
Max pedal load ....118 (kg)
Braking power.......................1812 (kw)

11

Initial speed ....200 (km/h)
Final speed ....138 (km/h)
Stopping distance ....23 (m)
Braking time ....0.46 (sec)
Max deceleration...................2.79 (g)
Max pedal load ....68 (kg)
Braking power.........................695 (kw)

13

Initial speed ....216 (km/h)
Final speed ....135 (km/h)
Stopping distance ....47 (m)
Braking time ....0.98 (sec)
Max deceleration...................3.07 (g)
Max pedal load ....76 (kg)
Braking power.........................841 (kw)

14

Initial speed ....169 (km/h)
Final speed ....79 (km/h)
Stopping distance ....41 (m)
Braking time ....1.15 (sec)
Max deceleration...................2.29 (g)
Max pedal load ....56 (kg)
Braking power.........................479 (kw)

* Turn 01 is considered the most demanding for the braking system.

Monaco Grand Prix May 2015
Circuit de Monaco (Monte Carlo)

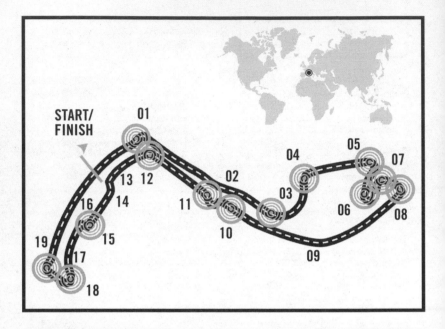

Circuit Braking Data

CIRCUIT DE MONACO (Monte Carlo)

This is a historic city circuit that winds through the streets of the Principality and can create many problems for the single-seater brakes. In fact, the winding track with poor grip often means that the drivers need to control the car often using the brakes, with negative reflexes on the caliper and brake fluid temperature. In the past this event has often resulted in a catalogue of problems connected to overheating and vapour lock of the braking system (a phenomenon in which the brake fluid reaches the boiling point inside the caliper). This has resulted in a lengthening of the pedal in braking, which has many times caused drivers to retire, if not crash.

Contemporary progress made in brake cooling has held these problems at bay, although particular attention still needs to be given to managing temperatures during the race weekend. The braking sections are not particularly dramatic, but the time spent on the brakes here is among the highest of the season at 21 per cent.

- **Length:** 3,340 m
- **Number of laps:** 78
- **Type of circuit:** Hard
- **Number of brakings:** 13
- **Time spent under braking per lap:** 21%

01

Initial speed .... 283 (km/h)
Final speed .... 110 (km/h)
Stopping distance .... 94 (m)
Braking time .... 1.88 (sec)
Max deceleration .... 4.45 (g)
Max pedal load .... 108 (kg)
Braking power ····················· 1581 (kw)

02

Initial speed .... 280 (km/h)
Final speed .... 167 (km/h)
Stopping distance .... 60 (m)
Braking time .... 1.00 (sec)
Max deceleration .... 4.37 (g)
Max pedal load .... 106 (kg)
Braking power ····················· 1535 (kw)

04

Initial speed .... 173 (km/h)
Final speed .... 126 (km/h)
Stopping distance .... 15 (m)
Braking time .... 0.34 (sec)
Max deceleration .... 2.39 (g)
Max pedal load .... 57 (kg)
Braking power .... 508 (kw)

05

Initial speed .... 221 (km/h)
Final speed .... 63 (km/h)
Stopping distance .... 79 (m)
Braking time .... 2.09 (sec)
Max deceleration .... 3.21 (g)
Max pedal load .... 79 (kg)
Braking power .... 891 (kw)

06

Initial speed .... 145 (km/h)
Final speed .... 44 (km/h)
Stopping distance .... 45 (m)
Braking time .... 1.68 (sec)
Max deceleration .... 2.02 (g)
Max pedal load .... 48 (kg)
Braking power .... 345 (kw)

07

Initial speed .... 102 (km/h)
Final speed .... 89 (km/h)
Stopping distance .... 2 (m)
Braking time .... 0.08 (sec)
Max deceleration .... 1.57 (g)
Max pedal load .... 27 (kg)
Braking power .... 162 (kw)

08

Initial speed .... 124 (km/h)
Final speed .... 105 (km/h)
Stopping distance .... 10 (m)
Braking time .... 0.33 (sec)
Max deceleration .... 1.77 (g)
Max pedal load .... 38 (kg)
Braking power .... 237 (kw)

10 *

Initial speed .... 287 (km/h)
Final speed .... 70 (km/h)
Stopping distance .... 115 (m)
Braking time .... 2.66 (sec)
Max deceleration .... 4.53 (g)
Max pedal load .... 110 (kg)
Braking power .... 1624 (kw)

11

Initial speed .... 77 (km/h)
Final speed .... 59 (km/h)
Stopping distance .... 6 (m)
Braking time .... 0.33 (sec)
Max deceleration .... 1.39 (g)
Max pedal load .... 39 (kg)
Braking power .... 149 (kw)

12

Initial speed .... 229 (km/h)
Final speed .... 174 (km/h)
Stopping distance .... 30 (m)
Braking time .... 0.54 (sec)
Max deceleration .... 3.36 (g)
Max pedal load .... 83 (kg)
Braking power .... 961 (kw)

* Turn 10 is considered the most demanding for the braking system.

Canadian Grand Prix June 2015
Circuit Gilles Villeneuve (Montréal)

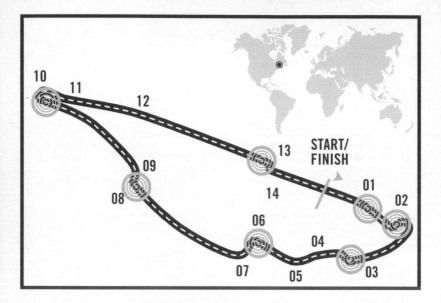

Circuit Braking Data

CIRCUIT GILLES VILLENEUVE (Montréal)

Montreal is without a shadow of a doubt the most demanding test bench for the single-seater braking systems. It is a 'stop and go' type circuit characterized by sudden braking sections and acceleration. The braking sections, all hard and very close together, result in an extremely high operating temperature for the discs and pads, which do not have time to cool sufficiently in the short straight stretches. These characteristics, combined with a significantly high percentage of time spent on the brakes, make for a very hard mix for the braking systems; another factor is that the aerodynamic load (in other words, the resistance to forward progress) is not one of the highest. The scenario can get even worse when there is a tail wind on the two main straight stretches which can significantly increase the straight line speed, putting even more onus on the brakes. A critical point is the chicane before the famous 'wall of champions' where having control going into the turn is a must to avoid hopping the kerb. On this turn being in tune with the brakes can make all the difference between having a good race and retiring following a crash!

- **Length:** 4,361 m
- **Number of laps:** 70
- **Type of circuit:** Hard
- **Number of brakings:** 7
- **Time spent under braking per lap:** 13%

01

Initial speed .... 301 (km/h)
Final speed .... 145 (km/h)
Stopping distance .... 78 (m)
Braking time .... 1.45 (sec)
Max deceleration .... 4.89 (g)
Max pedal load .... 119 (kg)
Braking power ····················· 1834 (kw)

02

Initial speed .... 135 (km/h)
Final speed .... 72 (km/h)
Stopping distance .... 29 (m)
Braking time .... 0.99 (sec)
Max deceleration .... 1.91 (g)
Max pedal load .... 49 (kg)
Braking power ····················· 334 (kw)

03

Initial speed .... 255 (km/h)
Final speed .... 138 (km/h)
Stopping distance .... 55 (m)
Braking time .... 0.98 (sec)
Max deceleration .... 3.91 (g)
Max pedal load .... 96 (kg)
Braking power .... 1255 (kw)

06

Initial speed .... 271 (km/h)
Final speed .... 103 (km/h)
Stopping distance .... 88 (m)
Braking time .... 1.78 (sec)
Max deceleration .... 4.24 (g)
Max pedal load .... 104 (kg)
Braking power .... 1440 (kw)

08

Initial speed .... 295 (km/h)
Final speed .... 120 (km/h)
Stopping distance .... 94 (m)
Braking time .... 1.76 (sec)
Max deceleration .... 4.74 (g)
Max pedal load .... 116 (kg)
Braking power .... 1748 (kw)

10

Initial speed .... 291 (km/h)
Final speed .... 57 (km/h)
Stopping distance .... 118 (m)
Braking time .... 2.74 (sec)
Max deceleration .... 4.67 (g)
Max pedal load .... 114 (kg)
Braking power .... 1706 (kw)

13 *

Initial speed .... 321 (km/h)
Final speed .... 137 (km/h)
Stopping distance .... 97 (m)
Braking time .... 1.64 (sec)
Max deceleration .... 5.43 (g)
Max pedal load .... 130 (kg)
Braking power .... 2134 (kw)

* Turn 13 is considered the most demanding for the braking system.

Austrian Grand Prix June 2015
Red Bull Ring (Spielberg)

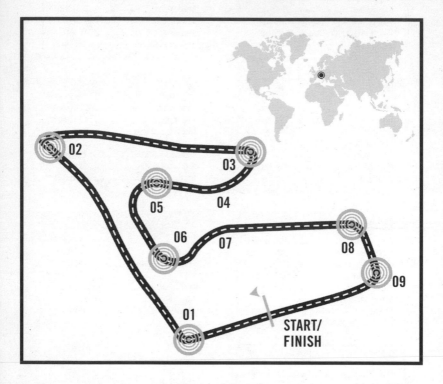

Circuit Braking Data

RED BULL RING (Spielberg)

The Spielberg circuit is a very hilly track, characterized by sharp bends with seven significant deceleration braking sections. The circuit is also quite short with little space for the system to cool between one braking section and another. As it is a circuit that has been reinstated to the Championship after several years, all teams will have to pay close attention to the temperature of brake discs and calipers.

- **Length:** 4,326 m
- **Number of laps:** 71
- **Type of circuit:** Hard
- **Number of brakings:** 7
- **Time spent under braking per lap:** 14%

01

Initial speed .... 302 (km/h)
Final speed .... 126 (km/h)
Stopping distance .... 91 (m)
Braking time .... 1.60 (sec)
Max deceleration .... 4.83 (g)
Max pedal load .... 118 (kg)
Braking power ···················· 182 (kw)

02 *

Initial speed .... 304 (km/h)
Final speed .... 67 (km/h)
Stopping distance .... 125 (m)
Braking time .... 2.80 (sec)
Max deceleration .... 4.89 (g)
Max pedal load .... 118 (kg)
Braking power ···················· 1849 (kw)

03

Initial speed .... 299 (km/h)
Final speed .... 98 (km/h)
Stopping distance .... 108 (m)
Braking time .... 2.10 (sec)
Max deceleration .... 4.78 (g)
Max pedal load .... 117 (kg)
Braking power .... 1788 (kw)

05

Initial speed .... 270 (km/h)
Final speed .... 181 (km/h)
Stopping distance .... 49 (m)
Braking time .... 0.80 (sec)
Max deceleration .... 4.11 (g)
Max pedal load .... 99 (kg)
Braking power .... 1392 (kw)

06

Initial speed .... 247 (km/h)
Final speed .... 188 (km/h)
Stopping distance .... 33 (m)
Braking time .... 0.56 (sec)
Max deceleration .... 3.61 (g)
Max pedal load .... 86 (kg)
Braking power .... 1115 (kw)

08

Initial speed .... 299 (km/h)
Final speed .... 210 (km/h)
Stopping distance .... 47 (m)
Braking time .... 0.67 (sec)
Max deceleration .... 4.77 (g)
Max pedal load .... 115 (kg)
Braking power .... 1783 (kw)

09

Initial speed .... 235 (km/h)
Final speed .... 158 (km/h)
Stopping distance .... 44 (m)
Braking time .... 0.82 (sec)
Max deceleration .... 3.38 (g)
Max pedal load .... 82 (kg)
Braking power .... 999 (kw)

* Turn 02 is considered the most demanding for the braking system.

Santander British Grand Prix July 2015
Silverstone Circuit (Silverstone)

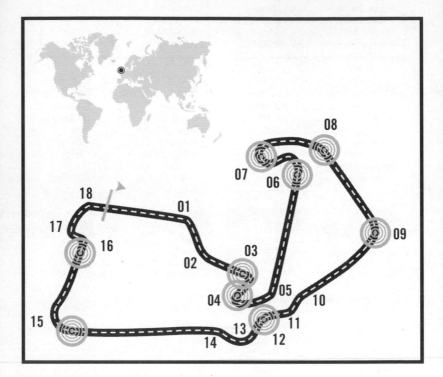

Circuit Braking Data

SILVERSTONE CIRCUIT (Silverstone)

This is perhaps the least demanding track for the braking system with just 8% of each lap spent on the brakes. In fact, it is a very 'driven' circuit where the long, fast turns generally translate into not-too-demanding braking sections. In the event of adverse weather conditions, given the low energy forces in play, there can be problems connected to excessive cooling and the 'glazing' of the friction material. In fact, the carbon from which the discs and pads are made do not guarantee correct friction generation if the operating temperatures are too low, thereby compromising braking performance.

- **Length:** 5,901 m
- **Number of laps:** 52
- **Type of circuit:** Light
- **Number of brakings:** 9
- **Time spent under braking per lap:** 8%

03 *

Initial speed..........................290 (km/h)
Final speed.............................105 (km/h)
Stopping distance..................90 (m)
Braking time..........................1.68 (sec)
Max deceleration...................4.70 (g)
Max pedal load.....................115 (kg)
Braking power ·····················1709 (kw)

04

Initial speed..........................160 (km/h)
Final speed.............................96 (km/h)
Stopping distance..................27 (m)
Braking time..........................0.72 (sec)
Max deceleration...................2.21 (g)
Max pedal load.....................50 (kg)
Braking power ·····················414 (kw)

06

Initial speed..........................310 (km/h)
Final speed.............................176 (km/h)
Stopping distance..................70 (m)
Braking time..........................1.07 (sec)
Max deceleration...................5.19 (g)
Max pedal load.....................124 (kg)
Braking power......................1990 (kw)

07

Initial speed..........................181 (km/h)
Final speed.............................127 (km/h)
Stopping distance..................18 (m)
Braking time..........................0.39 (sec)
Max deceleration...................2.52 (g)
Max pedal load.....................58 (kg)
Braking power......................556 (kw)

08

Initial speed..........................299 (km/h)
Final speed.............................290 (km/h)
Stopping distance..................5 (m)
Braking time..........................0.05 (sec)
Max deceleration...................4.90 (g)
Max pedal load.....................82 (kg)
Braking power......................190 (kw)

09

Initial speed..........................281 (km/h)
Final speed.............................254 (km/h)
Stopping distance..................14 (m)
Braking time..........................0.18 (sec)
Max deceleration...................4.49 (g)
Max pedal load.....................109 (kg)
Braking power......................1594 (kw)

12

Initial speed..........................227 (km/h)
Final speed.............................202 (km/h)
Stopping distance..................14 (m)
Braking time..........................0.24 (sec)
Max deceleration...................3.35 (g)
Max pedal load.....................81 (kg)
Braking power......................959 (kw)

15

Initial speed..........................314 (km/h)
Final speed.............................230 (km/h)
Stopping distance..................42 (m)
Braking time..........................0.56 (sec)
Max deceleration...................5.29 (g)
Max pedal load.....................127 (kg)
Braking power......................2051 (kw)

16

Initial speed..........................274 (km/h)
Final speed.............................106 (km/h)
Stopping distance..................91 (m)
Braking time..........................1.83 (sec)
Max deceleration...................4.34 (g)
Max pedal load.....................106 (kg)
Braking power......................1490 (kw)

* Turn 03 is considered the most demanding for the braking system.

Hungarian Grand Prix July 2015
Hungaroring (Budapest)

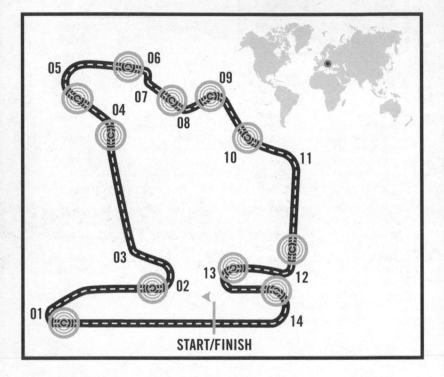

01 02 03 04 05 06 07 08 09 10 11 12 13 14

START/FINISH

Circuit Braking Data

HUNGARORING (Budapest)

This winding circuit is characterized by the high aerodynamic load put on the cars. Most of the circuit is quite driven, but it has a rather demanding braking section right after the main straight stretch. This track is among the most demanding for braking systems, even if friction material temperature management on this track is the key to managing the race along with ensuring consistent performance and wear kept under control.

- **Length:** 4,381 m
- **Number of laps:** 70
- **Type of circuit:** Hard
- **Number of brakings:** 11
- **Time spent under braking per lap:** 15%

01 *

Initial speed.........................322 (km/h)
Final speed.........................90 (km/h)
Stopping distance.................122 (m)
Braking time.........................2.37 (sec)
Max deceleration..................5.31 (g)
Max pedal load ....121 (kg)
Braking power ··················2112 (kw)

02

Initial speed.........................280 (km/h)
Final speed.........................118 (km/h)
Stopping distance.................87 (m)
Braking time.........................1.64 (sec)
Max deceleration..................4.31 (g)
Max pedal load ....99 (kg)
Braking power ··················1520 (kw)

04

Initial speed.........................295 (km/h)
Final speed.........................223 (km/h)
Stopping distance.................108 (m)
Braking time.........................0.53 (sec)
Max deceleration..................4.63 (g)
Max pedal load....................108 (kg)
Braking power.....................1729 (kw)

05

Initial speed.........................247 (km/h)
Final speed.........................157 (km/h)
Stopping distance.................51 (m)
Braking time.........................0.93 (sec)
Max deceleration..................3.62 (g)
Max pedal load....................83 (kg)
Braking power.....................1129 (kw)

06

Initial speed.........................248 (km/h)
Final speed.........................103 (km/h)
Stopping distance.................81 (m)
Braking time.........................1.77(sec)
Max deceleration..................3.64 (g)
Max pedal load....................81 (kg)
Braking power.....................1137 (kw)

08

Initial speed.........................209 (km/h)
Final speed.........................158 (km/h)
Stopping distance.................30 (m)
Braking time.........................0.58 (sec)
Max deceleration..................2.94 (g)
Max pedal load....................67 (kg)
Braking power.....................777 (kw)

09

Initial speed.........................183 (km/h)
Final speed.........................154 (km/h)
Stopping distance.................17 (m)
Braking time.........................0.35 (sec)
Max deceleration..................2.49 (g)
Max pedal load....................56 (kg)
Braking power.....................565 (kw)

10

Initial speed.........................260 (km/h)
Final speed.........................221 (km/h)
Stopping distance.................21 (m)
Braking time.........................0.31 (sec)
Max deceleration..................3.88 (g)
Max pedal load....................89 (kg)
Braking power.....................1281 (kw)

* Turn 01 is considered the most demanding for the braking system.

Shell Belgian Grand Prix July 2015
Circuit de Spa-Francorchamps (Spa-Francorchamps)

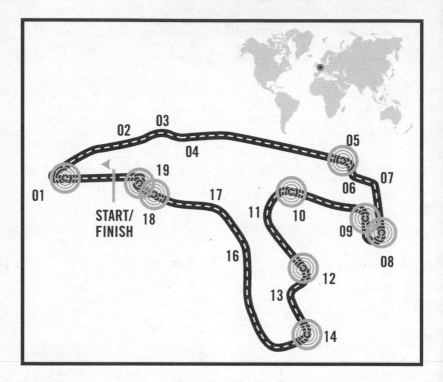

Circuit Braking Data

CIRCUIT DE SPA-FRANCORCHAMPS (Spa-Francorchamps)

At just over seven kilometres, this is the longest track of the season. Despite the presence of two braking sections ('Les Combes' at the end of the Kemmel straight and the 'Bus Stop' chicane right before the finish line) that are characterized by extremely high energy forces, the rest of the track is rather light on the braking system. Fast turns translate into not-so-demanding braking and so ensure excellent cooling of the system itself. This is especially so in adverse weather conditions, quite common in this region — in fact, problems connected to excessive cooling can occur.

- **Length:** 7,004 m
- **Number of laps:** 44
- **Type of circuit:** Light
- **Number of brakings:** 9
- **Time spent under braking per lap:** 11%

01

Initial speed ....287 (km/h)
Final speed ....73 (km/h)
Stopping distance ....108 (m)
Braking time ....2.24 (sec)
Max deceleration....................4.42 (g)
Max pedal load ....108 (kg)
Braking power ·····················1622 (kw)

05

Initial speed ....327 (km/h)
Final speed ....155 (km/h)
Stopping distance ....94 (m)
Braking time ....1.50 (sec)
Max deceleration....................5.41 (g)
Max pedal load ....131 (kg)
Braking power ·····················2207 (kw)

08

Initial speed ....272 (km/h)
Final speed ....122 (km/h)
Stopping distance ....61 (m)
Braking time ....1.03 (sec)
Max deceleration....................4.09 (g)
Max pedal load....................102 (kg)
Braking power......................1412 (kw)

09

Initial speed ....208 (km/h)
Final speed ....171 (km/h)
Stopping distance ....21 (m)
Braking time ....0.41 (sec)
Max deceleration....................2.82 (g)
Max pedal load....................69 (kg)
Braking power......................736 (kw)

10

Initial speed ....298 (km/h)
Final speed ....259 (km/h)
Stopping distance ....20 (m)
Braking time ....0.26 (sec)
Max deceleration....................4.68 (g)
Max pedal load....................115 (kg)
Braking power......................1774 (kw)

12

Initial speed ....306 (km/h)
Final speed ....176 (km/h)
Stopping distance ....72 (m)
Braking time ....1.11 (sec)
Max deceleration....................4.87 (g)
Max pedal load....................118 (kg)
Braking power......................1884 (kw)

14

Initial speed ....251 (km/h)
Final speed ....158 (km/h)
Stopping distance ....54 (m)
Braking time ....0.96 (sec)
Max deceleration....................3.63 (g)
Max pedal load....................90 (kg)
Braking power......................1171 (kw)

18 *

Initial speed ....313 (km/h)
Final speed ....82 (km/h)
Stopping distance ....116 (m)
Braking time ....2.22 (sec)
Max deceleration....................5.04 (g)
Max pedal load....................122 (kg)
Braking power......................1984 (kw)

19

Initial speed ....97 (km/h)
Final speed ....74 (km/h)
Stopping distance ....11 (m)
Braking time ....0.46 (sec)
Max deceleration....................1.47 (g)
Max pedal load....................36 (kg)
Braking power·····················189 (kw)

* Turn 18 is considered the most demanding for the braking system.

Italian Grand Prix September 2015
Autodromo di Monza (Monza)

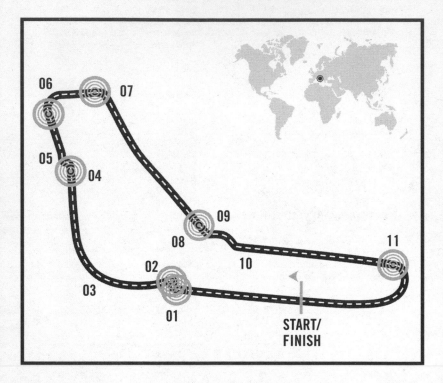

06 07

05 04

09

08

10

02

03

01

11

START/
FINISH

Circuit Braking Data

AUTODROMO DI MONZA (MONZA)

Known by fans as the 'Temple of Speed', the Monza track is extremely demanding and significantly tests the braking systems of the single-seater cars. The circuit's long straight lines and the lack of aerodynamic load, which reduces the possibility of efficiently offloading braking torque to the ground, makes the braking sections extremely tough and demanding to manage.

- **Length:** 5,793 m
- **Number of laps:** 53
- **Type of circuit:** Hard
- **Number of brakings:** 7
- **Time spent under braking per lap:** 11%

Racing legend Niki Lauda was instrumental in attracting Lewis Hamilton to Mercedes and the two would develop a close working relationship during 2014.

The addition of Paddy Lowe to the Mercedes pit wall as Executive Director (Technical) completed the team restructuring under Team Principal Toto Wolff.

The technical challenges of the new regulations for 2014 gave Mercedes the opportunity to create the all-conquering PU105A Hybrid Power Unit.

Nico Rosberg would have his best chance yet to challenge for the Driver's World Championship and he did not disappoint by winning the opening race in Australia.

Malaysia would be the first of five 1-2 finishes in a row for the team out of a total of eleven for the season.

The incredibly close wheel to wheel racing had to be kept in check by Toto Wolff both from the pit wall and in the drivers' debrief sessions.

The closeness of the battle meant that pit stops took on an added dimension as every advantage was sought on and off the track.

Hamilton took an important victory at Silverstone in front of his home crowd to cut the deficit to Rosberg to a mere four points.

The fight between the two world championship contenders intensified when they collided controversially on lap two in Belgium. Rosberg managed to hold on to second place, extending his lead over Hamilton who was forced to retire.

The incident at Spa heightened the rivalry between the two drivers with a more determined Hamilton taking the win at Monza as he began a streak of five consecutive victories.

The night race in Singapore became the turning point of the season as Hamilton was finally able to overhaul Rosberg in the standings with an emphatic victory.

Hamilton continued his great form with another victory in a rain- and accident-affected Japanese Grand Prix.

Hamilton won the inaugural race in Russia ahead of Rosberg which was enough to secure for Mercedes their first World Constructors' Championship.

Hamilton took his fifth consecutive win in the United States, again ahead of Rosberg.

Rosberg took his fifth win of the season in Brazil to ensure that the title would go down to a decider in Abu Dhabi at the final race of the season.

With double points on offer, both drivers sought out any tiny advantage before the race that would give them victory, but in the end it was Hamilton who triumphed, winning the race and the title.

The 2008 champion became only the fourth Briton to win two world titles. He moved level with Jim Clark and Graham Hill on two titles and is one behind Sir Jackie Stewart.

Hamilton became Mercedes' first champion since Juan Manuel Fangio's back-to-back titles in 1954 and 1955.

Hamilton said
in the podium
interview "This is
the greatest day
of my life."

At the start of
the 2015 season,
Mercedes
unveiled their new
AMG PETRONAS
F1 W06 Hybrid,
the next chapter
in their pursuit of
excellence.

01 *

Initial speed...........................341 (km/h)
Final speed............................85 (km/h)
Stopping distance.................139 (m)
Braking time...........................2.60 (sec)
Max deceleration....................5.54 (g)
Max pedal load......................135 (kg)
Braking power ·················· 2387 (kw)

02

Initial speed...........................85 (km/h)
Final speed............................75 (km/h)
Stopping distance.................2 (m)
Braking time...........................0.10 (sec)
Max deceleration....................1.37 (g)
Max pedal load......................26 (kg)
Braking power ·················· 128 (kw)

04

Initial speed...........................324 (km/h)
Final speed............................109 (km/h)
Stopping distance.................110 (m)
Braking time...........................1.87 (sec)
Max deceleration....................5.10 (g)
Max pedal load......................125 (kg)
Braking power.......................2096 (kw)

06

Initial speed...........................261 (km/h)
Final speed............................186 (km/h)
Stopping distance.................44 (m)
Braking time...........................0.71 (sec)
Max deceleration....................3.65 (g)
Max pedal load......................91 (kg)
Braking power.......................1225 (kw)

07

Initial speed...........................261 (km/h)
Final speed............................166 (km/h)
Stopping distance.................56 (m)
Braking time...........................0.97 (sec)
Max deceleration....................3.64 (g)
Max pedal load......................91 (kg)
Braking power.......................1223 (kw)

08

Initial speed...........................335 (km/h)
Final speed............................178 (km/h)
Stopping distance.................87 (m)
Braking time...........................1.26 (sec)
Max deceleration....................5.36 (g)
Max pedal load......................131 (kg)
Braking power.......................2267 (kw)

11

Initial speed...........................331 (km/h)
Final speed............................194 (km/h)
Stopping distance.................76 (m)
Braking time...........................1.07 (sec)
Max deceleration....................5.27 (g)
Max pedal load......................129 (kg)
Braking power.......................2205 (kw)

* Turn 01 is considered the most demanding for the braking system.

Singapore Grand Prix September 2015
Marina Bay Street Circuit (Singapore)

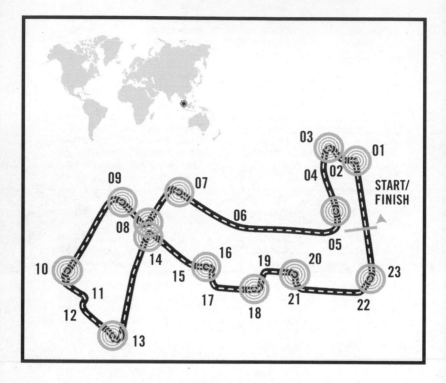

Circuit Braking Data

MARINA BAY STREET CIRCUIT (SINGAPORE)

As they pick their way through the turns and chicanes on the Singapore Street Circuit in their single-seaters, the drivers are well aware that they will put a lot of stress on their brakes with almost a quarter of the time spent on them. None of the thirteen braking sections that characterize this circuit are particularly demanding, but the hot pace and the lack of adequate opportunities for cooling make it one of the toughest on the braking systems. Friction material wear is one of the factors that needs to be monitored constantly by telemetry during each lap of the race.

- **Length:** 5,073 m
- **Number of laps:** 61
- **Type of circuit:** Hard
- **Number of brakings:** 13
- **Time spent under braking per lap:** 20%

01

Initial speed..........................304 (km/h)
Final speed............................145 (km/h)
Stopping distance.................87 (m)
Braking time..........................1.46 (sec)
Max deceleration...................4.79 (g)
Max pedal load......................116 (kg)
Braking power ··················1815 (kw)

03

Initial speed..........................142 (km/h)
Final speed............................89 (km/h)
Stopping distance.................26 (m)
Braking time..........................0.78 (sec)
Max deceleration...................1.93 (g)
Max pedal load......................48 (kg)
Braking power ··················350 (kw)

05

Initial speed..........................257 (km/h)
Final speed............................138 (km/h)
Stopping distance.................67 (m)
Braking time..........................1.27 (sec)
Max deceleration...................3.79 (g)
Max pedal load......................93 (kg)
Braking power......................1237 (kw)

07 *

Initial speed..........................311 (km/h)
Final speed............................109 (km/h)
Stopping distance.................108 (m)
Braking time..........................2.00 (sec)
Max deceleration...................4.98 (g)
Max pedal load......................119 (kg)
Braking power......................1917 (kw)

08

Initial speed..........................205 (km/h)
Final speed............................70 (km/h)
Stopping distance.................66 (m)
Braking time..........................1.68 (sec)
Max deceleration...................2.84 (g)
Max pedal load......................69 (kg)
Braking power......................724 (kw)

09

Initial speed..........................192 (km/h)
Final speed............................130 (km/h)
Stopping distance.................32 (m)
Braking time..........................0.72 (sec)
Max deceleration...................2.61 (g)
Max pedal load......................61 (kg)
Braking power......................608 (kw)

10

Initial speed..........................271 (km/h)
Final speed............................130 (km/h)
Stopping distance.................79 (m)
Braking time..........................1.50 (sec)
Max deceleration...................4.07 (g)
Max pedal load......................100 (kg)
Braking power......................1384 (kw)

13

Initial speed..........................170 (km/h)
Final speed............................123 (km/h)
Stopping distance.................27 (m)
Braking time..........................0.67 (sec)
Max deceleration...................2.27 (g)
Max pedal load......................55 (kg)
Braking power......................476 (kw)

14

Initial speed..........................279 (km/h)
Final speed............................82 (km/h)
Stopping distance.................97 (m)
Braking time..........................1.97 (sec)
Max deceleration...................4.25 (g)
Max pedal load......................104 (kg)
Braking power......................1484 (kw)

16

Initial speed..........................236 (km/h)
Final speed............................92 (km/h)
Stopping distance.................71 (m)
Braking time..........................1.55 (sec)
Max deceleration...................3.40 (g)
Max pedal load......................80 (kg)
Braking power......................1009 (kw)

* Turn 07 is considered the most demanding for the braking system.

Japanese Grand Prix September 2015
Suzuka Circuit (Suzuka)

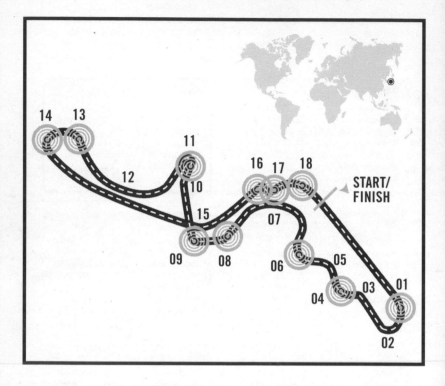

Circuit Braking Data

SUZUKA CIRCUIT (SUZUKA)

As with the other superquick tracks, the long, fast turns at Suzuka also mean that braking is relatively light. In fact, the single-seaters do not face any particularly sudden braking sections except for the 130R turn where they go from more than 300 km/h to about 120 km/h in less than 100 metres.

- **Length:** 5,807 m
- **Number of laps:** 53
- **Type of circuit:** Light
- **Number of brakings:** 11
- **Time spent under braking per lap:** 10%

01

Initial speed .... 306 (km/h)
Final speed .... 130 (km/h)
Stopping distance .... 53 (m)
Braking time .... 1.10 (sec)
Max deceleration .... 4.22 (g)
Max pedal load .... 90 (kg)
Braking power ····················· 1124 (kw)

04

Initial speed .... 224 (km/h)
Final speed .... 218 (km/h)
Stopping distance .... 4 (m)
Braking time .... 0.06 (sec)
Max deceleration .... 2.86 (g)
Max pedal load .... 53 (kg)
Braking power ····················· 641 (kw)

06

Initial speed .... 202 (km/h)
Final speed .... 179 (km/h)
Stopping distance .... 15 (m)
Braking time .... 0.28 (sec)
Max deceleration .... 2.52 (g)
Max pedal load .... 68 (kg)
Braking power .... 655 (kw)

08

Initial speed .... 279 (km/h)
Final speed .... 195 (km/h)
Stopping distance .... 51 (m)
Braking time .... 0.79 (sec)
Max deceleration .... 3.82 (g)
Max pedal load .... 105 (kg)
Braking power .... 1321 (kw)

09

Initial speed .... 214 (km/h)
Final speed .... 120 (km/h)
Stopping distance .... 59 (m)
Braking time .... 1.31 (sec)
Max deceleration .... 2.71 (g)
Max pedal load .... 75 (kg)
Braking power .... 712 (kw)

11

Initial speed .... 254 (km/h)
Final speed .... 65 (km/h)
Stopping distance .... 89 (m)
Braking time .... 2.67 (sec)
Max deceleration .... 3.37 (g)
Max pedal load .... 89 (kg)
Braking power .... 1071 (kw)

13

Initial speed .... 293 (km/h)
Final speed .... 173 (km/h)
Stopping distance .... 73 (m)
Braking time .... 1.16 (sec)
Max deceleration .... 4.08 (g)
Max pedal load .... 111 (kg)
Braking power .... 1491 (kw)

14

Initial speed .... 190 (km/h)
Final speed .... 162 (km/h)
Stopping distance .... 16 (m)
Braking time .... 0.32 (sec)
Max deceleration .... 2.35 (g)
Max pedal load .... 63 (kg)
Braking power .... 563 (kw)

16 *

Initial speed .... 295 (km/h)
Final speed .... 89 (km/h)
Stopping distance .... 123 (m)
Braking time .... 2.53 (sec)
Max deceleration .... 4.12 (g)
Max pedal load .... 112 (kg)
Braking power .... 1512 (kw)

17

Initial speed .... 87 (km/h)
Final speed .... 77 (km/h)
Stopping distance .... 3 (m)
Braking time .... 0.09 (sec)
Max deceleration .... 1.35 (g)
Max pedal load .... 38 (kg)
Braking power .... 151 (kw)

* Turn 16 is considered the most demanding for the braking system.

Russian Grand Prix October 2015
Sochi Autodrome (Sochi)

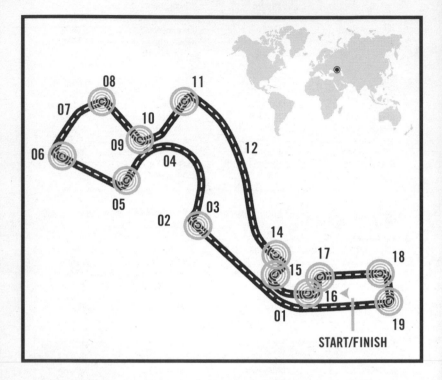

START/FINISH

Circuit Braking Data

SOCHI AUTODROME (SOCHI)

The F1 circus came to Sochi for the first time in 2014 and the teams will still need to pay close attention to the temperature of brake discs and calipers. Sochi is not one of the most challenging circuits for the braking system, even if the management of the friction material temperature is the key to managing the race with the guarantee of consistent performance and controlled wear. The most critical aspect, with regard to the braking system, is linked to the correct sizing of air intakes that ensure the optimum operating temperature for the brakes.

- **Length:** 5,853 m
- **Number of laps:** 53
- **Type of circuit:** Hard
- **Number of brakings:** 12
- **Time spent under braking per lap:** 10%

02 *

Initial speed .... 312 (km/h)
Final speed .... 107 (km/h)
Stopping distance .... 105 (m)
Braking time .... 1.94 (sec)
Max deceleration .... 5.27 (g)
Max pedal load .... 126 (kg)
Braking power ···················· 2046 (kw)

05

Initial speed .... 282 (km/h)
Final speed .... 130 (km/h)
Stopping distance .... 73 (m)
Braking time .... 1.29 (sec)
Max deceleration .... 4.53 (g)
Max pedal load .... 108 (kg)
Braking power ···················· 1599 (kw)

06

Initial speed .... 270 (km/h)
Final speed .... 136 (km/h)
Stopping distance .... 71 (m)
Braking time .... 1.29 (sec)
Max deceleration .... 4.28 (g)
Max pedal load .... 104 (kg)
Braking power .... 1453 (kw)

08

Initial speed .... 264 (km/h)
Final speed .... 141 (km/h)
Stopping distance .... 60 (m)
Braking time .... 1.12 (sec)
Max deceleration .... 4.14 (g)
Max pedal load .... 94 (kg)
Braking power .... 1245 (kw)

09

Initial speed .... 244 (km/h)
Final speed .... 146 (km/h)
Stopping distance .... 53 (m)
Braking time .... 1.00 (sec)
Max deceleration .... 3.70 (g)
Max pedal load .... 89 (kg)
Braking power .... 1130 (kw)

11

Initial speed .... 260 (km/h)
Final speed .... 128 (km/h)
Stopping distance .... 68 (m)
Braking time .... 1.28 (sec)
Max deceleration .... 4.05 (g)
Max pedal load .... 97 (kg)
Braking power .... 1327 (kw)

14

Initial speed .... 306 (km/h)
Final speed .... 116 (km/h)
Stopping distance .... 101 (m)
Braking time .... 1.86 (sec)
Max deceleration .... 5.11 (g)
Max pedal load .... 122 (kg)
Braking power .... 1960 (kw)

15

Initial speed .... 146 (km/h)
Final speed .... 139 (km/h)
Stopping distance .... 4 (m)
Braking time .... 0.10 (sec)
Max deceleration .... 2.03 (g)
Max pedal load .... 39 (kg)
Braking power .... 337 (kw)

16

Initial speed .... 225 (km/h)
Final speed .... 112 (km/h)
Stopping distance .... 55 (m)
Braking time .... 1.16 (sec)
Max deceleration .... 3.32 (g)
Max pedal load .... 80 (kg)
Braking power .... 946 (kw)

17

Initial speed .... 135 (km/h)
Final speed .... 192 (km/h)
Stopping distance .... 3 (m)
Braking time .... 0.10 (sec)
Max deceleration .... 1.91 (g)
Max pedal load .... 38 (kg)
Braking power .... 294 (kw)

* Turn 02 is considered the most demanding for the braking system.

United States Grand Prix November 2015
Circuit of the Americas (Austin)

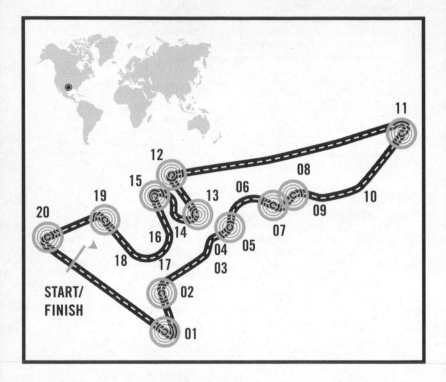

Circuit Braking Data

CIRCUIT OF THE AMERICAS (AUSTIN)

The Austin track can be considered to have a medium demand on the braking system with the drivers using the brakes for about 15 per cent of the time on each lap. However, the circuit is characterized by two very sudden braking sections. Turn Twelve is worth a mention. It is one of the most demanding of the season in terms of dissipated energy and one of the most sudden for the driver with a G-force of 5.5 Gs.

- **Length:** 5,513 m
- **Number of laps:** 56
- **Type of circuit:** Medium
- **Number of brakings:** 11
- **Time spent under braking per lap:** 14%

01

Initial speed .... 310 (km/h)
Final speed .... 75 (km/h)
Stopping distance .... 102 (m)
Braking time .... 1.86 (sec)
Max deceleration .... 5.15 (g)
Max pedal load .... 134 (kg)
Braking power .... 1990 (kw)

02

Initial speed .... 280 (km/h)
Final speed .... 271 (km/h)
Stopping distance .... 4 (m)
Braking time .... 0.05 (sec)
Max deceleration .... 4.44 (g)
Max pedal load .... 79 (kg)
Braking power .... 1181 (kw)

05

Initial speed .... 260 (km/h)
Final speed .... 231 (km/h)
Stopping distance .... 15 (m)
Braking time .... 0.22 (sec)
Max deceleration .... 4.02 (g)
Max pedal load .... 105 (kg)
Braking power .... 1324 (kw)

07

Initial speed .... 220 (km/h)
Final speed .... 209 (km/h)
Stopping distance .... 6 (m)
Braking time .... 0.10 (sec)
Max deceleration .... 3.19 (g)
Max pedal load .... 74 (kg)
Braking power .... 849 (kw)

08

Initial speed .... 202 (km/h)
Final speed .... 147 (km/h)
Stopping distance .... 31 (m)
Braking time .... 0.65 (sec)
Max deceleration .... 2.87 (g)
Max pedal load .... 74 (kg)
Braking power .... 729 (kw)

11

Initial speed .... 287 (km/h)
Final speed .... 80 (km/h)
Stopping distance .... 108 (m)
Braking time .... 2.33 (sec)
Max deceleration .... 4.61 (g)
Max pedal load .... 121 (kg)
Braking power .... 1672 (kw)

12 *

Initial speed .... 323 (km/h)
Final speed .... 81 (km/h)
Stopping distance .... 104 (m)
Braking time .... 1.81 (sec)
Max deceleration .... 5.51 (g)
Max pedal load .... 142 (kg)
Braking power .... 2198 (kw)

13

Initial speed .... 199 (km/h)
Final speed .... 103 (km/h)
Stopping distance .... 42 (m)
Braking time .... 0.96 (sec)
Max deceleration .... 2.81 (g)
Max pedal load .... 72 (kg)
Braking power .... 706 (kw)

15

Initial speed .... 210 (km/h)
Final speed .... 79 (km/h)
Stopping distance .... 31 (m)
Braking time .... 0.62 (sec)
Max deceleration .... 3.01 (g)
Max pedal load .... 76 (kg)
Braking power .... 793 (kw)

19

Initial speed .... 277 (km/h)
Final speed .... 202 (km/h)
Stopping distance .... 40 (m)
Braking time .... 0.60 (sec)
Max deceleration .... 4.38 (g)
Max pedal load .... 112 (kg)
Braking power .... 1526 (kw)

* Turn 12 is considered the most demanding for the braking system.

Brazilian Grand Prix November 2015
Autodromo Jose Carlos Pace (São Paulo)

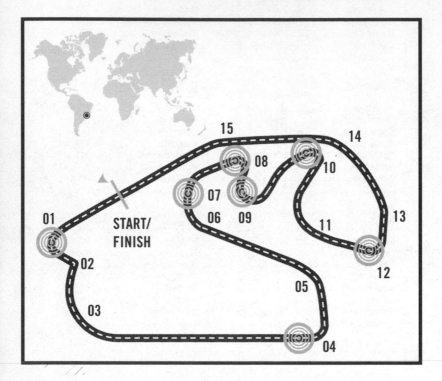

Circuit Braking Data

AUTODROMO JOSE CARLOS PACE (SÃO PAULO)

This is a very 'driven' track with long, fast turns that translate into not-so-demanding braking sections. Of the track's seven braking sections, none are particularly difficult for the braking system, which has plenty of time to cool down despite the fact that the drivers have a foot on the brake pedal for about 14 per cent of the time.

- **Length:** 4,309 m
- **Number of laps:** 71
- **Type of circuit:** Medium
- **Number of brakings:** 7
- **Time spent under braking per lap:** 14%

01 *

Initial speed .... 328 (km/h)
Final speed .... 106 (km/h)
Stopping distance .... 115 (m)
Braking time .... 2.00 (sec)
Max deceleration .... 5.33 (g)
Max pedal load .... 126 (kg)
Braking power ···················· 2170 (kw)

04

Initial speed .... 326 (km/h)
Final speed .... 152 (km/h)
Stopping distance .... 97 (m)
Braking time .... 1.56 (sec)
Max deceleration .... 5.26 (g)
Max pedal load .... 125 (kg)
Braking power ···················· 2123 (kw)

07

Initial speed .... 291 (km/h)
Final speed .... 224 (km/h)
Stopping distance .... 104 (m)
Braking time .... 0.52 (sec)
Max deceleration .... 4.40 (g)
Max pedal load .... 104 (kg)
Braking power .... 1616 (kw)

08

Initial speed .... 229 (km/h)
Final speed .... 96 (km/h)
Stopping distance .... 63 (m)
Braking time .... 1.34 (sec)
Max deceleration .... 3.15 (g)
Max pedal load .... 74 (kg)
Braking power .... 915 (kw)

09

Initial speed .... 142 (km/h)
Final speed .... 110 (km/h)
Stopping distance .... 14 (m)
Braking time .... 0.38 (sec)
Max deceleration .... 1.87 (g)
Max pedal load .... 44 (kg)
Braking power .... 315 (kw)

10

Initial speed .... 220 (km/h)
Final speed .... 96 (km/h)
Stopping distance .... 67 (m)
Braking time .... 1.52 (sec)
Max deceleration .... 2.99 (g)
Max pedal load .... 70 (kg)
Braking power .... 839 (kw)

12

Initial speed .... 269 (km/h)
Final speed .... 124 (km/h)
Stopping distance .... 85 (m)
Braking time .... 1.64 (sec)
Max deceleration .... 3.94 (g)
Max pedal load .... 94 (kg)
Braking power .... 1333 (kw)

* Turn 01 is considered the most demanding for the braking system.

Etihad Airways Abu Dhabi Grand Prix
September 2015
Yas Marina Circuit (Yas Island)

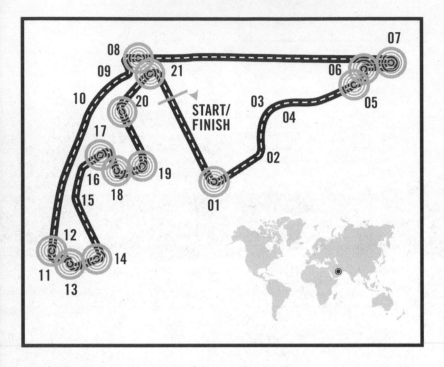

Circuit Braking Data

YAS MARINA CIRCUIT (YAS ISLAND)

Rather like the Bahrain circuit, the same considerations apply here, although the make-up of the track leads to lower speeds and therefore less stress on the brakes. On this track the stress the braking system is subjected to is in any case quite significant and above average: here the drivers spend more than 18 per cent of each lap with a foot on the brake. The thirteen braking sections are quite demanding and the hot pace and torrid climate, with its knock-on effects of increased grip and stress, can create thermal dissipation problems as well as problems with friction material wear.

- **Length:** 5,554 m
- **Number of laps:** 55
- **Type of circuit:** Hard
- **Number of brakings:** 13
- **Time spent under braking per lap:** 18 %

01

Initial speed................284 (km/h)
Final speed................147 (km/h)
Stopping distance................80 (m)
Braking time................1.40 (sec)
Max deceleration................4.08 (g)
Max pedal load................106 (kg)
Braking power ....1447 (kw)

05

Initial speed................290 (km/h)
Final speed................116 (km/h)
Stopping distance................84 (m)
Braking time................1.46 (sec)
Max deceleration................4.22 (g)
Max pedal load................109 (kg)
Braking power ....1521 (kw)

06

Initial speed................121 (km/h)
Final speed................89 (km/h)
Stopping distance................18 (m)
Braking time................0.60 (sec)
Max deceleration................1.66 (g)
Max pedal load................46 (kg)
Braking power................269 (kw)

07

Initial speed................150 (km/h)
Final speed................62 (km/h)
Stopping distance................38 (m)
Braking time................1.21 (sec)
Max deceleration................1.95 (g)
Max pedal load................47 (kg)
Braking power................342 (kw)

08 *

Initial speed................326 (km/h)
Final speed................62 (km/h)
Stopping distance................133 (m)
Braking time................2.58 (sec)
Max deceleration................5.09 (g)
Max pedal load................129 (kg)
Braking power................2026 (kw)

11

Initial speed................316 (km/h)
Final speed................84 (km/h)
Stopping distance................126 (m)
Braking time................2.45 (sec)
Max deceleration................4.82 (g)
Max pedal load................123 (kg)
Braking power................1866 (kw)

13

Initial speed................125 (km/h)
Final speed................115 (km/h)
Stopping distance................6.0 (m)
Braking time................0.17 (sec)
Max deceleration................1.70 (g)
Max pedal load................41 (kg)
Braking power................261 (kw)

14

Initial speed................171 (km/h)
Final speed................102 (km/h)
Stopping distance................40 (m)
Braking time................1.07 (sec)
Max deceleration................2.21 (g)
Max pedal load................57 (kg)
Braking power................466 (kw)

17

Initial speed................253 (km/h)
Final speed................91 (km/h)
Stopping distance................86 (m)
Braking time................1.80 (sec)
Max deceleration................3.49 (g)
Max pedal load................89 (kg)
Braking power................1113 (kw)

18

Initial speed................142 (km/h)
Final speed................109 (km/h)
Stopping distance................16 (m)
Braking time................0.44 (sec)
Max deceleration................1.87 (g)
Max pedal load................45 (kg)
Braking power................315 (kw)

* Turn 08 is considered the most demanding for the braking system.

DATA

THE FASTEST MEN IN F1—THE IT MEN

Data is key to the modern Formula One team and, with the difference between winning and losing being so small, the handling of data represents an opportunity to gain vital fractions of a second. Each lap more than 4 MB of data relating to speed, g-forces, pressures, temperatures and engine performance are transmitted from the race car to the pit system via real-time radio transmissions adding up to a massive 25 GB of telemetry data collected per three-day race weekend. Here we look at how two of the teams and their IT partners meet the data challenge to gain that crucial advantage.

THE DATA WRANGLERS

Graham Hackland, formerly of Lotus F1 Team and now at Williams, explains the role of data during the race weekend: "All of the teams arrive at the circuit with a base configuration for the car that they've done in simulation back at the factory, or from their experience of their previous years at the track, and very quickly on the Friday they have to get as close as they can to the set-up for the car that they think they're going to qualify with. Getting the data up to speed so it can be fed back into the car and analyzed to get the most out of the set-up is absolutely paramount.

"There is a large team back in the factory working on the real-time data that's coming off the car so that by the Saturday, it becomes even more critical that the performance of your data and the applications that we develop are spot on. Being able to access the right data at the right time means that engineers can be faster in getting new and effective ideas into the car and onto the track, and therein lies the competitive edge. By the time of the Grand Prix on Sunday, they have a strategy system that, at the press of a button and within three minutes, can run 10,000 race simulations based on the current situation; for instance it's started raining, there is a crash, track temperature, your competitors, the tyres you've got, the tyres your competitors have got. All of that data collected from the sensors on the car is factored in. This can be done continually at the press of a button, based on current real-time data."

The data wranglers in F1 may very well be the fastest men in the sport as they collect, process and disseminate vast amounts of crucial data at unheard of speeds. According to Hackland, "In less than a second (absolute real-time terms), we're getting the data, processing it and delivering it to the team sitting on the pit wall. Lap by lap, corner by corner, sector by sector, they are getting the data they need in realtime so as it's happening on the car, it's on the pit wall or at the mobile data centre in the back of the garage."

At any point in time, up to thirty engineers may be accessing a single file from either of the two live or the two back-up storage clusters. While the Race Team at the track are analyzing the data to improve performance in minuscule increments, there is an even bigger team back at the factory that are hard at work on larger sets of data to analyze performance in terms of the bigger, long-term picture. Speed, capacity and availability are the key indicators, but these are nothing without reliability. Teams will expect 100 per cent reliability from the car and they expect the same from the IT as well. For the past seven seasons, Lotus has run NetApp at the track with zero downtime; that is, a 100 per cent record for seven consecutive seasons for five days at a time, Wednesday to Sunday. At the end of the five days, the whole system is shut down and either flown or trucked back to the factory. The system has to maintain reliability and performance despite the rugged trackside and transport environments.

There are many in Formula One who talk about how important data is to the sport and to the business and even some who believe that no racing could take place without it. Hackland is quick to provide some perspective. "Sometimes, we IT people think they can't do anything without what we provide, and the reason is that the engineers absolutely need IT for the running of their jobs. The reality is that we should just be in the background providing the race team with the services they need to engineer the car. At the end of each race weekend, my hope is that they don't mention IT at all, then I know we have done our job."

These days, Formula One teams are looking for much more agility in terms of the way they handle data—they need to provide the capability to consume the resources at business level rather than having to refer to IT. Lawrence James, Alliances and Solutions Manager at NetApp, explains: "What we do for F1 is simple: the ability to provision, expand and scale seamlessly. For instance, they've got elastic demand coming through where they need to provision more capacity, and they can do that really quickly. And it's the guys that actually need that capacity who can instigate the commands rather than having to go through IT all of the time. It's automated and self-provisioning, which is key for agility."

He continues: "What is expected these days is that the business and IT are working as a unit and the IT is enabling the business. IT is used in Formula One to effectively *enable* their engineers, both trackside and back at the factory, so that when they need capacity trackside or when they need to get data back to HQ quickly, they can do that. They can provision it themselves in a future-proof, self-sufficient way."

NetApp, in its partnership with Lotus, has moved away from storage islands towards a converged infrastructure so that rather than being a modular approach, it is an integrated approach to both networking and storage so that the two levels work in tandem with each other effectively, virtualizing each layer. James poses the question: "So how do you retire the old, introduce the new, but make it totally transparent to the applications that are running? Our agile data infrastructure does just that: it collapses all of those layers, it virtualizes those layers. When I want to retire an old piece of storage, I can transparently migrate all the data. The application doesn't see anything and isn't disrupted. I can bring in the new. I can rebalance. I can replicate so as to protect the data as well but totally transparently to the server layer and to the application layer."

Agile data is crucial in Formula One. The need to restore data from a few days ago (or even earlier in the season) so that it can be analyzed again and again is vital in the search for any performance advantage. James elaborates: "NetApp intelligently manages the Lotus data by protecting it very efficiently. They use a lot of our ability to copy data in the form of snapshots but with a difference in that the snapshots are created locally but don't consume any more physical storage. We provide the ability to grab data and to grow it efficiently by the use of 'snap mirror'; i.e. replicating from A to B and 'snapshotting' at A or B to create point-in-time copies which are spatially efficient. So, in other words, they are not consuming two, three or four times the capacity with every copy that they make."

Formula One provides NetApp with an opportunity to demonstrate what it does and the capability of the company's gear in a number of different environments from business applications to fluid dynamics through to the race situation. As James adds: "Let's face it; it's a great talking point. It's very positive in terms of talking about it in front of our clients. This technology would work wherever you put it. It's proved itself. There are no excuses for it; it's been there, it's done the job. It's been taken from one side of the world to the other. It never fails."

EVERY MILLISECOND COUNTS
Thanks to increasing high-speed connectivity the back-end is the new front line. When Tata Communications first entered Formula One the Indian Telco giant became Formula One's official connectivity provider. Here was a company that owned the world's largest (and only) round-the-world sub-sea cable network—over 20,000 route kilometres of sub-sea and terrestrial network fibre—allowing a data transfer capacity of one terabit per second. Back then, it was clear that Tata Communications was one of the few companies in the world with the technology to cater for and improve upon Formula One's colossal, split-second connectivity requirements.

Mehul Kapadia, Managing Director of Formula One Business at Tata Communications, and his team were looking forward to showcasing their talents on a global platform by providing the infrastructure for Formula One's connectivity and communications in all twenty race locations and offices worldwide. Melbourne 2012 was Tata Communication's first race, and it was here, and almost immediately, that Kapadia bumped into former CEO of the Mercedes Team, Nick Fry.

From that meeting came an agreement that would deepen the company's involvement in the race: to provide MERCEDES AMG PETRONAS with trackside connectivity, enabling the team to transfer vital real-time data from the Silver Arrow cars at any Grand Prix location to its headquarters in the UK. The relationship with Mercedes puts Tata in the front-line, although, Kapadia explains, thanks to modern high-speed connectivity, the "front-line" can now be several hundred miles from the track.

"[On race days] we run a pit crew of about twenty people who from across the globe are actually running the operation," he says. "It's quite a task, but it's about practising what you preach. So if we are telling the industry that through Telecom you can save on operational cost by doing stuff backstage rather than having to send people to a lot of places, we also optimise it that way. We send these two people to the races; we don't send a battery of thirty people out there.

"There are specialists working out of London, out of India and also, at times, out of the US, depending on the need. We'd never operated within a level of a sport like Formula One before. It is a place for us to demonstrate our capability in delivering solutions and services across the world in crazy timelines, to modify our processes to suit our customers' needs, create response teams, which are so much in the backend but still able to deliver at the front-end."

For Mercedes, the data has to come back to Team HQ in Brackley where it's analyzed and then sent back to the racetrack. The goal is to decrease latency—the split-second time delay in transferring data—and, in the pursuit of efficiency and a competitive edge, every millisecond counts. Through its global network, Tata Communications was able to upscale Mercedes' connectivity to make it three times faster.

"The laws of physics mean latency can't completely go away," says Kapadia. "There will always be latency. Right now, when I'm talking to you, you can immediately hear me but if you were at a distance, the voice would take more time to travel.

It's sheer physics. How do I ensure that even when you are in Australia I can minimize that latency? If there are x-hundred milliseconds happening, I reduce a few tens of milliseconds out of it. That also helps because if you can get it faster, if you can analyze it faster, you are going to make your decision a split second faster.

"It seems simplistic, but when you scale it up that much—by three times—imagine the amount of potential you've opened for those engineers. Let's not kid ourselves; we are a telecom player, we are not going to change a particular component on the car. But it can enable the people who *are* doing that to do their work with more flexibility, and three times faster."

Tata Communications hosts technical workshops with Mercedes during which the company explains the depth of its own services and listens to the needs of its client; although the relationship with Mercedes is that of service provider and customer, as opposed to an in-kind technical partnership or sponsorship. There is, however, a marketing partnership in existence. The opportunity to align itself with the Mercedes brand and its assets is just too attractive, says Kapadia, but he hastens to add that this agreement is independent of the company's role as a service provider. "This is truly a success, where you can have win-win between partners. It's not actually about sponsorship, I think. Formula One is way beyond that."

The deal with Mercedes is a multi-year contract and while immediate plans are under wraps, Kapadi sees an opportunity for all of the teams to improve their connectivity. "The way I look at it is that this is an opportunity to look at all the teams. Mercedes is definitely the team that we started work with, but all the other teams also need to look at getting into a strong relationship with a strong Telco that has the capability and the commitment to the sport.

"I'm sure what people are getting today is good. It's not about bad becoming good, it's about how you can better what is already there. Three times capacity is not to say that one times capacity was bad. It is just that three gives you the opportunity to do more. Our theme for the year is to provide to our customers the speed to lead. This is our way of saying that we truly believe that, with our kind of fibre capability globally, our kind of telecom experience, and our kind of data centre technology, we can help teams to look at telecommunications in a different way."

FUEL AND LUBRICANTS

FUEL

Two new restrictions were introduced at the start of last season that placed the spotlight on the role of fuels in Formula One. 100 kg maximum mass (around 35 per cent less than in 2013), and a maximum fuel flow limited to 100 kg/h (no limits in 2013). F1 fuels contain more than 200 ingredients providing 100 ignitions per cylinder per second, and last season more than twenty F1 fuel formulas were produced by Total as part of their long-term partnership with Renault Sport F1 and their partner teams. This partnership has produced 158 victories and twenty-one world titles over a twenty-seven-year period, so it is to them that we turn to examine how they embrace the new philosophy—optimization of energy efficiency—that will bring Formula One ever closer to the car industry and the everyday motorist.

FUELS: AN ESSENTIAL CONTRIBUTION TO ENERGY SAVINGS

The power of an engine is down to the quality of its combustion. Optimizing the performance of a turbo engine also comes from maximizing the flexibility of the engine over a wide rev band and a large number of set-ups. In F1, igniting the air-fuel mixture and burning it as quickly as possible is an on-going preoccupation. If fuel pump petrol can be likened to ready-to-wear, that used in Formula 1 is a bespoke product. Total formulates it, and as soon as it has been tested and validated by Renault Sport F1 and homologated by the FIA, produces and transports it to the circuits.

The Formula 1 fuel regulations are strict. They control the formulation by limiting the Hydrocarbons content that go into the final product. An F1 car must race on fuel that is very similar to the unleaded super at the pump and complies with draconian restrictions: except for one point—the octane rating is not limited. Thus, despite an apparent rigour, there is room for manoeuvre which allows the researchers to make the difference. As long as the ingredients that make up the F1 fuel are regulated and their proportions controlled then the recipe is *almost* open to interpretation.

OPTIMIZING THE INJECTION

The new power units introduced in 2014 have direct injection so the fuel is injected directly into the combustion chamber as in a modern diesel engine: maximum revs are limited to 15,000 rpm. In reality, the combination between the search for reliability, the fuel flow limit and the way the lean mixture works imposes a working rev band of around 11,000 rpm. Nonetheless, because of direct injection the new power unit should be fed by a fuel that vaporizes very quickly. The injection is triggered during the compression phase just before the ignition and the power stroke. It is managed electronically to be as quick as it is accurate. The high-pressure injection favours vaporization while the spray of petrol goes into fresh gases and the vaporization can also be enhanced by the quality of the fuel. The sprays should be very short to vaporize quickly without wetting the surface of the pistons or the cylinder sleeves sleeves too much. In the past, indirect injection above the valve took more time and the temperature and pressure were less critical.

STABILIZING COMBUSTION

Running an electric motor on the shaft of the turbocharger by the exhaust gases to generate electricity is an innovative technology to harvest energy. It is efficient, however, only when it can pass the largest quantity of gas at the highest possible speed over the turbo's hot turbine placed in the stream of the exhaust gases. The solution? Make the petrol engine work like a diesel with a considerable excess of air thus creating a very lean fuel/air mixture. In addition to rapid ignition, Total formulates a fuel that provides stable combustion in lean mixture conditions. It must also ensure the cooling and the behaviour of the fuel pumps including the feeding pump in the fuel tank, which raises the pressure by a few bars, and feeds the high pressure pump generating the injection pressure.

MASTERING KNOCKING

The role of the fuel can be summed up in just a few words. To liberate the power it has to combust very quickly and thus it needs to be vaporized very rapidly in a homogenous manner. After ignition, the spread of the flame must be almost instantaneous without generating knocking. Turbo engines are particularly prone to this phenomenon as they have a higher thermal load than normally aspirated motors. This was one of the challenges of the 2014 season that needs to be met in 2015 and beyond. Knocking is uncontrolled combustion causing explosions, which spread at high speed in the combustion chamber. This explosive phenomenon produces high-speed pressure waves that bounce off the walls causing oscillation with large amplitude in pressure. The consequences are: sharp sounds, destruction of the internal dynamics of gases in the combustion chamber, and heat transfer between the burnt gases and the metal. Thus, in a few seconds a piston can be holed or destroyed by fusion. The aerodynamics of the combustion chamber, the flow from the injectors, the design of the plugs, the electronics and the quality of the fuel all have to be taken into account.

ENERGY CONTENT AND DENSITY

Our researchers supply a fuel with a much higher octane rating. The 100 kg of fuel allowed for the race has to be formulated in such a way that it contains the maximum amount of energy that the engine can use. This bespoke formulation for the Renault Energy 2015-F1 has a few secrets. Out of the 100 kg fuel load a 0.025 difference in the density leads to a reduction of 5 litres in the contents and the size of the fuel tank. Thus, Total formulates its fuels to meet the requirements of the Renault Energy 2015-F1 power unit without penalizing the design of the chassis. A single aim: find the best possible compromise in terms of density to meet the needs of the chassis (size of the fuel tank) and of the engine while ensuring bulletproof resistance to the temperature. This is a crucial point as the storage batteries are installed under the fuel tank, which they heat.

INGREDIENTS OF NON-FOSSIL ORIGIN

Since 2008, the law imposes ingredients of non-fossil origin in fuels—the bio-fuels. One of the fundamental elements in the recipe is the obligation to incorporate 5.75 per cent (in mass) of renewable ingredients in the fuels. These non-fossil molecules may contain oxygen as well as ethanol and hydrocarbons obtained from decomposition by bacteria of vegetable waste, which are not part of the food consumption chain. They are the second-generation or renewable bio-fuels. However, the oxygen sometimes found in these bio-ingredients limits the calorific power, hence the energy available in the fuel, but it provides ingredients like alcohol, which give a high octane rating. Analyzing the combustion is vital to find the best compromise on the turbo engine between reliability provided by the high octane rating and the maximum power available.

ADDITIVES

Additives are crucial for the reliability of the engine, both to avoid failure and also the progressive loss of power during its life cycle. Our work on the resistance to knocking and in the field of lubrication answers these problems. While Renault Sport F1 has acquired in-depth knowledge about the optimization of parts, Total knows how to mix additives with the fuel to reduce friction on the rings, prevent deposits forming on the piston crowns and the hot areas of the engine, lubricate the high-pressure petrol pumps and prevent the injectors from cocking. These active molecules have different properties: cleaning, anti-corrosion, anti-oxidation, reduction of emulsion with the air, modification of the frictions, amelioration of the combustion and the octane rating.

PRACTICALLY SPEAKING: ON-GOING DIALOGUE

Formulating the best possible F1 fuel means knowing all the secrets of the engine. The history of the partnership between Total and Renault Sport F1 is part of this synergy. There is complete openness between the Total Research Centre (CReS), Total ACS and the Renault Sport F1 factory and the engineers are in permanent contact. Certain parts of the engine are tested in the CReS. Today, our shared experience means we can start collaborating on the design of the combustion chamber to optimize its functioning whether through simulation or tests on real parts. It's a major advantage.

THE FINAL RESULT

It is a fuel comprising dozens of ingredients specially formulated for the Energy 2015-F1 power unit guaranteeing top-class performance. The V6 turbocharged Renault engine is fed by a lead-free super petrol with a high octane rating (antidetonation properties and power gain), excellent combustion stability (anti-knock), and a high energy content.

DEVELOPMENT

Numerous evolutions of F1 fuels are tested every season. Once the optimum formulation has been chosen and validated by Renault Sport F1, Total homologates the product with the FIA (Fédération Internationale de l'Automobile). It's the birth certificate of the fuel. Some ten litres are sent to the Federation's laboratory in Great Britain to receive the legislator's agreement. The analysis of this sample by gas chromatography gives the genetic code of the fuel, which will serve as a reference until the homologation of the next batch. Another barrel of twenty-five litres is sent to one of the FIA's service providers for the calibration of the cars' fuel flow metre. Each time the formulation changes a new validation is required. The homologation process usually takes three to four weeks.

PRODUCTION

Total produces its F1 fuels in different volumes from 200 litres for specific tests to large volumes for Grand Prix. Twenty to 100 cubic metres are produced per cycle, which guarantees an autonomy of eight to ten race weekends for teams as well as for the Renault Sport F1 test beds. Production takes several days, and putting the fuel into barrels is done quickly to ensure the homogeneity of their contents. In the middle of production a barrel is divided up into several samples for the FIA and for our own references. Finally, each barrel is numbered, a crucial stage that requires precision. Certain ingredients represent less than 1 per cent of the mixture. Total may have to bring several specific fuels to the circuits depending on the choice of the teams.

ANALYZING CONFORMITY AT THE GRAND PRIX

Despite the multiple hurdles (customs, sales and safety regulations, etc), which differ from country to country, the Total ACS
logistics team cannot make mistakes when it comes to transport-
ing the fuels to the Grands Prix. The FIA takes fuel samples on a
random basis during race weekends and also takes a sample from
the cars of the drivers who finish in the top three, meaning that
Total fuels have undergone the highest number of checks over the
past few seasons. Numerous parameters may alter a fuel compared
to its homologated content: a solvent still present after cleaning
the fuel pump or the fuel tank and evaporation of the lightest
molecules if a car has run with a low fuel load. The alteration of
the formulation compared to its reference may lead to the disqualification of a driver or his team. To ensure that the fuel used
always complies with the registered genetic code Total brings a
chromatograph to the circuits. This machine used for checking
can identify the molecules present in the fuel on a daily basis and
ensures that the fuel complies in real time.

LUBRICANTS

Unlike the restrictions placed on fuels, the FIA technical regulations allow considerable freedom in the field of F1 lubricants where
the challenge is to ensure reliability while searching for performance. A wide range of ingredients and additives are used to
formulate the most suitable products that contribute to power and
energy saving, protect the mechanical parts, cool the engine, evac-
uate air and obtain a high level of thermal stability. The influence of
lubricants is not just limited to the engine (lubricants and cooling
liquid) but also plays a role in terms of wheel bearings (grease),
power steering (hydraulic fluid), brake liquid, KERS (cooling
liquid), gearbox (lubricant), transmission (grease) and the DRS
operation system (hydraulic fluid).

POWERTRAIN

THE POWER UNIT ARRIVES—NO LONGER JUST AN "ENGINE"

There is no doubt that in 2014 Mercedes got it more right than their competitors when it came to responding to a set of revolutionary technological challenges. The team's solution to the new powertrain regulations was an integrated approach that focused on maximizing the lap time of the overall technical package. Over the three-year lead time each trade-off was discussed and debated in detail in order to find the optimum overall solution. The result was a power unit and chassis designed in parallel for the full works advantage to be realized. Here they place the challenge in context, as well as providing the technology, terminology and specifications of the triumphant PU106A Hybrid Power Unit.

FORMULA ONE: A HISTORY OF INNOVATION

The technical revolution of 2014 can be expressed in one simple phrase: the engine is no more, long live the Power Unit!

The idea of the engine as a standalone source of propulsion in Formula One was consigned to history several years ago through the introduction of KERS Hybrid power in 2009 and from 2011 through 2013. That said, the change for 2014 was altogether more far-reaching.

Out went the 2.4-litre, normally aspirated V8 power plants used over the past eight years. In came a 1.6-litre, turbocharged V6 configuration with integrated Hybrid Energy Recovery System (ERS) to form the Power Unit. Each driver is limited to a maximum of five Power Units per championship; three fewer than the allocation of eight in 2013.

This latest amendment to the powertrain regulations draws on a long line of similar—if sometimes less far-reaching—regulation changes dating back to the very beginnings of Formula One. In the early years of the competition from 1950 to 1953, 4.5-litre normally aspirated and 1.5-litre supercharged engines were permitted (although races were run to Formula 2 regulations in 1952 and 1953), before the introduction of a restricted 2.5-litre maximum capacity in 1954; the same year in which Mercedes first entered the championship with its W196 R.

For 1961, maximum engine capacity was reduced to 1.5 litres, at the same time as the "rear-engine revolution" took hold of chassis technology. Although initially underpowered, the units quickly grew in power output; eventually resulting in faster lap times than those seen under the previous regulations, and setting a trend that has continued throughout each era of Formula One powertrain technology to date.

With Formula One beginning to fall behind the more powerful sports cars in the mid-1960s, maximum engine capacity was raised to 3.0 litres with 1.5-litre compression charged formats also permitted. The 3.0-litre format was the norm until, in 1977, Renault exploited the opportunity of turbocharging for the first time. Where the French manufacturer led others soon followed, with every championship from 1983 to 1988 won by turbo power until the technology was outlawed at the end of the year.

The ban on pressure charging led to larger capacity engines being reintroduced as the sport's governing body sought to allay fears that Formula One would once again fall behind sports cars as the world's fastest racing category. Between 1989 and 1994, a mandatory 3.5-litre maximum capacity was set in place, before being reduced to 3.0 litres in 1995 as constant development of the unit began to produce ever-higher levels of power. The high-revving, high-pitched screaming era of the V10 was thus born; recognized by many as a peak of uniformly regulated Formula One engine performance.

Fast forward to 2006 and the latest incarnation of regulations—driven by the twin objectives of capping performance and controlling costs—introduced a 2.4-litre, normally aspirated V8 configuration of minimum 95 kg weight. The reduction in capacity was designed to give a power reduction of around 20 per cent from the three-litre engines, however constant development meant that performance consistently improved. Further restrictions introduced in 2007 saw engine specification homologated in order to contain development costs.

The new rules for 2014 marked a watershed for the sport of Formula One, with a set of regulations written to encourage and promote the development of advanced new technologies with which efficiency and performance will become synonymous. Where previous revolutions were prompted by engineers identifying and exploiting opportunities in the regulations, this step change in Power Unit technology has been applied across the board from 2014.

These rules position Formula One firmly at the cutting edge of automotive technology, redefining what's possible in the field of engineering and actively encouraging innovation to stretch technological boundaries—in other words, exactly what Formula One has been about since its early days.

The Technological Revolution
V8 to V6

	V8 (2006–13)	V6 (2014)
Capacity	2,400cc V8	1,600cc V6
Maximum RPM	18,000	15,000
Fuel Mass Flow	Unlimited	Max 100kg/hr
Admission	Normally Aspirated	Single-stage Compressor & Exhaust Turbine
Minimum Weight	ICE = 95 kg KERS = Unlimited	Power Unit = 145 kg

KERS to ERS

	KERS (2009, 2011–13)	ERS (2014)
Components	MGU-K Power Electronics Energy Store	MGU-K, MGU-H Power electronics Energy Store
Power	MGU-K, Max 60 kW	MGU-K, Max 120 kW MGU-H, Unlimited

Energy Input	No Maximum	Max 2 MJ per lap MGU-K Unlimited MJ per lap MGU-H
Energy Output	400 kj	4 MJ to MGU-K
Weight	No Regulation	ES between 20–25 kg (must be contained within Survival Cell)

Technology and Terminology

Power Unit

In regulatory terms, the Power Unit comprises six different systems:
Internal Combustion Engine (ICE)
Motor Generator Unit-Kinetic (MGU-K)
Motor Generator Unit-Heat (MGU-H)
Energy Store (ES)
Turbocharger (TC)
Control Electronics (CE)

The change in terminology reflects the fact that this new power-train is far more that simply an internal combustion engine. Where the previous V8 format utilized a KERS hybrid system, which was effectively "bolted on" to a pre-existing engine configuration, the Mercedes-Benz PU106A Hybrid has been designed from the outset with hybrid systems integral to its operation.

ICE

The Internal Combustion Engine (ICE) is the traditional fuel-powered heart of the Power Unit previously known simply as the engine. From 2014 this took the form of a 1.6-litre turbo-charged V6 configuration with direct fuel injection up to 500 bar of pressure. Where the V8 engines could rev up to 18,000 rpm the ICE is limited to 15,000 rpm from 2014 onwards. This reduction in crankshaft rotational speed coupled with the reduction in engine capacity and number of cylinders reduces the friction and thus increases the total efficiency of the Power Unit. This down-speeding and down-sizing approach is the key technological change at the heart of the ICE structure.

Turbocharger

The turbocharger is an energy recovery device that uses waste exhaust energy to drive a single stage exhaust turbine that in turn drives a single stage compressor via a shaft thereby increasing the pressure of the inlet charge (the air admitted to the engine for combustion). The increased pressure of the inlet charge offsets the reductions in engine capacity and RPM when compared to the V8, thus enabling high power delivery from a down-speeded, down-sized engine. The turbocharger is the key system for increasing the efficiency of the ICE.

ERS

From 2014 the notion of hybrid energy shed a letter with KERS becoming ERS, but with a significant increase in sophistication. Energy could still be recovered and deployed to the rear axle via a Motor Generator Unit (MGU) however this is now termed the MGU-K (for "Kinetic") and is permitted twice the maximum power of the 2013 motor (120 kW or 161 hp instead of 60 kW or 80.5 hp). It may recover five times more energy per lap (2 MJ) and deploy ten times as much (4 MJ) compared to its 2013 equivalent equating to over thirty seconds per lap at full power. The rest of the energy is recovered by the MGU-H (for "Heat") an electrical machine connected to the turbocharger. Where the V8 offered one possible "energy journey" to improve efficiency via KERS there are up to seven different efficiency enhancing energy journeys in the ERS system.

MGU-K

The Motor Generator Unit-Kinetic (MGU-K) has double the power capability of the previously used KERS motors and operates in an identical way. Some of the kinetic energy that would normally be dissipated by the rear brakes under braking is converted into electrical energy and stored in the Energy Store. Then, when the car accelerates, energy stored in the Energy Store is delivered to the MGU-K which provides an additional boost up to a maximum power of 120 kW (approximately 160 hp) to the rear axle for over thirty seconds per lap.

MGU-H
The Motor Generator Unit-Heat (MGU-H) is a new electrical machine that is directly coupled to the turbocharger shaft. Waste exhaust energy that is in excess of that required to drive the compressor can be recovered by the turbine, harvested by the MGU-H, converted into electrical energy and stored in the Energy Store. Where the MGU-K is limited to recovering 2 MJ of energy per lap, there is no limit placed on the MGU-H. This recovered energy can be used to power the MGU-K when accelerating or can be used to power the MGU-H in order to accelerate the turbocharger, thus helping to eliminate "turbo lag". This new technology increases the efficiency of the Power Unit and most significantly provides a method to ensure good driveability from a boosted, down-sized engine.

ES
The Energy Store (ES) does exactly what it says on the tin; storing the energy harvested from the two Motor Generator Units (MGUs) for deployment back into those same systems. It is capped in terms of maximum and minimum weight: the maximum (25 kg) setting engineers an aggressive target while the minimum (20 kg) means weight reduction will not be chased at all costs.

kJ, MJ and kW
A joule (J) is a unit of energy; (kinetic, heat, mechanical, electrical etc.) A kilojoule (kJ) is equal to one thousand joules while a megajoule (MJ) represents one million joules. To put this into context, kJ is a unit often used to describe the energy present in nutritional goods while 1 MJ represents the approximate kinetic energy of a one-tonne vehicle travelling at 160 km/h. Meanwhile, a watt (W) is a unit of power that quantifies a rate of energy flow; with a kilowatt (kW) equal to one thousand watts. This unit is commonly used to express the power output of an engine, where 1 kW is equal to 1.34 horsepower.

Mercedes-Benz PU106A Hybrid Technical Specification

Power Unit Specification

Type:	Mercedes-Benz PU106A Hybrid
Minimum weight:	145 kg
Power Unit Perimeter:	Internal Combustion Engine (ICE)
	Motor Generator Unit—Kinetic (MGU-K)
	Motor Generator Unit—Heat (MGU-H)
	Energy Store (ES)
	Turbocharger (TC)
	Control Electronics (CE)
Power Unit Allocation:	Five Power Units per driver per season

Internal Combustion Engine

Capacity:	1.6 litres
Cylinders:	Six
Bank angle:	90
No of valves:	24
Max rpm ICE:	15,000 rpm
Max fuel flow rate:	100 kg/hour (above 10,500 rpm)
Fuel injection:	High-pressure direct injection (max 500 bar, one injector/cylinder)
Pressure charging:	Single-stage compressor and exhaust turbine on a common shaft
Max rpm exhaust turbine:	125,000 rpm

Energy Recovery System

Architecture:	Integrated Hybrid energy recovery via electrical Motor Generator Units
Energy Store:	Lithium-Ion battery solution, between 20 and 25 kg
Max energy storage/lap:	4 MJ
Max rpm MGU-K:	50,000 rpm
Max power MGU-K:	120 kW (161 hp)

Max energy recovery/lap
MGU-K: 2 MJ
Max energy deployment/
lap MGU-K: 4 MJ (33.3 s at full power)
Max rpm MGU-H: 125,000 rpm
Max power MGU-H: Unlimited
Max energy recovery/
lap MGU-H: Unlimited
Max energy deployment/
lap MGU-H: Unlimited

Fuel & Lubricants
Fuel   PETRONAS Primax
Lubricants   PETRONAS Syntium
Gearbox & hydraulic oil PETRONAS Tutela

SAFETY

When Kimi Räikkönen hit the wall head-on at Silverstone in 2014 at 240 km/h, he experienced a 47 g impact. The fact that he was able to walk away from the accident with minor injuries is a testament to the level of safety that exists in modern Formula One. From track safety, cockpit safety, crash tests, medical provision, helmets, HANS (Head & Neck Support), clothing and deployment of the Safety Car, Formula One sets the highest standards in safety and continually strives to improve every aspect taking nothing for granted. For the most comprehensive exploration of safety in F1 please go to www.formula1.com/inside_f1/safety where F1 Global Partner Allianz in association with the official Formula One Website examine each aspect in extensive detail. Below we take a look at two aspects of safety in F1, the Safety Car and Track Design, from the perspective of two key Formula One experts: Bernd Mayländer and Philippe Gurdjian.

BERND MAYLÄNDER—THE SAFETY CAR DRIVER
Always in the background and still a vital part of Formula One: Bernd Mayländer, the Official Formula One Safety Car driver. The FIA is responsible for the deployment of the Official Formula One Safety Car and sends it onto the track in hazardous situations. Bernd Mayländer gives exclusive insights into his unique role.

The Car
Tell us about the Official Formula One Safety Car, Bernd:
"The Safety Car is a 525-HP Mercedes Benz SL 63 AMG. It is not a standard road car. In order to ensure the best possible reliability even in tropical temperatures, the vehicle incorporates large additional coolers for engine oil, transmission oil, cooling water and power steering. Despite a larger braking system including brake cooling, additional cooling measures, lighting system and communications equipment the Safety Car is significantly lighter than the road version. It has been possible

to dispense with all the mechanical and hydraulic components of the convertible roof. Also, the sound-absorbing materials, which serve primarily to improve the driving comfort, have been removed. The result is a weight reduction of 220 kg and a Safety Car that only weighs 1750 kg."

When are you deployed onto the track?
"According to the official regulations of the Fédération Internationale de l'Automobile (FIA), the car is deployed 'if competitors or officials are in immediate physical danger but the circumstances are not such as to necessitate stopping the race', for example after an accident or in severe rain showers."

What's the role of the Official Formula One Safety Car?
"It takes up its position at the front of the field and leads the Formula One cars around the track at reduced speed until the dangerous situation has passed. All the cars, beginning with the race leader, must line up behind the Safety Car."

Who decides when you should take to the track?
"The decision is made by the FIA race director, Charlie Whiting. He also decides when the Safety Car phase is finished."

How are the drivers informed about the Safety Car phase?
"The drivers are notified by the marshals and light-panels that show yellow flags together with the letters "SC". Additionally the driver is informed via radio by the team and a warning light inside the cockpit flashes until the Safety Car phase is over."

How long does the Safety Car remain on the circuit?
"It will remain until the hazardous situation is under control and the FIA feel that it is safe to resume. The laps completed during the Safety Car phase count as normal race laps. If the specified number of laps is completed, a race can also come to an end behind the Safety Car."

What influence does a Safety Car phase have on the race strategy?
"As a rule, the teams use a Safety Car phase for an unscheduled

pit stop, because it involves a much smaller loss of time than if the field is racing at full speed. If a team manages to bring its driver into the pits at exactly the right time, it can result in a crucial advantage. Because the field is pressed up close together during a Safety Car phase, it also increases the excitement for the spectators."

Since when has the Safety Car been used in Formula One?
"Its first introduction was in 1973 at the Canadian Grand Prix. However, the FIA laid down clear guidelines for the role of the Safety Car in 1992."

The Driver

In 2000 the FIA entrusted the task of driving the Official Formula One Safety Car to Bernd Mayländer, a successful touring-car driver. He knows how to keep the pace during the safety period just high enough so that the race-car tyres do not cool down too much. Bernd started his career in karting at the end of the 1980s. In the following years he progressed to Formula Ford, the Porsche Carrera Cup, the FIA GT Championship and the Deutsche Tourenwagen Meisterschaften (DTM) before becoming the Official Formula One Safety Car driver in 2000.

A part of Formula One—Bernd Mayländer outlines his work environment

"The Safety Car is deployed on Formula One race weekends, so my working week generally starts on Thursday and ends on Sunday. Besides Formula One, the car is also used in other racing series over the course of the weekend, such as the GP2 Series, Porsche Supercup, etc. The same rules and regulations apply as in Formula One. Even if this means additional work for me, it ensures the same safety standard in all series and this is our main objective.

"Our safety team consists of two Safety Cars and a pair of medical cars—we transport approximately 3.5 tons of material from race to race, which we store in a small area on the pit lane or next to it. Along with my co-driver Peter Tibbetts, I am responsible for the Official Formula One Safety Car. The medical car is staffed by an additional driver and the F1 rescue coordinator of the FIA. Two mechanics take care of our cars, and remain present at all times. My co-driver and I have been working together since 2000—we always work as a team! Our small team is part of the FIA and is coordinated by the Race Director Charlie Whiting, and FIA observer Herbie Blash in Race Control. During a race, I am permanently in radio contact with Herbie, who keeps me up-to-date on all the racing action."

A typical Formula One race weekend
Thursday
"Generally I arrive on Wednesday evening, at European races early on Thursday morning. We meet at the racetrack at around 10 a.m. First, I go to the FIA office, where we have a short meeting and go through the important documents for the race weekend, such as race schedules, circuit maps, rules and regulations etc. Then I get changed into my race overalls and I am in the Safety Car at 1.35 p.m. Between 2 p.m. and 3 p.m. the first circuit test takes place. The Safety Car is therefore the first car that enters the circuit each race weekend.

"The track test is very important because both the car and the track are being tested, also the radio system, the GPS systems, as well as the cameras. Then, I forward our test results to Charlie Whiting, change into my official FIA clothing and attend the Drivers Briefing for the GP 2 series. The meeting takes roughly ten to thirty minutes, depending on the topics and how much needs to be discussed. After that I return to the hotel. I usually spend the evening exercising or I go for dinner with the teams and sponsors. It depends on which city we are in—cities like Melbourne and Istanbul obviously offer more possibilities to go out in the evening than others."

Friday
"Friday morning we leave the hotel at around 7 to 7.30 a.m. After arriving at the circuit we have a brief meeting with Charlie Whiting, the press, the technical and software department and with my team. Afterwards we perform another GPS test of the circuit—this test is performed from Thursday through to Sunday.

This is very important because a track system is built into the Safety Car and all other Formula One race cars, which not only provides an exact location of the vehicles but also transmits the flag signals on the side of the track to the display in the car. Tests have shown that the driver can see them much better on the display—this goes a long way towards increasing safety at the circuit. I follow the practice session on the monitor in the FIA trucks, but I am not in my car because there is no Safety Car during the practice session. I can always be reached though, in case the Safety Car is needed.

"The Formula One drivers briefing takes place at 5 p.m. and during the driver briefing the previous race and the current race are analyzed and we discuss what can be improved and how. The meeting is led by Charlie Whiting, and all Formula One drivers and test-drivers, FIA race stewards and I are present. After the meeting I return to the hotel. I try to exercise or go for dinner with friends—it is completely up to me how I plan my evening."

Saturday
"Saturday morning the FIA arrive at the track very early and have another meeting and this is followed by a GPS test, then I watch the third practice session. Directly after Formula One qualifying, the countdown starts for the first GP2 race. The GP2 race on Saturday afternoon is principally the same as the Formula One event, just at a different time and with different cars. Saturday evening we usually socialize a little, but we tend not to stay up that late—because Sunday is the all-important race day."

Sunday
"On Sunday we arrive even earlier at the track. It is one of the greatest moments of the weekend—to see how the circuit and the people slowly awake and embrace the exciting day ahead. The toughest part of the day for me begins right after the GPS test. After the second GP2 race, the Porsche Supercup takes place and at 1 p.m. the showdown for the Formula One race begins. At 1.10 p.m. my boss Charlie Whiting brings the Official Formula One Safety Car to the starting grid and hands it over to me. I check again whether the camera and the radio function properly and I

make sure I get the most recent weather update—which is a very important part of my race preparation.

"At 1.50 p.m. I join my co-driver in the car. We adjust our helmets, buckle up and check the radio frequency. At 1.55 p.m. we leave the starting grid and park the car in the agreed parking position for the first lap. As soon as all cars have completed the first curve, I am told to move the Safety Car to the parking position for the rest of the race by my colleagues in Race Control. I observe the race on the TV monitor in the car and I also watch the weather. I usually communicate two to three times with Race Control to check whether all frequencies function properly and to receive further weather updates.

"When I get a command, I always have to confirm it by stating what I am currently doing. It is like the relationship between an airplane pilot and air traffic control: when the pilot receives the order he then confirms it and also re-confirms the update of his/her new position. In addition, our mechanics also follow the radio on the pit lane. If the race is finished without a Safety Car phase—which is thankfully usually the case—I wait for the last race car and I follow it. With this, I notify the marshals that there are no other cars coming behind me and that they are able to enter the circuit. If no support race takes place, my day usually ends there. There are races after which I leave straight away and others where we depart on Monday morning. When my work is done it is nice to spend Sunday evenings at home."

Bernd Mayländer on track

"During the race I am constantly in the Safety Car and I follow the race on the monitor in my car and also listen to the radio, which connects me to Race Control. If the weather conditions worsen, or an accident occurs, I communicate with Race Control several times. I give them my opinion of the situation and I wait for their feedback. Race Control then decide whether I will be deployed or not. Along with the information I provide, Race Control takes the information of the weather station and the teams into consideration.

"If Race Control sees potential for the deployment of the car, I get the command—'Safety Car stand-by!' I prepare for deployment and wait for further commands. If I receive a 'Safety Car stand-down!' the dangerous situation no longer exists, and there is no need for me to go on the track. If I hear 'Safety Car GO!' I immediately drive onto the circuit and try to quickly go in front of the leading car, so that the race cars can line up behind me.

"During a Safety Car phase, safety is the most important element; however I still need to maintain a certain level of speed. This is so the race cars do not overheat from the lack of cooling air or that their tyre pressure does not decrease. The teams have also an impact on the velocity. They inform Race Control if they want me to speed up or slow down.

"I tend to drive at my limit during the Safety Car phase—the Safety Car often seems slower than it is. Just to give an example: A Formula One race car is on average thirty-five to fifty-five seconds faster with every lap it completes, depending on the length of the track. This means that a Formula One car can overtake the Safety Car every three laps. It is incredible how fast these cars are. I then stay on the circuit until the hazardous situation has been overcome. This is the decision of Race Control. At the end of the second section I switch off the warning lights. Before taking the next possible exit, I turn into the pit lane and the grid is released. Overtaking is only permitted after having crossed the start/finish line."

Safety is essential—Bernd Mayländer about safety in Formula One

"In the field of safety, a lot has changed over the past few years. Regardless of which Formula One topic you speak about, whether it is the race car, the circuit, or the procedure of a Safety Car deployment—everything is being actively thought about and there is always room for improvements. There are different departments within the FIA that are responsible for safety in Formula One and there are many regulations that are created in co-operation with the Formula One teams. These intensive exchanges with the drivers ensure constant improvement. Currently, we already are at a very high level and strive to continuously enhance safety standards in the future."

SAFETY BY DESIGN—PHILIPPE GURDJIAN

Philippe Gurdjian (1945–2014) was one of Formula One's most gifted practitioners. The organizer and facilitator of almost thirty Grands Prix across five different countries during his four decades in the sport, including the inaugural races in Malaysia, Bahrain and Abu Dhabi, Gurdjian was most closely associated with his role at the heart of the quest for improved circuit safety. Writing shortly before his untimely passing, Gurdjian reflected on the improvements in trackside safety over the past thirty years, and considered where future innovation can lead to even safer racing environments.

THE SAFETY IDEAL

Since I organized my very first Grand Prix at France's Paul Ricard Circuit in 1985, I have always concentrated my attention towards safety. When I first came to Formula One, at the height of the turbo age in the era of Jean-Marie Balestre, concerns about safety were rising as the speed of the cars grew ever faster. Paul Ricard, where I organized six Formula 1 Grands Prix, was notorious as one of the world's fastest and most dangerous circuits. To improve safety, we created "red zones" (no spectators and no marshals), removed the "fences system" inside the gravel traps and tripled the height of the safety rail over the Mistral Straight— as well as adding three further layers of guard rails. Following Elio de Angelis' fatal accident while testing at Paul Ricard in 1986, we took the decision to modify the circuit layout, removing the high-speed "S" of the Verrerie complex to better control the speed of the cars.

After the French Grand Prix moved to Magny-Cours in 1991, I remained the race promoter, and once again innovated improvements at the circuit to improve safety. The most significant of these came in 1994 when, as Formula One collectively focused on ways to make the cars and circuits safer after the tragedy of Imola, we introduced two new systems at Magny-Cours: the "AVD" (Visual Starting Aid) system, which saw twelve quadri-flashes placed along the grid to indicate the exact location of a problem at the start of a race, and the "AVP" (Visual Driving Aid) system, where flashing lighted signs backed up the flags used by the stewards in the event of an incident. Furthermore, we increased the size of the gravel trap at Turn One and pushed the guardrails back by 100 m—all executed in less than eight weeks at a then substantial cost of 6 million Francs. A year later, we were able to overhaul the pit lane and starting grid areas, installing armoured glass windows above the pit lane wall to protect the teams and marshals, introducing electronic control of the safety lights from the race manager's post, and adding flying start control and speed monitoring systems in the pit lane. Our work at Magny-Cours set a standard that was successfully applied to a new generation of racetracks.

DEVELOPING NEW SOLUTIONS

Working at Paul Ricard and Magny Cours was about upgrading and modernizing safety at existing facilities, and allowed me to develop solutions that could subsequently be fed into the design of new circuits and facilities. Bernie Ecclestone consequently gave me total freedom to execute my vision with the development of Paul Ricard into a High Tech Test Track (HTTT). We needed to create new solutions for the track layout, the pit lane and pit garages, the race control and the medical centre. In addition, the safety solutions had to satisfy the needs of all types of vehicles, from touring cars to Formula 1. With the freedom and trust granted by Mr Ecclestone, we were able to implement the development of three significant safety mechanisms at HTTT Paul Ricard: TECPRO barriers, a video tracking system, and asphalt run-off.

TECPRO barriers were developed in order to replace the archaic system of tyre barriers. Today, racetracks such as Monaco, Circuit of the Americas and Buddh International Circuit are using the TECPRO technology. Yas Marina Circuit in Abu Dhabi, which I created from scratch, and Singapore are the only racetracks equipped entirely with TECPRO barriers, but the safety, efficiency and lifespan of these new barriers means they are being seen as not only viable, but an essential component of new circuit design.

A unique Lighting System replacing flag marshals was also developed at HTTT Paul Ricard and applied in Abu Dhabi. The video tracking being linked to the race control room allowed the stewards to instantly follow the progress of each car around the circuit. In addition, I worked closely with COLAS [Group] to create and implement 25 hectares of run-off made of three different types of asphalt abrasiveness; replacing all the gravel traps on the new HTTT Paul Ricard. The old gravel traps used to cost a lot of money and required regular maintenance at that time in order to remain efficient and safe. They were also fundamentally flawed, as accidents such as Michael Schumacher's crash at Silverstone in 1999 showed, because cars were unable to brake on the gravel and couldn't reduce their speed before impact.

FROM THE PAST TO THE MODERN AGE

The solutions developed at Paul Ricard were extremely efficient and are now not only set as safety standards for new circuits, but have also been used to update historic tracks such as Silverstone, the Nürburgring and the Circuit de Catalunya. With HTTT Paul Ricard, I was able to demonstrate that before becoming the most modern and safest racetrack in the world, it was an obsolete and unusable facility. In Barcelona, we decided to implement some ECPRO barriers around the racetrack and, most importantly, we created a chicane between Turn Thirteen and Fourteen to reduce the speed of the cars entering the finish line. Thus, it is becoming crucial that old racetracks such as Spa, Monza, Le Mans and Montreal start investing more money and significantly update the safety for the drivers. I strongly believe that TECPRO barriers should become mandatory by the FIA in 2015 replacing all the existing tyre barriers which are extremely dangerous and displace energy inefficiently—as we have seen during enormous accidents at the 24 Hours of Le Mans in the past two years.

SAFETY FROM SCRATCH

Of course, updating historic circuits to make them safer is crucial, but it is a different consideration entirely to design a completely new circuit and implement the safety features as a key design component rather than an afterthought. With innovations in racetrack safety, so new possibilities have emerged. The most compelling of these has been the development of night races.

When Bernie Ecclestone decided to introduce the first night race in Singapore, he asked me to initiate the testing at HTTT Paul Ricard. An Italian company was chosen to find and develop the best solution. Such a task required us to take the drivers' visibility and the quality of the images on TV into consideration. Even though the drivers' visibility was an easy task, finding the most efficient solution for TV images was substantially more difficult. After several weeks of testing, we finally came up with the solution that allowed me to call Bernie and tell him he would host the first night race in Singapore in the safest conditions. The success of Singapore pushed other tracks such as Abu Dhabi and now Bahrain to replicate it. Talking about Bahrain, even though I had only ninety days to complete the racetrack constructions for the inaugural race ten years ago, I took an important decision to push back the guardrails by an additional fifteen metres. Today, I am extremely satisfied to see that this circuit has kept improving its safety and replaced most of the gravel traps with run-off areas.

Combining all of the lessons we have learned through the years, the most developed circuit in terms of design and safety is undoubtedly the Yas Marina Circuit in Abu Dhabi. I have to pay tribute to Khaldoon Al Mubarak and the Royal Family of Abu Dhabi for entrusting me. I was provided with all the resources I needed to build the most advanced, developed, modern, safest and unique racetrack in the world.

In terms of architecture, it is by far the most accomplished racetrack in the world, symbolizing the image of Abu Dhabi and featuring creative touches such as the tunnel at the pit lane exit, the new generation of garages, the team houses in the paddock and Paddock Club hospitalities, the media centre and the medical centre. From a spectator's point of view, it is the only track that has 55,000 covered seats, and where the positioning of the grandstands has been designed to bring the public closer to the action. Above all, the Yas Marina Circuit is the safest racetrack in the world featuring the entire range of solutions developed at HTTT Paul Ricard, including asphalt run-off areas, TECPRO barriers, lighting systems and painting solutions for the run off areas. New safety solutions are yet to be developed and "risk zero" will never exist. The investment behind each solution is definitely significant, however protecting the lives of the drivers, the crews, the marshals, the media and the spectators around the track remains the primary obligation, no matter what the costs.

Finally, I would like to thank the drivers, Charlie Whiting, Herbie Blash, Jean-Louis Piet and Professor Sid Watkins for their strong collaboration through the development of safety solutions during my work and over the last thirty years.

TYRES

Sole F1 tyre supplier Pirelli expects the 2015 cars to make a good improvement in performance especially in terms of race pace, which will impact on tyre behaviour. It is unlikely that the same compound choices as 2014 will be made as what was conservative last season may become aggressive this year, particularly towards the end of the season. Below are Pirelli's technical notes as well as their circuit-by-circuit tyre guide. As ever their objective for the 2015 tyres remains the same as previous seasons—to provide between two and three pit stops per race. At the end of the chapter are the complete FIA 2015 Tyre Regulations which govern the supply, quantity, control, use and testing of tyres throughout the race weekend.

MEET THE 2015 COMPOUNDS
Dry Compounds
Supersoft (red)

The softest compound in the range is ideal for slow and twisty circuits, especially in cold weather, when maximum mechanical grip is needed. The supersoft benefits from an extremely rapid warm-up time, which makes it ideal in qualifying as well, but the flip side to that important characteristic is of course increased degradation. This is a low working range compound. The structure of the rear tyre has been altered in order to distribute heat more evenly which will lead to more consistent performance and better traction.

Soft (yellow)

This is one of the most frequently used tyres in the range, striking a very good balance between performance and durability, with the accent on performance. It is still biased towards speed rather than long distances, but is nonetheless capable of providing teams with a competitive advantage both at the beginning of the race on full fuel and when used as a "sprint" tyre at the end. This is a high working range compound.

Medium (white)
Theoretically this is the most perfectly balanced of all the tyres, with an ideal compromise between performance and durability. As a result, it is extremely versatile, but it often comes into its own on circuits that tend towards high speeds, temperatures and energy loadings. This is a low working range compound.

Hard (orange)
The toughest tyre in Pirelli's range is designed for the circuits that put the highest energy loadings through the tyres, with fast corners or abrasive surfaces, and are often characterized by high ambient temperatures. The compound takes longer to warm up, but offers maximum durability—which frequently means that it plays a key role in race strategy. This is a high working range compound.

Wet Compounds

Intermediate (green)
The intermediates are the most versatile of the rain tyres, dispersing approximately 25 litres of water per second at full speed. They can be used on a wet as well as a drying track. This is the only tyre that has not changed compared to last year.

Wet (blue)
The full wet tyres can disperse up to sixty-five litres of water per second at full speed (increased from sixty litres last year), making them the most effective solution for heavy rain. The latest evolution of the Cinturato Blue means that it is also effective on a drying track, with increased durability. The full wet tyre has a new compound and a redesigned rear tread pattern to further reduce aquaplaning.

2015 FORMULA ONE TRACKS AND THEIR TYRE CHARACTERISTICS

Australian Grand Prix
The Australian Grand Prix is a race loved by teams and fans alike. The atmosphere is great, the weather is usually warm and

the track is close to the vibrant city of Melbourne. All the dry compounds will be a little bit more conservative compared to previous years but they should still provide interesting racing, with a window of two pit stops per car and race. With the Australian circuit being a semi-permanent facility, there is a high level of track evolution as the weekend goes on.

Malaysian Grand Prix

Sepang is hot, humid and frequently home to tropical downpours. The track itself is very abrasive and, in combination with the high ambient temperatures, this circuit is one of the toughest all year for the tyres. In order to master this difficult track, high-speed stability from the tyre is particularly important. The chances of seeing the latest generation of our Cinturato Green intermediate and Cinturato Blue full wet compounds in action here is one of the highest all year.

Bahrain Grand Prix

The Bahrain International Circuit is very tough on the rear tyres in particular, and a high level of traction and grip is necessary for a representative lap time. As it is located in the desert, it is quite common to have sand on the track, which can lower the grip levels, in particular at the start of a session. The temperatures can be very high, a fact that increases thermal degradation: the real limiting factor at this track. This is a track where tyre management is important, with rear traction in particular being the key to a strong qualifying and race pace.

Chinese Grand Prix

A wide variety of pit stop strategies can work at this track. Although ambient temperatures can be quite low and the surface is relatively smooth, tyre degradation can be high due to the unique track layout. Overall, it is a demanding circuit for the tyres, in particular the front left tyre, and the heavy braking that is a characteristic of this track tends to put more strain on the front tyres than the rears.

Spanish Grand Prix

Barcelona has several medium- to high-speed corners that test every aspect of a tyre's performance and can lead to a notable degree of wear and degradation. The asphalt is rather abrasive and temperatures can be high. It is one of the most challenging circuits of the season for tyres, also due to the high lateral loads it puts on them, particularly on the left-hand side.

Monaco Grand Prix

Monaco is the slowest and least abrasive circuit of the year. Tyre wear is low; mechanical grip and high levels of low-speed downforce are very important here. Overtaking is nearly impossible, putting the emphasis on qualifying. With limited run-off areas and a high risk of incidents the Safety Car has often influenced the race outcome at this track in the past. Nevertheless, a one-stop strategy can work here under the right circumstances.

Canadian Grand Prix

The iconic Circuit Gilles Villeneuve is a semi-permanent track, infrequently used during the year, which means that at the beginning of the weekend the track is very "green" and slippery. However, there is plenty of track evolution as more rubber gets laid down throughout the weekend. This track puts a lot of longitudinal stress through the tyres, and is very demanding in terms of braking and traction. The surface provides little grip, which can lead to the cars sliding more. The rear tyres are particularly stressed at this track, due to the traction required out of the low and medium speed corners.

Austrian Grand Prix

With two main straights and mostly sharp corners keeping average speeds generally low, the cars had to rely on mechanical grip from the tyres more than aerodynamic downforce at the refurbished Red Bull Ring. Despite not having held a race since 2003, tyre performance in Austria was very much as expected with the two softest compounds delivering the best compromise between grip and performance. The race became a straightforward two-stopper with teams generally starting on the supersoft then changing to the soft compound before changing again to finish the race on the soft.

British Grand Prix

Silverstone is another challenging track for the tyres. Here, aerodynamics play a more important role than mechanical grip, due to a high average speed. The circuit features a technical layout, taking in a variety of very high-speed corners with some slower and more technical sections. This, together with an abrasive surface, tends to lead to a high wear rate. Temperatures in Silverstone can often be rather low and there is always a risk of rain. But when dry, temperatures can equally be quite high, which means that a very versatile compound choice is required for this track.

Hungarian Grand Prix

Hungary is the slowest permanent circuit of the year, which does not make it any easier on the tyres. It is a very twisty track and often slippery, in particular at the start of the weekend. This means that much more heat is put through the tyres than on a fast and flowing layout because the tyres are moving around more: particularly when it is hot, which is often the case in Hungary. Therefore, balancing the demands of speed and durability is key to getting the most out of the compounds, in order to keep degradation under control. Overtaking is traditionally difficult at this track, so drivers have an opportunity to use strategy to gain track positions.

Belgian Grand Prix

Spa-Francorchamps is the track that puts the highest vertical loads on the tyres during the season: often in excess of 1,000 kg per tyre in the Eau Rouge and Radillon section. The undulating surface at Spa also means that the tyres work hard as part of the car's suspension, placing heavy demands on the structure, while the sustained high speeds mean that the drivers have to be careful not to overheat the rubber. On top of this, the weather can be very changeable—and because the track is so long, it is not unusual for it to be raining on one part of the circuit but dry in another.

Italian Grand Prix

The track is characterized by very high speeds, putting a lot of energy through the tyres, so the cars run the lowest downforce levels of the year. Consequently, the tyres tend to slide more. There are also a number of high kerbs, which mean that the tyres have to constantly absorb heavy structural impacts. Overall, this track puts some of the highest demands on the tyres that they will face all year.

Singapore Grand Prix

Singapore is another street circuit but unique as it is the season's only night race, run under powerful spotlights. The circuit itself is low-grip and slippery, evolving considerably as more rubber gets laid down. Average speeds are contained, so degradation is not usually an issue provided that wheelspin—which can lead to overheating and blistering—is controlled out of the slower corners. The softest compounds are usually well suited for this race, and the fact that the high temperatures tend to fall as the race goes on can put an interesting spin on strategy.

Japanese Grand Prix

Suzuka was resurfaced a few years ago, so it is a little less aggressive on the tyres than in the past, but the circuit is still a difficult one for the rubber because of the track layout. It contains a wide variety of fast and flowing corners that place high lateral loads on the tyre but there are also some heavy braking areas and tighter corners. The first half of the lap is essentially a non-stop series of corners, which put plenty of heat through the tyres with no significant straight where they can cool down.

Russian Grand Prix

Wear and degradation were extremely low at the brand new track in Sochi last year with most drivers being able to complete a one-stop strategy by starting on the soft compound and changing to the medium compound to finish the race. Sochi is a medium speed circuit with a variety of corners that test all aspects of performance. Overall tyre energy demands are average and the surface measures in the low-abrasion range.

United States Grand Prix
The track contains a variety of elevations, plus some slow and technical sections alternating with very fast parts. It's a good test of a tyre's all-round ability, with traction demands out of slow corners just as important as lateral grip through the high-speed changes of direction that are another key characteristic of the 5.513-km Circuit of the Americas.

Brazilian Grand Prix
The track is one of the shortest but also one of the most challenging of the year. There are some big elevation changes and it is notably bumpy, which makes it hard for the tyres to find traction and increases the physical demands on the drivers. The track surface is on the rough side but as the corners are not very fast, degradation is not usually an issue.

Abu Dhabi Grand Prix
The track surface is generally smooth, so degradation is low. The circuit provides a variety of speeds and corners, so the tyres have to withstand a wide range of different demands. As the race starts late in the afternoon and continues into dusk, ambient and track temperatures tend to fall as the race goes on, which alters the usual strategy calculations.

SUPPLY OF TYRES IN THE CHAMPIONSHIP AND TYRE LIMITATION DURING THE EVENT
Supply of Tyres
A single tyre manufacturer has been chosen by the FIA for the 2014, 2015 and 2016 seasons. The appointed tyre supplier must undertake to provide:

a) Two specifications of dry-weather tyre at each Event, each of which must be visibly distinguishable from one another when a car is on the track. At certain events one additional specification of dry-weather tyre may be made available to all teams for evaluation purposes following a recommendation to the FIA from the appointed tyre supplier. Teams will be informed about such an additional specification at least one week before the start of the relevant event.

b) One specification of intermediate tyre at each event.

c) One specification of wet-weather tyre at each event.

Quantity of Tyres During an Event

a) Except under Article 25.1(a) and (d) below, no driver may use more than thirteen sets of dry-weather tyres, seven of "prime" specification and six of "option" specification.

b) Except under (e) below, no driver may use more than four sets of intermediate tyres and three sets of wet-weather tyres.

c) A set of tyres will be deemed to comprise two front and two rear tyres all of which must be of the same specification.

d) Following a recommendation to the FIA from the appointed tyre supplier, one additional set of either "prime" or "option" specification tyres may be made available to all drivers. Teams will be informed about such an additional set at least one week before the start of the relevant event.

e) If either P1 or P2 are declared wet, one additional set of intermediate tyres will be made available to all drivers. Under such circumstances, one used set of intermediate tyres must be returned to the tyre supplier before the start of P3.

Control of Tyres

a) The outer sidewall of all tyres which are to be used at an event must be marked with a unique identification.

b) Other than in cases of *force majeure* (accepted as such by the stewards of the meeting), all tyres intended for use at an event must be presented to the FIA technical delegate for allocation prior to the end of initial scrutineering.

c) At any time during an event, and at his absolute discretion, the FIA technical delegate may select alternative dry-weather tyres to be used by any team or driver from among the stock of tyres the appointed supplier has present at the event.

d) A competitor wishing to replace one unused tyre by another identical unused one must present both tyres to the FIA technical delegate.

e) The use of tyres without appropriate identification may result in a grid position penalty or exclusion from the race.

25.4 Use of Tyres

Tyres will only be deemed to have been used once the car's timing transponder has shown that it has left the pit lane.

a) Thirteen sets of dry-weather tyres will be allocated by the FIA technical delegate to each nominated driver, seven of "prime" specification and six of "option" specification. One set of "prime" specification tyres may only be used during the first thirty minutes of P1 and must be returned to the tyre supplier before the start of P2. One further set of "prime" specification tyres must be returned to the tyre supplier before the start of P2 and one further set of "prime" specification tyres and one set of "option" specification tyres before the start of P3. If P1 and P2 are both declared wet, one set of the tyres normally returned before the start of P3 may be retained by each driver but must be returned to the tyre supplier before the start of the qualifying practice session. If Article 25.2(d) is invoked an additional set of either "prime" or "option" specification tyres will be available to each nominated driver for use during P1 and P2. This set of tyres must be returned to the tyre supplier before the start of P3. One set of "option" specification tyres may only be used during Q3, by those cars that qualified for Q3, and must be returned to the tyre supplier before the start of the race. One set of "option" specification tyres, which were allocated to cars that did not qualify for Q3, may only be used during the race. If an additional driver is used (see Article 19.1(b) he must use the tyres allocated to the nominated driver he replaced.

b) If an additional specification of dry-weather tyre is made available in accordance with Article 25.1(a) two sets of these tyres will be allocated to each driver for use during P1 and P2. Any such tyres must be returned to the tyre supplier before the start of P3.

c) From the remaining dry-weather tyres one set of each specification must be returned to the tyre supplier before the start of the qualifying practice session.

d) Prior to the start of the qualifying practice session intermediate and wet-weather tyres may only be used after the track has been declared wet by the race director, following which intermediate, wet or dry-weather tyres may be used for the remainder of the session.

e) At the start of the race each car which qualified for Q3 must be fitted with the tyres with which the driver set his fastest time during Q2. This will only be necessary if dry-weather tyres were used to set the fastest time in Q2 and if dry-weather tyres are used at the start of the race. Any such tyres damaged during Q2 will be inspected by the FIA technical delegate who will decide, at his absolute discretion, whether any may be replaced and, if so, which tyres they should be replaced with. A penalty under Article 16.3(c) will be imposed on any driver whose car is not fitted with the tyres with which he set his fastest time in Q2 (except if damaged tyres have been replaced with the approval of the FIA technical delegate).

f) Unless he has used intermediate or wet-weather tyres during the race, each driver must use at least one set of each specification of dry-weather tyres during the race. If the race is suspended and cannot be restarted, thirty seconds will be added to the elapsed time of any driver who was unable to use both specifications of dry-weather tyre during the race. However, any driver who completes the race without using both specifications of dry-weather tyre will be excluded from the race results.

g) If the race is started behind the Safety Car because of heavy rain (see Article 40.17), or resumed in accordance with Article 42.5(a), the use of wet-weather tyres until the Safety Car returns to the pits is compulsory. A penalty under Article 16.3(c) will be imposed on any driver who does not use wet weather tyres whilst the Safety Car is on the track at such times.

25.5 Testing of Tyres

a) Tyres supplied to any competitor at any time may not be used on any rig or vehicle (other than an F1 car on an F1 approved track, at the exclusion of any kind of road simulator), either team-owned or rented, providing measurements of forces and/or moments produced by a rotating full size F1 tyre, other than uniquely vertical forces, tyre rolling resistance and aero-dynamic drag.

b) Tyres may be used on a test rig providing forces control and monitoring by F1 rim manufacturers for the sole purpose of proof testing their products

ANTHOLOGY

ANTHOLOGY INTRODUCTION

Formula One is a completely strange, somewhat mad world populated by completely strange, somewhat mad people. While the first part of this book seeks to unravel the whys and wherefores of the sport it falls to the following collection of writing to illuminate the complex characters, personalities and relationships that keep the global spectacle at the forefront of both the sporting and business worlds.

Beginning with *The Piranha Club*, Timothy Collings presents an incisive and entertaining look behind the scenes with the most fascinating and influential figures that wield the real power within the Formula One paddock.

Steve Matchett's *The Mechanic's Tale* is an eyewitness and very personal account of life in the pit lanes of Formula One where there is little glamour, long hours, hard work, unbelievable pressure and great camaraderie. It is often labelled "a must read" for its unique and compelling perspective.

The Limit by Michael Cannell is a remarkable work on the inspirational and deadly rivalry between the American Phil Hill and the German Count Wolfgang von Trips in a lost era as they hurtle towards the ultimate accolade—the Formula One Drivers' Championship.

Mark Gallagher is an industry insider who in *The Business of Winning* examines the business lessons to be learnt from Formula One and in particular the leadership qualities of those who succeed. He uncovers those attributes that make successful business leaders in a highly competitive global industry.

While most F1 books claim to be an insider's guide, *Tales from the Toolbox* by Michael Oliver truly provides an absorbing glimpse into the world of the Grand Prix mechanic with insightful stories and illuminating anecdotes from a rarely heard about side of the sport in the golden days of the fifties, sixties and seventies.

Performance at the Limit is a scholarly examination of the role that innovation plays in Formula One in the quest for continual change and improvement to outpace the competition. It shows how the hard-won lessons from the track can be applied to the ever-challenging world of business.

Overdrive from Clyde Brolin is an esoteric journey into "the Zone". Drawing on exclusive interviews with the world's quickest drivers, Brolin uncovers the notion of reaching a higher plane to unlimited performance.

Among the hundreds of driver biographies I have carefully chosen an excerpt from David Tremayne's marvellous *Jochen Rindt: Uncrowned King* not only because it is a remarkable study of Formula One's only posthumous World Champion but also because it gets to the essence of a complex and restless character. It is also a fascinating insight into Jochen's relationships with Bernie Ecclestone and Jackie Stewart during a period in Formula One when drivers and their partners raced, partied and holidayed together.

Finally, in *Bernie's Game*, Terry Lovell, using multiple first-hand accounts, meticulously documents the extraordinary efforts of the Commercial Rights Holder to gain a permanent foothold in the United States, so crucial to Formula One becoming a truly global marketing platform.

This eclectic collection of writing is designed to delve into the innermost workings of the complex organism that is Formula One. I trust that the quality of the selection will inspire you to seek out the titles in their entirety and I have no doubt that they offer a unique opportunity to understand a fascinating entity.

BERNIE'S GAME by Terry Lovell, 2003

Until November 1998, when a thirty-million-pound deal with Tony George, the billionaire president of the Indianapolis Motor Speedway heralded its return to America after an absence of almost ten years, Ecclestone had failed with distinction in making Formula One a truly global marketing platform. Ceaseless efforts during the eighties had been largely frustrated by inconsistencies of venue and television coverage. Its total absence during most of the nineties prompted Jacques Nasser, the then president of Ford to remark: "It seems to me it's difficult to portray Formula One as a world sport without having an event in the world's largest market."

It was an absence that plagued Ecclestone. In his mind there was no question where the problem lay. It was "because the people who do business there seemed to have failed to understand what a contract is, and to respect it." (*Autosport*, 1 August 1985.) To others, however, there was a more familiar reason. Demanding contractual terms, particularly in respect of television rights, that circuit owners and promoters were not prepared to accept. In a country where there were more than sixty major motorsport events at forty-five tracks attracting twelve million spectators, he was discovering that he was not dealing with a European organizer or promoter in fear of his every word and move. For the first time he was dealing with equally hardheaded businessmen whose commercial success did not depend on the patronage of Formula One.

Ecclestone's problems in America began when, in pursuit of bigger bucks, he turned his back on the venue where Formula One had been consistently successful for almost twenty years— Watkins Glen, a purpose-built circuit situated about 300 miles northwest of New York in a small town of about 3,000 people. It had been built in 1956 on 680 acres of land—later extended to 1,180 acres—acquired by the Watkins Glen Grand Prix Corporation to replace the 6.6-mile course on which the first American road race was staged in 1948, founded and organised until 1970 by a young lawyer, Cameron Argetsinger.

The first two United States Grands Prix in Sebring, Florida in 1959 and Riverside, a suburb east of Los Angeles, in 1960, proved financial disasters. But Watkins Glen, from its debut event in October 1961, won by Innes Ireland in a Lotus, proved an immediate hit with the teams and public alike. In 1966 it put up a record prize purse of $100,000, increased four years later to $250,000. However, its popularity made it a victim of its own success. Following discussions with FISA (Federation Internationale du Sport Automobile), the Corporation had to find $3.5 million to extend the circuit by a mile and improve team and public facilities. To raise the money, a high-risk, quasi-municipal board was floated with the Bank of New York as trustee, securing the mortgage on the land as security. It was an ambitious sum, incurring a debt service charge of $1,000 a day, payments that the organisers found increasingly difficult to meet.

According to hotelier Vic Franzese, chairman of the Industrial Development Agency, responsible for issuing the quasi-municipal bond and then-president of the Watkins Glen Chamber of Commerce, of which the Watkins Glen Grand Prix Corporation was a subsidiary, the causes of the problem were safety demands by, initially, the Grand Prix Drivers' Association for safety improvement and latterly, Ecclestone's financial demands on behalf of the Formula One Constructors Association (FOCA). "The new track was very smooth and much faster. This enabled the cars to sustain much higher speeds—by the mid-seventies they were completing the three-and-a-half-mile laps in about 1 minute 34 seconds—but they were too high for the safety factor in the construction of the cars. We tried slowing down the circuit by adding, for example, chicanes, but it didn't work because the cars were still getting faster. So each year the Grand Prix Corporation was forced to add or remove from the track whatever the Formula One powers decided they wanted to try next. To cover the cost we were shelling out, perhaps, $200,000 every year. That was in addition to about $3 million the Grand Prix was costing towards the end of the seventies. Much of that went on flying the teams in, and the prize money, which by then was a closely-guarded secret between the Pied Piper [Ecclestone] and one or two others."

No longer able to service the debt, the bank of New York fore-closed on the property, forcing the Grand Prix Corporation to declare itself bankrupt in 1981 with assets of $3.5 million against debts totally $5.9 million. Of that sum, $800,000 was owed to the FOCA. It could have been paid, says Franzese, if Ecclestone had responded more positively to efforts to refinance the Grand Prix Corporation. "The Formula One Group was prepared to swallow that loss because they were committed to go elsewhere for more money. He [Ecclestone] was looking for millions of dollars, which he thought couldn't come out of here. He figured there were cities that were going to throw millions at them. Watkins Glen no longer fitted in with Ecclestone's plan for the future of Formula One."

The last Grand Prix at Watkins Glen, won by Alan Jones in a Williams-Ford, took place in 1980. The economy of the community of Watkins Glen, which over nearly two decades had invested a good deal of effort and money in the Grand Prix, was hit badly by Ecclestone's decision, which also hurt fan support of all American road race sports, said Franzese. "Only now is the support for road racing starting to pick up a little." But he insists it was Ecclestone and the teams who were the losers. "Watkins Glen hosted twenty successful Formula One Grands Prix that were basically financially profitable. Bernie and company had ten years of failures and ten years of no races in this country after leaving Watkins Glen. They didn't understand what American fans will support and where they will spend their money. The new track owners are enjoying a very profitable business. Bernie should have taken a second look before jumping ship. He and the Formula One Group miscalculated badly."

A few years later, Ecclestone blamed the backwoods image of Watkins Glen for the teams' departure. "It was in the wrong place," he said. "Everything was wrong with it. It wasn't the image we're trying to project. Didn't do anything for us." (*Autosport*, 1 August 1985.) Its image was seriously tarnished, said Max Mosley, by a series of incidents involving locals alleg-edly setting light to a bus containing the luggage or Argentinian tourists and a car being set alight, which caused its two sleeping occupants to be burnt to death. They were incidents that "dear old Vic" had forgotten about. "It got to the point where it really wasn't good for the image of Formula One," said Mosely. And Ecclestone commented, "If that wasn't bad enough, they forgot to pay us $800,000."

Cameron Argetsinger, who left the Grand Prix Corporation in 1970 and became president of the Sports Car Club of America in 1974 expressed a different point of view. "That's bullshit. Watkins Glen was the right place at the right time, as its success over almost twenty years proved. It is the only circuit in America where a United States Grand Prix has been successful. I think his comments might be coloured by the fact that he got stiffed for $800,000."

In an era where the global monetary system was less efficient and cheques were not always to be trusted, the financial arrangements between the Grand Prix Corporation and the teams were, as elsewhere, strictly cash only. Michael Tee, then vice chairman of Champion Sporting Specialists International, which was extensively involved in promoting Grands Prix, was present when a FOCA associate of Ecclestone's made a payment in cash to one of the team bosses. He said; "He came in and said, 'Look, can I just have a moment?' I got up [to leave] and he opened a case and it was a mass of dollar bills. He said, 'I think you will find everything is there.' He just handed over a wad of notes, closed the case and walked off. He was doing the rounds of the various teams and handing over their 'start' and prize money. It was entirely up to them how they put it through their books. In those days everything was cash."

Such deals, he confirmed, were an intrinsic part of the Formula One culture. He was paid $7,000 cash to produce a series of promotional articles and programmes. "That, in a minor way, was what was happening [throughout Formula One]," he said. "Now, of course, it doesn't have to be that way, because money can be put in and moved about so easily from overseas banks . . . but the amounts now are ridiculous."

In moving out of Watkins Glen, Ecclestone began looking for somewhere like Long Beach, California, where since 1976 he had also been staging the USA West Grand Prix. He believed he had found the ideal venue in Nevada when he struck a deal with Clifford Perlman, the chairman of Desert Palace Inc., owners of

Caesar's Palace in Las Vegas. Billy Weinberger, the president of the casino, believed the glamour of the Grand Prix would help pull in the gamblers, and he drove Ecclestone to a board meeting where the deal was agreed in record time. Ecclestone was escorted to a vault-like, soundproofed room deep below ground level where he was introduced to members of the board.

"This is Bernie Ecclestone," Perlman announced. "We want to have a Grand Prix here at Caesar's Palace and this is the guy who is going to fix it. Are you in favour?" A voice to Perlman's right broke the silence with a solitary yes. "Congratulations, gentlemen. Done." A Las Vegas Grand Prix was staged in 1981 and 1982 in the casino's huge car park but both events proved financial disasters. They failed to attract the crowds or the interest of the major television companies on which their success had essentially depended. They heralded, in fact, the beginning of a series of similar disasters.

In the same year that Las Vegas was hosting its second and last Grand Prix, Ecclestone had moved east to stage a Grand Prix in the streets of Detroit—an unusual choice for someone keen on an image more glamorous than Watkins Glen—which, with Long Beach, meant that during 1982 there were three Grands Prix in America, a number which Ecclestone believed the size of the country could support. The deal that he cut with the organisers—four years and three on option—was a confident one. The length of the contract could be guaranteed, he said, because, rather than a flat fee, the teams would be taking a percentage of the race earnings. It was structured to give the FOCA the first $500,000 with the next 1.2 million dollars going to the organizers and the next $500,000 to the FOCA; after that, the receipts would be split down the middle.

However, long before the contract had expired, his relationship with the organizers of the Detroit Renaissance Grand Prix, so named because the street circuit looped around the city's huge Renaissance Center, had turned sour. Ecclestone claimed the organizers had failed to provide the standard of facilities necessary to maintain a successful Grand Prix including adequate garage facilities for the teams, media and sporting control centres, and on-track medical units, while the organizers reportedly claimed that Ecclestone made so many last-minute financial demands to increase his share of profits that budgeting became a nightmare.

By the time the Grand Prix in Detroit was in its second year, the USA West Grand Prix was bringing the shutters down for good due to the financial demands made by Ecclestone of organizer and promoter Chris Pook, a Somerset-born businessman who had arrived on the Pacific Coast in 1963. Ten years later Pook began discussions with the City of Long Beach to permit a Grand Prix street race, which led, to the first of eight Grands Prix. However, despite every effort by Pook—not averse to rolling his sleeves up, as witnessed at the 1977 Grand Prix, when, with the installation of safety fencing behind schedule, he assisted crews in its construction, and, during the second practice session, manned one of the pit access gates to check credentials—the event became so unprofitable that he was forced to drop Formula One for the more popular Championship Auto Racing Teams (CART) series.

He said, "We were running this huge race every year and only making $80–100,000 profit, and if we had a bad, rainy weekend our company would have been upside down. We could not see the light at the end of the tunnel to be able to build our company, and that is why we had to change." It was brought about, he said, by the size of the prize money expected by Ecclestone and the cost of the teams' transportation. Advertising and hospitality rights had also been a factor. "F1 was growing in a certain direction and wishing to maintain control over certain assets that it saw [for] itself, and, quite candidly, in the US the market is so fierce for similar assets that F1 wanted . . . that those of us who were promoting it [Formula One] couldn't see a way through the maze."

Coverage of the last Long Beach Grand Prix, produced by Grand Prix Teleproductions, was syndicated through cable television. Network television companies, in a country where Formula One attracted a quarter of the television audience of IndyCar racing and up to eight times fewer than NASCAR (North American Sports Car Racing), argue that viewing figures simply didn't justify Ecclestone's financial terms. For Pook, as with others, it meant only one recourse—to hike up ticket prices, which did nothing to improve the prospects of Formula One's success. Said Pook, "There is huge competition [in the USA] for the leisure dollar, and if you start to charge $150–200 for a ticket, your market diminishes very, very fast. It wasn't that Americans didn't embrace Formula One. It was a commercial issue."

Lest he should appear too critical of Ecclestone, Pook, who, like many others in Formula One, has a healthy respect for his power and favour, later insisted that his comments should include reference to Ecclestone's "incredible negotiating ability, his brilliance, his quickness and [that his] word is his bond." He failed to mention though, that Ecclestone's recall of detail apparently falls somewhat short of his. The demise of Formula One at Long Beach, according to Ecclestone, had nothing whatsoever to do with any commercial factors, but "because we were told by Chris Pook that there was going to be major developments in the city which would necessitate that shortening of the circuit, which would not thereafter comply with the FISA regulations." (*Autosport*, 28 June 1990.) Following the departure of the teams, Pook nevertheless continued to use the appeal of Formula One to pull in the crowds, said Ecclestone. "We built up Formula One and it was called the Toyota Long Beach Grand Prix. It never mentioned Formula One. When we left, and they went to CART cars, Chris still called it [Toyota Long Beach Grand Prix]. In fact, so many people complained to me that they had gone to see Formula One and it wasn't Formula One."

With the split at Long Beach, and Detroit falling short of expectations, Ecclestone attempted to strike oil in Texas. After much haggling in a deal brokered by Chris Pook, the financial terms were finally agreed with Dallas real-estate developer Donald R. Walker, who ran Dallas Motorsports Inc. from a plain, unmarked warehouse north of the city near LBJ Freeway and Interstate 35. The Dallas Grand Prix, which cost co-promoter Walker and his associates an estimated six million dollars, proved to be one of the most short-lived US Grands Prix. Ecclestone's negotiating tactics, according to Walker, caused an intensity of conflict that seriously threatened cancellation of the race. He complained that Ecclestone avoided agreeing specifics until he had had become committed to heavy investment. "Then, once he knows you're committed and have a heavy investment, he'll then use the leverage of cancelling your race to get the specifics the way he wants them." (*Dallas Morning News*, 4 August 1985.)

It was a charge rejected by Ecclestone, who denied that Walker had invested heavily. "The only investment he had . . . it was a street circuit." He added, "The only complaint we ever had with him was that, firstly, he would have to respect FOCA passes, and, [secondly], the track broke up in the heat. All the drivers wanted to cancel the race. I pushed them to make the race go on." Italian Michele Alboreto, who drove for Ferrari, was particularly alarmed, said Ecclestone. "On the starting grid, he came to me and said, 'I'm going to die in this race, and you are going to be the one that's murdered me.' I said, 'Well, the easiest thing is don't race, if that's what you believe.'"Tragically, Alboreto, who spun off during the fifty-fourth lap, did die on the track—he was killed on 25 April 2002 at the Lausitz Circuit, Germany, when his Audi test car had a puncture at 200 mph, causing it to hit a crash barrier.

For all his criticism of Ecclestone, there were others who had cause to complain about Walker's business methods. By the following March he was forced to file for bankruptcy proceedings following a $17 million tax dispute with the Internal Revenue Service, who claimed that between 1982 and 1984 he had understated tax liabilities to partners involved in real-estate developments totalling $165 million. He and his wife, Carol, were later charged with owing $85 million in taxes, interest and penalties dating back to 1981. In January 1988, Walker was jailed for seven years after pleading guilty to three counts of felony tax fraud related to three limited partnerships in 1982.

The 1984 Dallas Grand Prix, won by Keke Rosberg in a Williams-Honda car on his way to a World Championship title, proved to be the first and last. Attempts by other businessmen to fulfil the four-year contract with Ecclestone were thwarted by the sheer cost of the event and legal difficulties, including a lawsuit for damages by residents living near the Fair Park Circuit against the City of Dallas and Walker. It was Chris Pook's opinion that the Dallas Grand Prix would have proved to have been "an incredibly successful venue" but for the downfall of Walker. But to seasoned sports reporter Gene Wojciechowski, with the *Dallas*

Morning News at the time of the Grand Prix, "it was the wrong place and the wrong time. It was novelty, something different. Dallas considers itself world-class but it is very provincial. They don't go to the sports pages looking for the results of Formula One. Bernie Ecclestone's style didn't go down too well either. It was 'my way or the highway', and with so many other motor sports around that wasn't the smartest approach with people who wanted to see Formula One succeed."

Ecclestone's attempts to establish a long-term Grand Prix venue in America proved no more successful when the Formula One big top rolled on to Phoenix, Arizona, several years later. In 1989, the city council members gave their seven-to-one backing to the plan, along with a one-off $2.9-million grant towards the cost of building and maintaining a 2.2-mile, thirteen-turn street circuit and an annual $1.6-million grant thereafter. But, three years later, the Phoenix Grand Prix was no more. The first race was staged in June, a time of blistering heat when the city attracts few visitors. It was a disaster with no more than an estimated 13,000 spectators going through the gate. It proved particularly costly for Ecclestone, who, as the promoter, ended up with costs totalling £12 million. He decided to stage the following year's race in March, but, while the attendance figure marginally improved, the layout of the track was widely criticized by spectators who had paid more than $200 for grandstand seats and could see little more than the drivers' helmets.

The third race, in 1991, proved no more appealing. Won by Ayrton Senna, who dominated the race from start to finish to take his fifth US Grand Prix win, it turned out to be the last Grand Prix to be held in America for the next nine years. Describing the Phoenix Grand Prix as "a debacle", Chris Pook said that Ecclestone "really did himself some harm, or Formula One did themselves some harm, because it just didn't work. The circuit wasn't compatible, the spectators couldn't see. It was a non-event. Unfortunately, Bernard received some very bad advice, very bad advice. The dilemma was that Formula One came to be seen as an anchor around the neck [of promoters] . . . trying to swim."

Ecclestone attempted that year to set up his own Grand Prix base in America by making a $4-million offer for the twelve-turn,

2.52-mile Road Atlanta track in Georgia, which had hit financial difficulties. He said he agreed its purchase on a handshake, but when another interested party offered $100,000 more, he was invited to top it. "The guy came back to me and said, 'Do you want to give another $500,000?' I said no. You can't sell it twice. You'd be put in prison for that." Twelve months later the owners filed for bankruptcy. Ecclestone and a New Hampshire businessman bid $2.6 million against bids of $3.2 million from a tyre corporation and an Atlanta-based group of companies. Ecclestone's bid was unsuccessful; the track was sold but reportedly failed to reach its investment potential.

By February 1996, its new owner, multi-millionaire Don Panoz, who had made a fortune in pharmaceutical research, was talking to Ecclestone about hosting a Formula One Grand Prix. Widely considered to be one of the best tracks in America, Panoz, the kind of businessman who established a successful vineyard in Georgia when the experts said it couldn't be done, spent $30 million improving the circuit to Formula One's standards and facilities. But it all came to nothing. Although Panoz had committed $30 million to bringing Formula One to Georgia, he was not prepared to back down over the financial terms. Ecclestone insists the issue was safety not money. "There was a bridge with concrete on either side of the track, and when I agreed to buy it I agreed to knock it down. It was dangerous. There was no way we would race there."

Ecclestone became so desperate to break into the American market that he was willing to do so on the back of the hugely popular Indy 500 series. He supported an FISA proposal, mooted by Max Mosley in 1993, that oval races of up to 500 miles be allowed into the Formula One Championship series, with slightly modified Formula One cars competing against IndyCars. As the United States Auto Club, the then organisers of the Indy, is sanctioned by the FIA, it was theoretically possible. But, across the Atlantic, it was rejected out of hand. The technical differences between the two cars apart—the lighter, nimble Formula One cars with their carbon brakes would have had an overwhelming advantage over the heavier, turbo-charged IndyCars and their steel brakes—it made no sense for the owners of the top-class ovals, strongly committed to the IndyCar and the NASCAR series, to threaten that profitable relationship by getting involved with a formula that had little following. The proposal, criticised by the American motorsport media as "silly and irrelevant", was dismissed as a headline-grabbing stunt to raise the profile of Formula One in America. More crucially, the teams themselves were far from keen.

A few months later similar judgement was accorded a challenge Ecclestone issued to IndyCar team owners: a £5-million showdown between Formula One cars and IndyCars on the streets of Adelaide the week before the Australian Grand Prix. Probably to his surprise, the challenge was accepted. Roger Penske, owner of Marlboro Team Penske, and the Michigan International Speedway, which accepted on behalf of another team owner, between them agreed to supply eight IndyCars to race against four Williams and four McLarens in a road race one day and on an oval circuit the next. Fine, responded Ecclestone. The race was on. But there was a qualification. The Formula One cars, he insisted, would race only on an FIA-approved track—and when one was built in Europe. Ecclestone had grabbed the headline and his £5 million was safe.

He had consistently refused to have a US Grand Prix as a separate event on an IndyCars programme. It was either a head-to-head, he insisted, or nothing, otherwise the television rights would be devalued along with the level of his profits. "Television coverage would be confused," he said. "No mileage in that for anyone." It had probably proved the stumbling block when Ecclestone and Tony George had talks at the last US Grand Prix at Phoenix in 1991. But that was then, in the days when Ecclestone probably still believed that America was there for the taking. Now, after several years of varying degrees of failure trying to replicate the long-term success of Watkins Glen, he was ready to meet again with George—after talks, which involved Chris Pook, with Willy Brown, the controversial mayor of San Francisco, to stage a street race similar to Long Beach floundered.

During the weekend of 13–14 September 1997, George, accompanied by his executive vice president, Leo Mehl, flew to London to discuss once more the obvious synergy of a commercial alliance at the Indianapolis Motor Speedway. George, from one of the richest families in the USA, said to be worth more than $9 billion, guaranteed a level of professionalism to ensure a stable Grand Prix home for Formula One. A further three meetings took place during 1998 to see a conclusion to the negotiations that had stalled seven years earlier in Phoenix. On 2 December it was announced that a $30-million multi-year contract had been agreed to stage a US Grand Prix beginning in 2000 at the Speedway.

It took place on a 2.606-mile circuit combining an infield loop with a mile-long section of the storeyed oval and before a record crowd of more than 250,000. The US Grand Prix, won by Ferrari's Michael Schumacher had, in a sense, returned home after a forty-year absence. Until 1960, the Indy 500 had been included in the Formula One calendar to justify the World Championship series title. The Indianapolis circuit had been bought by George's grandfather, Tony Hulman, as a derelict speedway in 1945 for $500,000 to save it from housing development, and now George described the return of Formula One to the venue as the "realization of my dream". After nine years in the American wilderness, it was probably no less a dream come true for Ecclestone.

Before his link-up with Tony George, there was one other Stateside deal that went sour for Ecclestone. This time it was with Hollywood star, Sylvester Stallone, who was keen to direct and star in a film on Formula One. At the Italian Grand Prix at Monza in 1997 Ecclestone signed, for the benefit of Stallone's backers, an agreement giving him the film rights.

In an interview with *Indianapolis Star* motor sport journalist Robin Miller, Stallone claimed that for the next two years he worked hard with lawyers and producers to formalize the financial arrangements and complete the script, based on the fictional story of a retired driver coaxed out of retirement to help a Brad Pitt-type character become world champion. "I thought we had a deal and every time I went to sign, Bernie would raise the price. I apologize to the very dedicated F1 fans, but there's a certain individual who runs the sport who basically has his own agenda. He just made it impossible to work."

Ecclestone blamed its failure on Hollywood studios unable to raise the necessary finance. After signing the agreement with Stallone, he had been advised that the terms were too generous and did not reflect the true value of granting a Formula One franchise for a film production. The project went no further. But even if Ecclestone had been willing to proceed, it seems it would have run into a major obstacle with the FIA. Stallone wanted the rights to the film for infinity, something that Ecclestone couldn't guarantee once the commercial rights were returned to the FIA after 2010, when the contract in existence at the time between Ecclestone and the FIA terminated. Mosley had no problem in obliging Stallone, as long as he was given the opportunity to check the script. He explained he had to be sure that Formula One would not be portrayed in a sensational plot detrimental to its image. Stallone refused point blank, and phoned Ecclestone to express his anger at that "fugger Mosley".

THE LIMIT by Michael Cannell, 2011

1961

One cold night in the first week of January 1961, Wolfgang von Trips stood at a doorway, suitcase in hand, and rang the bell. He lingered in the chill air, listening for signs of stirring within.

Three years earlier he had hired a former tax advisor named Elfriede Flossdorf to type his diary entries and manage the details of his endorsement deals and appearances. He teasingly called her "the racing secretary" because she dashed to keep up with the growing workload. Flossdorf was so busy, in fact, that von Trips installed her and her husband Willi in a small house on the entrance road to Burg Hemmersbach so that she would be close at hand. He had come to their door in the middle of the night to pick up his key. "Guess where I've been," von Trips asked when they had roused themselves.

They looked puzzled. "Well," she said, "in South Africa . . ."

"I was walking with the Almighty," von Trips interrupted.

"Where?"

"I celebrated the new year in absolute heaven," he said. "Astounding, isn't it?"

He was making a facetious reference to the Aga Khan's home in South Africa where he had spent the holiday with Gabriella, Princess of Savoy. In the coming months he faced a murderous campaign for the championship, but he began the year in the arms of a blue-eyed princess.

The next morning von Trips unwrapped a set of plates sent by Huschke von Hanstein of Porsche, a steadfast friend throughout his rise and fall and eventual redemption. Hanstein had given the plates as a Christmas gift and to show that he accepted von Trips' decision to leave Porsche and return to Ferrari at the end of the 1959 season.

"I've just unpacked your Christmas present and note to my great regret that the large plates have broken into three clean pieces," von Trips wrote to Hanstein. "My next difficulty is to interpret what this oracle could mean. Shards can mean luck, or it could be a wink at my decision to go with the Reds."

As von Trips noted, broken plates are considered propitious. He may have carried the luck with him to Sicily for the Targa Florio, a rugged mountain race preceding the Grand Prix season. [Phil] Hill chased von Trips through the harsh countryside on rutted roads caked with dung and caught up to him on a rocky plateau 2,000 feet above the coast. He tried to pass but the road— no more than a paved track for donkey carts—was too narrow. Von Trips might have edged over to let Hill pass, but he was unaware that Hill had pulled so close; the engine howl drowned out the sound of Hill's car. Hill announced himself with a series of bumps and shoves delivered from behind at 120 mph. One rough nudge sent both whirling off the road and into the weeds. After a heated exchange, they pulled back onto the road, with von Trips leading again. Now driving in a fit of anger, Hill muscled past von Trips and skidded his way down a long switchbacked descent to the sea, flashing by ancient stone churches and old men riding mules. At the bottom of a steep stretch he hit a blind bend and lost control, sliding through concrete guardposts and crumpling his Ferrari in a ditch. He crawled out of the wreck in time to see von Trips rumble by on his way to a win. It was a taste of the dogfight to come.

Sicily was a prelude to Monaco, the first of nine Grands Prix that would decide the championship. Von Trips glowed with confidence at the pre-race galas and the house party thrown by Bernard Cahier at his home in Villefranche-sur-Mer, six miles down the Riviera coast. He was a bit shy, as always, with a hint of sadness in his eyes, but his charm and good humor seemed to inoculate him against misfortune. He was too nice to die.

Von Trips had the winner's demeanour, but a betting line would probably have favoured Hill due to his consistent record. "This is Hill's year," Moss said. "He has the ability . . . and the car." Though you wouldn't have known it to see Hill on practice days. As usual, he was a knot of nerves, pacing the pits with a cigarette and wiping his goggles. At every turn he exchanged sharp words—punctuated by animated Italian gestures—with the Ferrari mechanics buzzing about the cars in buff-coloured coveralls. When he was not practising, Hill burned off nervous energy and built stamina with long swims off the Monaco beachfront.

The warm, wet Mediterranean was his antidote to days spent staring at blacktop.

Hill, the incessant worrier, did not accept the common view that a Ferrari championship was inevitable. "There was always an uncomfortable feeling in the team," he later wrote, "and while the car was very competitive, I never was convinced that the championship was going to be easy or even possible to win."

Among other things, Hill feared that Stirling Moss might steal the race, or the season, with one of his sensational dark-horse performances. Moss was flinty-eyed and muscled, with almost superhuman discipline. He once said that he abstained from sex for a week before each race so as not to soften his resolve. His eyesight was so acute that he could read newspapers from across a room and scan the crowd for pretty girls while entering a curve at 85 mph.

Moss had won fourteen Grand Prix races—more than any other active driver—but he had yet to win a championship. His shrewd, cold-blooded precision made him a perennial threat, despite his refusal to join a manufacturer's team. Aside from a stint with Mercedes, Moss mostly relied on privately owned cars a year or so out of date. He enjoyed the underdog role, just as he relished driving cars painted British racing green, but he knew that he might never win a championship that way. In one race after another the obsolete cars broke down under his punishment.

"I like to feel the odds are against me," he said. "That is one of the reasons why I do not drive for a factory. I want to beat the factories in a car that has no right to do so. If I had any sense I would have been driving for Ferrari all these years. Year after year Ferrari has the best car. But I want to fight against odds, and in a British car."

Moss took particular delight in beating Ferrari. Ten years earlier, when Moss was a twenty-two-year-old sensation, Ferrari had courted him. After some negotiation Moss had agreed to race a sleek new four-cylinder Ferrari with a tapered nose in a race at Bari, a port city on the heel of the Italian boot. Moss and his father, a prosperous dentist, made the long trip from London, only to be rebuffed at the Ferrari garage. "The mechanic said,

'Who are you and what are you doing?'" Moss recalled. "I said I was going to drive that car. He said, 'I'm afraid you're not.'"

Moss telephoned Ferrari, who unapologetically explained that he had changed his mind and given the car to the veteran Piero Taruffi. Ferrari was most likely punishing Moss for taking a tough stance in their negotiations. Moss took it as a barefaced insult—to England as much as to himself—and vowed revenge on the race-track. Indeed, the affront gave him extra incentive over the following decade. "It gave me great pleasure to beat Ferrari," he said.

If Moss could beat Ferrari anywhere, it was Monaco. In the narrow streets winding above the blue Mediterranean, the driver counted at least as much as the car. The course contained only one straight where the Ferraris could unleash their decisive horse-power. Racing a Ferrari against Moss in Monaco, Hill said, was "similar to seeing which is quickest round a living room, a race horse or a dog." Monaco was not considered dangerous. The cars reached only 120 mph or so as they twisted their way up and down the hilly principality. It had been nine years since a driver had died there. The circuit might not be fast, but it was demand-ing. One hundred laps of short bursts and tight hairpins punished the clutch—and the driver. They changed gears about once every five seconds.

Monaco was one of the few Grands Prix run in city streets, and the smallest slip could send a car into a storefront or street-lamp. Pedestrians stood unprotected on the kerb, almost within reach of passing cars. Six years earlier Alberto Ascari had driven his Lancia into the harbour, and a driver once missed a sharp left turn and rammed his hood into the ticket office of the train station. "To go flat-out through a bend that is surrounded by level lawn is one thing," Moss said, "but to go flat-out through a bend that has a stone wall on one side and a precipice on the other—that's an achievement."

Race day felt like a fresh start for a sport afflicted by so many recent deaths. As the Sunday church bells tolled, tanned specta-tors gathered on balconies and in open-air cafés. The city was bright with lilies and bougainvillea. White yachts bobbed at their moorings with pennants fluttering and girls sunbathing in bikinis. Many of the yachts had moved a safe distance from their quayside slips. Rescue divers waited on small boats in case a driver flew off the harbourfront promenade.

At 2.30 p.m. the mayor of Monte Carlo arrived at the grandstand in a chauffeured convertible, a sign that the race was about to begin. Mechanics rolled a pair of Sharknoses onto the third row of the starting grid for Hill and von Trips. They sat one row behind Richie Ginther, Hill's Santa Monica friend and a former mechanic.

At Hill's urging, Chiti had hired Ginther to test drive the Sharknose prototypes. He would now race the latest version with a new engine configured with V-shaped cylinder banks spread at 120 degrees. Ginther's cylinder banks were 55 degrees wider than those found in the other two Sharknoses, allowing the engine to sit lower in the chassis. With a lower centre of gravity, the 120-degree version should in theory handle better on Monaco's hairpins and switchbacks. Chiti wanted the new version assigned to Hill or von Trips, but Ferrari did not trust it. He insisted that it go to a second-tier driver. Hill and von Trips would drive the more thoroughly tested model. "It was a typical piece of Ferrari meddling—jealous of a possible project of which he hadn't approved," Chiti later said. "He was almost afraid of a good result, after having said, over and over, for months, that the future didn't lie with rear engines."

Ferrari's reservations proved unfounded. The 120-degree engine was blindingly fast over three days of practice, earning Ginther a place beside Moss on the front row.

As usual, Moss was badly handicapped by an obsolete car, though it never seemed to bother him. The year-old Lotus was 20 mph slower than the Ferraris. It had undergone a series of hasty repairs, the last of which took place on the starting line. Minutes before the start Moss noticed a crack in the chassis. He stood by calmly sipping water as a mechanic covered the fuel tanks in wet towels and welded the fracture closed.

With five minutes to go, the drivers pulled their helmets on and lowered themselves into their seats. They stretched their legs inside the hollow cars and checked the alignment of rear-view mirrors. They twisted wax earplugs into place to muffle the engine noise and pulled on string-backed gloves dusted with talcum powder to absorb sweat. Up and down the starting grid drivers yawned in response to nervousness and swallowed to slake dry throats. Some spat for good luck. A helicopter buzzed overhead, ruffling palm fronds. Team managers leaned in for parting words. *Viel Glück. Bonne chance. Buona fortuna.* Good luck.

A minute to go. The drivers adjusted goggles, snapped helmet straps, and scanned gauges. The hornet shriek of exhaust reverberated off the tiers of hotels and grand homes ringing the semi-circular harbour. Hill and von Trips put their cars in gear and held the clutch to the floor, eyes fixed on the starting flag.

When the flag dropped, sixteen cars leapt from a billow of blue exhaust and the grey smoke of burning rubber. Ginther squeaked into the lead as the cars sprinted the first 300 yards and abruptly slowed to 30 mph for the first turn, a 180-degree hairpin known as the Gasometer. Moss and a young Scotsman named Jim Clark shadowed Ginther with the pack strung out behind. They moved like a herd of red, silver, and blue thoroughbreds, darting left and right but never touching. The drivers switched gears every three or four seconds, leaning out to watch as they placed the front wheels on a precise line.

They climbed the switchbacks to Casino Square, passing between the green-domed Casino and the Hotel de Paris, two Belle Époque wedding cakes at the clifftop. A wrong turn here and a car would fall eighty feet. The pack fought and wove for advantage as they funnelled downhill through zigzagging legs ending at a dark, curving tunnel that opened onto a long waterfront run back to the starting line.

Clark dropped back on lap eleven with a faulty fuel pump, leaving Moss to shadow Ginther. Moss pressed, hoping to force Ginther into a mistake. On lap fourteen Moss made his move. He slithered by Ginther coming out of the Gasometer to steal the lead. Eleven laps later Ginther yielded to Hill, who had worked his way up from seventh.

Von Trips never joined the lead pack. A faltering battery reduced his revs, causing him to miss a gear shift on the last lap and smack into a guardrail at Mirabeau, a hairpin leading down to the tunnel. He was unable to finish, but he would be awarded fourth place based on the number of laps completed behind the winner.

Now it was a three-car race: Moss chased by the Ferrari tandem of Hill and Ginther. Hill closed to within four-and-a-half seconds of Moss, waiting for the spot where he might pass. But he never found it. Moss was hitting his stride, smoothly shifting gears ninety times a lap. "What I remember about that race," Hill later said, "was the frustration of busting my ass and not being able to catch Moss."

By now Moss was lapping the slower cars, weaving among them for protection from his pursuer. "It was rather like a fighter plane being chased by a superior enemy and being saved by dodging into the clouds," he said.

With brakes worn by the punishing twists and turns and his carburettor faltering, Hill motioned to the pits that Ginther should take over the chase. "Richie was going much faster than I was," Hill said. "I kept waving for him to go by."

Tavoni agreed. On the seventy-fifth lap—the three-quarter mark—he flashed Ginther a sign marked "Go." It was his signal to move up into second place and assault Moss's seven-second lead. The two men flung themselves all over the road, swooping down the hills and peeling along the harbourfront, the Ferrari edging closer by tiny increments. Ginther jawed hard on a wad of gum, his face set with determination. Billows of exhaust hung in the narrow streets.

Moss kept the lead, but that was not necessarily an advantage. "It's much easier to chase a man than to be the one being chased," Moss said. "If a man is following you really closely—within a couple of yards—he's learning an awful lot about your techniques and where he might be able to pass you. If you have a particularly unusual line around a corner you immediately show it to him. So unless you can break away from the pack it's not a good thing to lead for ninety-nine out of a hundred laps and then have the bloke pull out and pass you on the last one."

It looked as if Ginther might do just that. With sixteen laps left he spat out his gum and stepped up the attack, closing to within four seconds. The Ferrari pit crew held out a sign marked "Bravo." In cafés and on verandas the spectators stood. All of Monaco watched to see if Ginther would pass Moss in the final minutes. "The last few laps I stopped watching," said Rob Walker, who owned the car Moss was driving. "I couldn't look anymore. I couldn't stand it."

Even in the midst of the race Moss held the steering wheel with his fingertips and lifted a hand to wave at the crowd, which gave him a deceptively nonchalant appearance. "At Monaco in 1961 I was on the limit," Moss later said. "One doesn't very often run a race flat-out ten-tenths. Nine-tenths, yes. But at Monte Carlo every corner, every lap as far as I can remember, I was trying to drive the fastest I possibly could, to within a hair's-breadth of the limit, for at least ninety-two of the hundred laps. Driving like that is tremendously tiring, just tremendously tiring, most people have no idea what it does to one."

Before the final lap Tavoni held out a pit sign that said "All". On the final lap Ginther mounted one last frantic blitz. He came within 2.8 seconds, but Moss held him off. He had snatched the win from the Sharknose pack in an ageing car. Ginther and Hill followed Moss in that order.

"Until I saw the checkered flag I wasn't sure what would happen," Moss said.

The British boats blasted their horns. Moss received congratulations from Prince Pierre in the royal box and held the trophy aloft while a band played "God Save the Queen". His mother tried to push her way in for a hug, but the police turned her away. She watched with tears as her son lit a cigarette and took a victory lap, waving to the crowd with that distinctive British parade gesture. Moss called the race "my greatest drive".

Hill was so exasperated as he talked with reporters that he looked as if he might break down. Ginther wiped his grime-streaked face with a rag and smiled ruefully. "Embracing him, it seemed to me that [Ginther's] overalls were completely empty," said Franco Gozzi, one of Ferrari's aides. "He was shattered, all in, he had really given everything."

Ferrari wouldn't have to wait long for revenge. Within days the teams packed up and moved to the Dutch Grand Prix at Zandvoort, a resort hard against the North Sea. The setting could not have been more different than Monaco. The Dutch track's broad, sweeping turns wound among massive sand dunes studded with scrub grass and scoured by cold sea winds. It was fast, with one side flattened to form a half-mile straight where the Ferraris could cut loose.

If they showed up. In the first practice session, held on a Saturday morning, the other teams whipped through biting wind and rain showers. There was no sign of Ferrari. The Lotus and Porsche crews conferred in hopeful tones—had the punishing maze of Monaco exposed a flaw in the Sharknose, sending Ferrari engineers back to the drawing board?

Minutes after the Monaco Grand Prix ended, Chiti had phoned Ferrari. Based on Ginther's performance he urged that 120-degree engines be installed in all three cars in time for the Dutch Grand Prix. After a short pause Ferrari said, "No, not even worth discussing." Chiti was livid. He went to Maranello and demanded an explanation. Ferrari convened a meeting to discuss the proposition. Not surprisingly, Ferrari's deputies unanimously sided with Ferrari. Chiti dug in. Either the new engines go in, he said, or he would quit. Ferrari backed down. "Do you as you like then," he said.

The Ferrari team missed the first morning of practice as the new engines were prepared and installed. At lunchtime a red Ferrari car transporter, a double-decker truck shaped like an oversized fire engine, pulled up to the paddock and unloaded three Sharknoses. No longer constrained by the tight Monaco turns, the Sharknoses put on a display of raw speed. They were easily fast enough to earn Hill, von Trips, and Ginther side-by-side positions on the front row—a blockade of red. Moss would line up directly behind them.

After practice Hill, as usual, dickered with the mechanics. Von Trips noticed a ten-year-old Dutch boy sitting on the roof overhanging the low pits. He stepped onto a bench and pulled himself up, tousling the boy's hair and sitting with him for a few minutes, trying to glean warmth from the wan sun glimmering through the cold fog. It was the kind of spontaneous gesture that came easily to him.

On the morning of the race Prince Bernhard and Princess Irene arrived by helicopter and shook hands with the drivers. It did not go unnoticed that the princess wore a headscarf with the Ferrari logo. During the final tune-ups Hill found that his car had developed a tendency to oversteer, which caused him to turn more sharply than intended. It was the sort of glitch that unhinged Hill, particularly when it arose during last-minute preparations. To make matters worse, his clutch conked out on the final warm-up lap. The mechanics swarmed, rushing to repair it five minutes before the start. Hill tried to "calm his palpitating heart with his right hand," *Motor Racing* reported, "while he unwrapped a stick of gum with his left."

Moss, who was not known for jokes, chose this moment to goad Hill. "Push that thing away," he said. "It will only jam up the rest of us."

"Oh, look Phil," he added. "They're taking the whole bloody back end out of your car." Hill had tried to ignore Moss, but now he turned in a panic. In the end the mechanics fixed the clutch and Hill got into his car on the front row. Moss chuckled just behind Hill, knowing that he had unsettled him.

After the start the cars vanished among the sand dunes. When they reappeared at the end of the first lap von Trips was leading. He had surged ahead while Hill tried to manhandle his errant steering through the rolling grey landscape. He was, by his own description, "all arms and elbows".

"I was going around oversteering all over the place," Hill added. "You can just oversteer your way around that course so long before you've done in your tyres—and run out of adrenaline."

With von Trips five seconds ahead, Hill sparred for second with Jim Clark, the Scot who had briefly challenged him in Monaco. Chiti and Tavoni were working their stopwatches when Clark ripped by an inch from the pits, forcing them to jump back. Clark's message was as unmistakable as it was intimidating. Even though Clark's Lotus handled the turns adroitly, it couldn't compete with a Ferrari whistling down the long straights. For the first time in the season, the Sharknose showed what it could do. With twenty laps left Hill pulled away. Chiti exhaled in relief. Il Commendatore could no longer question the 120-degree engine's capability. It looked unbeatable.

Now Hill was finding his rhythm, and he began eating into von

Trips' lead. Just as Hill drew up behind von Trips, the phone rang in the pits. It was Ferrari calling from his office. On his orders Tavoni held out a message board dictating the order of finish: "Trips-Hill". With the first Ferrari win of the season within reach, Tavoni would not risk any tangles. "They just didn't want us ripping each other up once the thing was stabilized," Hill said. Von Trips led all seventy-five laps, scoring his first Grand Prix win—the first by a German after the war—with a dominating performance. Cheers erupted in beer gardens and rathskellers all across Germany.

Prince Bernhard and his daughter, Princess Irene, were supposed to present the winner's trophy, but the crowd surged onto the field and formed an impenetrable throng around von Trips, who pulled Hill under the victory wreath with him. Von Trips had agreed to a post-race interview with Hermann Harster, the German journalist who was covering the race for a Hamburg tabloid, but von Trips had no way of getting to their meeting place. He borrowed a bicycle from a teenager and rode with the teen sitting on the handlebars. "It was a big day for him," Harster wrote. "How many kids are ridden around on their bicycle by a Grand Prix winner?"

Moss and von Trips had each won a race. The pressure was now on Hill to keep pace at the Belgian Grand Prix, held at Spa on 18 June. "Trips wanted it very much, and so did I," Hill said. "And despite the fact that we were members of the same team we each knew that we'd have to fight with everything we had to win the title. The tension continued to build from race to race."

If Hill were to win he would have to prevail on the fast and unforgiving track at Spa, where drivers rocketed through the Ardennes Forest at an average speed of 130 mph and slid around turns on asphalt softened to a slippery sheen by an early summer heat wave. As if to prove up front that he was in the fight for good, Hill scorched the practice laps and recorded the first lap under four minutes in track history. The fastest practice times earned him the right to start in the pole position, the inside spot on the front row that guaranteed an edge as the pack entered the first turn. Von Trips would line up beside him. Completing an all-Ferrari front row was Gendebien, who, as a Belgian, was driving a Sharknose painted yellow, the Belgian racing colour.

The lead switched back and forth between Hill and von Trips as they swung through curves at 150 mph. Hill eventually pulled away, three times recording track records. A heavy rain closed in and the pack slowed by about ten seconds a lap, impeding von Trips from challenging Hill, though he surely would have pressed had he known that Hill was vulnerable. A tiny pebble had lodged in Hill's eye on the twentieth lap. He drove the last third of the race half blind.

Once again Tavoni held out a sign freezing the positions, this time with Hill beating von Trips by barely a half-second. Hill looked embarrassed as the laurel wreath was laid on his shoulders. Victory displays always made him uncomfortable and he slipped away as fast he could. More than anything he was relieved to have done it—relieved to be back in the mix—but he called it "more of a joke than a race" because of Tavoni's orchestration.

The win in Belgium propelled Hill to the top of the standings. He now had nineteen points, one more than von Trips. He had shown his mettle by scrapping from behind in the tally, and it looked as if he would add to his lead two weeks later at the French Grand Prix in Reims, a race on public roads looping through the patterned geometric fields of Champagne country. His car ran particularly fast in warm-ups, giving him "a guilty surge of pleasure".

Hill's car was so fast that it became a point of contention in the elaborate game-within-a-game that preceded each race. Von Trips groused to Tavoni that Hill had an unfair advantage. Hill knew that von Trips was right. "God, my car was clearly superior to Trips'," he said. "I mean, my car was a full half-second, three-quarters-of-a-second faster."

Von Trips demanded that Hill take a few laps in his car to demonstrate its slowness. If Hill also logged sluggish lap times it would prove that the problem lay with the car, not the driver. "I didn't really want to," Hill said. "After all, it might not be running right and they might fix it so that it would be faster than mine."

Ferrari normally prohibited drivers from handling a teammate's car for fear of sabotage. In this case Tavoni relented under pressure from von Trips. "Sure," he told Hill, "take it around." Hill only agreed because there was an oil spill at Thillois, a hairpin turn just before the pits. If he drove slower than von Trips, he figured, he could blame it on the oil. In fact, he drove three blazing laps, beating von Trips' lap time by half a second, a generous margin by Grand Prix standards. "I really let it all hang out," he said. "I flew. When I came into the pits, Trips was the picture of gloom. I said, 'I'm sorry I wasn't able to turn in a good one with all that damn oil all over, but it doesn't feel half bad, to tell you the truth.'" Hill compounded the insult by winning the hundred bottles of champagne awarded to the driver with the fastest practice lap.

As if that wasn't humiliating enough, von Trips looked back on the next practice lap to see Moss riding an out-of-date Lotus in his slipstream so that the more powerful Ferrari would suck him along in its wake. Moss and other British drivers had perfected the technique during those many difficult years when their cars could not keep up with the Italians. When Tavoni waved in warning from the pits, von Trips sped up, hoping to shake Moss, but he succeeded only in pulling him along faster. Moss's practice times were consequently much faster than they would otherwise be—the speediest among non-Italian cars by two seconds—assuring that he would start near the front. On his way back into the pits he flipped the Ferrari crew the two-fingered up-your-arse gesture.

On the morning of the race the July sun scorched the blond wheat fields like a blowtorch. The thermometer touched 102° in the shade and 120° on the circuit. Women clutched glasses of cold champagne beneath broad-brimmed hats. The drivers lingered in the pits, dousing their coveralls with water. Ferrari mechanics wiped the cockpits down and removed body panels to allow cooling air to flow around the drivers' legs and bodies. Hill put a rubber bag of ice water on his floorboards with a jury-rigged hose snaking up his shoulders. When he stepped on the bag a cold spritz trickled down his back.

The heat softened the black tarmac, loosening bean-sized pieces of gravel. It was like "driving on a spill of ball bearings," Hill said. Two years earlier, on the same road, a bit of gravel had bloodied Hill's nose. "I was drinking blood for about five laps and couldn't feel a thing," he said.

As expected, the Ferraris jumped out front with Hill leading comfortably, followed by von Trips. After eight laps Hill eased back. Five laps later von Trips pulled up and Hill let him pass. It was still early and Hill could afford to sit back and spare his engine exertion and overheating.

On the other hand, Hill may have been ordered to give way to von Trips. Denis Jenkinson, a long-bearded correspondent for the British magazine *Motor Sport*, wrote that Tavoni had told Hill in a pre-race meeting that he would have to step aside for von Trips, despite Hill's faster practice times. Hill had accepted the order, according to Jenkinson, but with an eruption of anger. Hill's frustration is understandable. For five years he had accomplished everything asked of him, only to see Ferrari repeatedly favour von Trips. Hill's rage would only have pleased Ferrari. An angry driver, he knew, was a fast driver.

As it turned out, von Trips would not win, with or without Hill's help. He pulled into the pits on the eighteenth lap with his engine smoking and steam spewing from his right-hand exhaust. A piece of gravel had punctured his radiator.

Spectators saw a broad smile of vindication under the tinted visor of Hill's helmet as he reclaimed the lead. The race was now his to lose. With a sixteen-second lead he could afford to slow down and play it safe. He could coast home without incident. Ginther, in second, was too loyal to challenge him. Moss was laps behind. Hill was now all but assured of extending his lead in the Grand Prix tally and extinguishing some of the crushing pressure.

The race appeared locked up for Hill until he came out of a long downhill straight—among the longest and fastest on the entire Grand Prix circuit—at 160 mph and swung into the notorious Thillois hairpin expecting to drift his car around. Maybe he was too relaxed. Or maybe the heat had wilted his reflexes. Either way, he committed a rare miscue and skidded clockwise 180 degrees directly under the gaze of a grandstand. He might have quickly recovered if Moss, still struggling with his brakes, had not ploughed into the Ferrari's nose, spinning Hill another half turn. Hill then stood mid-track, cars whizzing by on either side, push-starting his car. He shoved it into motion with one hand and threw it into gear with the other. In the process the car ran over his foot. By the time the engine shook to life he had dropped to ninth place, which is where he finished.

Hill still led von Trips by a point, but he had fumbled his chance for an insurmountable lead. "My golden opportunity to make a decisive leap in the point standings was lost in one stupid move," he said, "but perhaps a certain Calvinist notion of retribution had been satisfied."

The charcoal skies hung low with a cold spit of rain when Phil Hill set out alone to walk the three-mile track at Aintree, site of the British Grand Prix and home of the Grand National, the country's most famous steeplechase. He had come a week early to inspect every detail—the pebbled texture of the road, the windy stretch near Bechers Bend, the sightlines into Cottage Corner. He had come to absorb the feel of the place, to visualize it. If he learned anything from the frustrating Reims episode of the previous week, it was that anything can happen. He wanted to eliminate uncertainties.

Hill walked along a dead-flat black asphalt road separated from the steeplechase course by high hedges and white fencing. It was a grim expanse, as featureless as a pool table, overlooked by a double-decked Victorian grandstand with a deep overhang to shelter spectators from persistent drizzle. Along with bland food, the cold Lancashire rain was Aintree's most distinctive aspect. Rain sprinkled on the practice runs and poured in sheets fifteen minutes before the start, sweeping the track with puddles. Dark roiling squall clouds mingled with factory smoke from Liverpool five miles to the south. Mechanics rushed to exchange smooth-tread tyres for wet-weather versions with deep grooves for traction. Hill and von Trips stood in drenched coveralls adjusting their helmet visors.

Spectators huddling under umbrellas hoped the rain would help Moss pull off another upset. He had grown up in the damp British weather, and his wet-road skills had earned him the nickname "Rain Master". But when the Union Jack dropped it was Hill, not Moss, who jumped out in front. He led for the first six laps with von Trips driving into a forty-foot plume of spray thrown up by his rival's Ferrari. Hill drove with great authority, safely negotiating flooded sections at Country Corner and Valentine's Way.

Before the race, while the track was still relatively dry, Hill had adjusted his car's brake balance so the front wheels had more braking force. He regretted it now as his front tyres began losing traction on the flooded tarmac. On the seventh lap he gained speed on the long backstretch, then braked in a tricky spot known as Melling Crossing, where a road crossed the track. He skated at 100 mph toward a massive solid oak gatepost, spinning first one way, then the other. His championship prospects seemed to vanish as he slid hopelessly closer to the post. He had unwillingly entered the state drivers feared most: he was a passenger without control, waiting for the slide to play itself out.

At the last second, as the gate loomed beside him, his front tyres found grip through the puddle. He engaged a low gear and dodged the post by an inch. It was a near miss, but the fright played on his nerves. "A few years earlier it would have been forgotten—like a letter dropped in a mailbox—the instant the wheel caught hold," he said, "but by 1961 it stayed with me."

Consciously or not, Hill drove more cautiously after his close call, allowing von Trips to nip by. Then Moss came up from behind to badger Hill. On the tenth lap Moss blew by him, drawing cheers from the grandstand. Badly shaken, Hill dropped to third.

Now it was von Trips' turn to see Moss looming in his rear-view mirror, waiting to pounce like a wolf on the smallest mistake. Moss was an endlessly resourceful fighter with British veins of ice. In the past, his unnerving presence, combined with the foul conditions, might have provoked von Trips into an impetuous blunder. But by now von Trips knew how to keep his composure, and he drove a nearly flawless race through unrelenting rain. "Sometimes I lost my Ferrari at 150 kilometers an hour and we skipped through the puddles like a stone thrown flat on the water," von Trips said.

Still, Moss kept turning up the heat. Lap after lap he hounded von Trips. By the twentieth lap Moss had closed to less than a second—an arm's reach away. He made his move at Tatts Corner, a ninety-degree turn at the far end of the backstretch, but von Trips closed the door on him.

Moss would have tried again, and might have succeeded, but he suffered his own skid at Melling Crossing, spinning an entire 360 degrees on the wet road. Five laps later he dropped out with a broken brake rod.

Half an hour later von Trips saw a tyre appear through the wall of rain ahead of him. A second tyre emerged along with the red body of a Ferrari. The rookie driver Giancarlo Baghetti had pulled close to another car to shield himself from the water spraying onto his windshield and goggles. When the other car unexpectedly braked, Baghetti went into a spin. "Suddenly I saw the eyes of the Italian coming out of the rain curtain toward me," von Trips said. "It was ghostly, eerie. He came closer and closer. Finally I flicked past him. On the next lap I saw Baghetti on the edge of the track. Without a car." Baghetti had spun five times and rammed into a wood fence.

By now the clouds had parted and the road dried. The sunlight shone on von Trips as he crossed the line in first, seven seconds ahead of Hill. More than any other, this race made von Trips look like a champion in waiting.

Minutes later the two men flanked Laura Ferrari on the podium. She smiled and linked her arms in theirs, a show of Ferrari unity. Von Trips wore a laurel wreath. He held his cup aloft and sipped champagne from it as the crowd whistled and clapped. Hill, who had led the standings until this moment, stood by in a suede windbreaker looking as if he had aged ten years.

Halfway through the Grand Prix season Hill and von Trips began wearing down. Neither could gain the upper hand, and the deadlock put added pressure on both men to crowd the limit. Von Trips called their contest "the highest test, the high wire". Their rivalry played large on the front pages of newspapers. Everywhere they went people asked about it. Teammates stole sideways glances to see how they were holding up.

"As I had feared, Trips and I became involved in an increasingly bitter competition for Championship Points," Hill later wrote. "Because the championship was at stake, I was not able to be reasonable and sensible about every race. After all these years I should have been, automatically, but there was this continual counting of points. By midseason my concentration was suffering."

Hill joined von Trips for a midseason break at Burg Hemmersbach, where the count lightened the mood with funny stories about his upbringing in the castle and the Americans stationed there. It is not known what else they discussed, but Hill clearly made a good impression on the countess. He was the sort of earnest and articulate young man parents approved of. Most of her son's friends cared only about speed and girls, but Hill's inquisitive mind took in everything European culture offered—wine, architecture, and history. The castle was like a museum, and she was his personal docent. In particular he shared her interest in music. It's easy to imagine him admiring the piano where she had played Chopin to von Trips as a boy. Hill and the countess agreed to attend the Salzburg music festival together later that month.

Back on the circuit, Hill and von Trips found it harder and harder to maintain a friendship under the strain of competition. In hotels and restaurants they were cordial, but they kept a protective distance. They greeted each other with tense smiles. "All year long it was him or me for the championship," Hill said. "It's not a normal situation race drivers are in: you try to beat the other guys all day, and then at night you're supposed to forget all that."

Their friendship was further strained by von Trips' driving. Too often he seemed oblivious to the subtle points of protocol that safeguarded drivers, and his tolerance for risk had an infectious quality that Hill struggled against. "I was very aware of staying within my limits, but Trips was unpredictable in this regard, and in a way I feared him," Hill said. "Racing against him, I soon learned that I was capable of being sucked into areas where I didn't want to be, even as I was having enough trouble knowing and sticking to my own limits."

A week before the German Grand Prix at the Nürburgring, von Trips found diversion in speed of a different magnitude. The 36th Fighter Wing of the US Air Force invited him to ride in a supersonic F-100. Just before the Ferrari transporter delivered the cars for practice, he drove an hour to the Bitburg air base where he was "passed around the officers' club like a champion cup," said Harster, who accompanied him. A small, wiry major named Charlie Davis fitted him into the cockpit with a flight suit, oxygen, and radio. After the sickening G-force of takeoff, Major Davis gave von Trips an aerial view of the Nürburgring. "There was the track," von Trips said. "There was the start and the finish. Charlie put the bird at an angle so I could see everything exactly. Then we were on the straightaway directly in front of the grand-stand, on it at more than 900 kilometres an hour."

"There was a green splash of colour in the pits," he added. "Probably the sleeping Lotus belonging to Stirling Moss." [The Lotus would actually have been blue, the Scottish racing colour.] "Thank goodness, no red," he said. "I didn't need a bad conscience. The Ferraris weren't there yet. My car wouldn't miss me."

Then the F-100 shot almost vertically through the clouds and into a dark blue sky. Major Davis rolled the jet and flicked on the afterburners. "About thirty seconds later there was a light tremor through the machine," von Trips said. "The needles with heights and speed began to dance like drunks." They broke the sound barrier over Burg Hemmersbach. Von Trips called it "a visit with the Gods".

He looked down from 40,000 feet at a serene and verdant countryside, but on the ground was a country in crisis. In August 1961 Germans fled East Berlin at a rate of fifteen hundred a day until Communist officials sealed the border with a barricade of barbed wire and jagged glass. Tanks patrolled the streets and police hauled thousands to detention camps. East and West faced off across the Berlin Wall, their fingers on the trigger.

Almost overnight von Trips' ambition to be the knight of a new Germany became closer to an imperative: He now carried the weight of national expectation. Divided and demoralized, Germany needed a uniting figure.

His countrymen had every reason to believe von Trips was on the verge of a Grand Prix title. His win at Aintree gave him nine points, leapfrogging him ahead of Hill in the championship standings. He arrived at the German Grand Prix with a two-point lead. He also had the comfort of racing on familiar ground. The Nürburgring was sixty miles from Burg Hemmersbach and practically within walking distance of the family hunting lodge. He was at home with the scent of pine needles and the short, steep mountains patrolled by wild boar and red deer. It was here that Bernd Rosemeyer had first roused his interest in racing.

Not that anyone would call the Nürburgring welcoming. On the contrary, the drivers nicknamed it the Green Hell. It was a fourteen-mile roller coaster created exclusively for racing, with narrow, winding turns without guardrails and abrupt plunges that pitched cars airborne. The road ran up and down hills that a civilian road would skirt. An atmosphere of foreboding hung over the course with Schloss Nürburg, a medieval ruin, overlooking Butcher's Field, Pick Axe Head, Enemy's Garden and other hazards. When an Australian driver was told (incorrectly) that Hitler himself had designed some of the features, he said, "I guessed as much!".

Von Trips arrived for practice a few days before the race and checked into the Sporthotel tucked under the grandstand. The hotel was reserved for teams on race weeks, and it took on a clubby aspect, with drivers wandering among rooms for meals and drinks. They could stroll to the pits in minutes. The drawback for von Trips was that thousands of admirers hovered, even on practice days. They encircled him at every turn, thrusting programmes for signatures and cornering him with questions. Girls lingered in the stairwell hoping to meet him.

The demands chafed on his goodwill. "I noticed that I had slowly but surely fallen into a state of nerves," he said. "I would like to leave the Nürburgring hotel and stay somewhere else, somewhere quiet, but I'm always there because it is practical . . . This time it was simply too dangerous for me. I had to find some peace at any price."

Von Trips had a reputation for conducting himself as a gentleman, but "sometimes it's just too much," he told a friend. "When I give a nice lad an autograph, I immediately have a whole tribe of people around me. So I'll go back to using my old method—I look at the floor, three feet ahead of me, and walk through the area, try not to make eye contact with anyone. It's the best way to get through." He thrust his hands in his pockets and nodded to avoid shaking hands.

With practice under way, word spread through the paddock that Enzo Ferrari had issued an edict from Modena that Hill should hold back, permitting von Trips to win the race and collect nine more points. According to an anonymous source interviewed by Robert Daley, the decision was based purely on business.

Von Trips is to win here and the world championship, too, in recognition of his splendid victory in the rain at the British Grand Prix in July and because this will sell many Ferrari touring cars in Germany. Ferrari's American market already is booming. He figures that a victory by Hill would not add anything in America.

At Reims, Hill had reluctantly complied with the order to step aside for von Trips, though fixing the finish had struck him as a shameful European arrangement determined by pedigree more than merit. But this time, with Hill trailing in points with three races to go, the suggestion that he take a dive stirred him to defiance. On the eve of the race he told Daley that he had not received any such instructions. Nor would he obey them.

"Ferrari can make all the decisions he wants to," he said. "But the decisions wouldn't mean anything unless I go along with them. I don't intend to go along with them."

As if to prove his resistance, Hill went out and drove a near-perfect practice lap in eight minutes and fifty-five seconds, with an average speed of 95.2 mph. One report described him looking "starry-eyed and trembling" when he returned to the pits. It was the fastest lap ever clocked at the 14.2-mile Nürburgring and the first to break the nine-minute barrier, which many considered inviolable. Hill called it a "freak" performance that he would not likely repeat.

"Out of the car he seems colder, more determined than ever before," Daley reported. "There is an icy edge to his apparently friendly banter with von Trips."

Meanwhile von Trips struggled through a series of mechanical setbacks. "Madre Mia," he said as he pulled in after a practice run. Tavoni came over with the conciliatory air of a physician. "First, the transmission is bad," von Trips told him. "Second, the chassis sits miserably on the course. And what about the time?"

The lap time was indeed laggard. At nine minutes and twenty-nine seconds, he was thirty-four seconds slower than Hill. Tavoni assured him that the mechanics would work through the night to get things right.

Von Trips' mood lightened on the short walk back to the Sporthotel. "If the engine cooperates," he said, "I will also come in under nine minutes." He paused to book a private hour of practice at 7 a.m. the next morning, for which he paid 100 deutsche marks.

The reworked engine proved surprisingly swift, allowing von Trips to break the nine-minute mark the next day, as he predicted. Later that morning a light rain fell, preventing Hill and von Trips from duplicating their earlier times. Hill clocked nine minutes and seven seconds; von Trips was two seconds faster. It was a good sign for von Trips. "Now I know that under the same conditions on the Ring I am probably just a tick faster than Phil," he said.

A pack of photographers encircled him, each calling for a different pose. *Sit behind the wheel. Look under the hood. Share a laugh with your teammates.* "If they only knew how tired I am," he whispered to a friend.

That night von Trips snuck away to an old hotel in a nearby village. He ate in a private dining room and borrowed a bed from a mechanic who had booked a double room above a stable. "I slept three feet above the manure pile," he said. "No one wanted anything from me. It was just wonderful."

Shortly after noon on race day von Trips ate lunch in his room at the Sporthotel, then napped briefly. His shirt was unbuttoned. Around his neck dangled a gold chain with the Ferrari insignia of the prancing stallion. Hill was staying a few rooms down the hall, and he dropped by. Von Trips could see that Hill was in the same state of nervousness. "Now we'll start together once more," said von Trips.

"To a good race," he added.

"We'll have a good race," Hill said.

Von Trips bathed with cold water and ate a final snack to fortify his blood sugar. At 1.57 p.m. he walked through the tunnel leading to the track. He was greeted by Fangio, who would drop the starting flag. The former champion embraced the man who hoped to succeed him.

Three hundred and fifty thousand Germans stood on the wooded hillsides to watch their national hero flash by in a red Ferrari. By most counts it was the largest crowd gathered at the Nürburgring since von Trips' hero, Bernd Rosemeyer, had driven there in the 1930s. The crowd didn't seem to care that von Trips was driving an Italian car. He wore a silver helmet in honour of Germany.

"There were so many people around, so much excitement in the air," he said. "I would've preferred to crawl into a hole, to get away. But I had to be there in the thick of it. There was no other way."

On the damp starting grid Tavoni questioned whether to begin the race on dry or wet tyres. Hill fretted along with him, pacing around the start area in his crewneck sweater.

With two minutes to go von Trips settled into his car and inserted his rubber earplugs. "The calm comes over me like redemption," he told Harster. "My helmet's strapped tight. The backup goggles are around my neck; the others are lowered over my eyes. The air valves are open. I slip on my gloves and turn the ignition. It sounds like Niagara Falls heard from far away."

Rumours had circulated all season that Britain's answer to the Sharknose would soon arrive on the Grand Prix circuit, diminishing the Ferrari advantage and possibly marking the start of a British comeback. At Nürburgring it finally happened, at least in part. A new V8 engine made by the British manufacturer Coventry Climax was fitted into a Cooper chassis for Jack Brabham, the defending world champion. "Doesn't it make a lovely noise?" asked John Cooper, the team owner. The new engine had 185 horsepower—enough to propel Brabham to the front row of the grid, alongside Hill and Moss. Von Trips started a row behind.

Von Trips had hoped to dart through the gap between Brabham and Hill, but Hill blocked his way. Instead, Brabham shot out to a quick lead as the front pack rounded the first curve and went up a straight behind the pits, around a left-hander, and through a serpentine series of downhill turns called the Hatzenbach. Then Brabham skidded on a damp, shaded patch and plunked his

Cooper through a thick hedge and into a ditch, disappearing in a cloud of leaves. Passing drivers were "smiling happily to themselves as they saw Brabham climb out of the V8 Cooper-Climax," *Motor Sport* reported.

With Brabham gone, Moss inherited the lead and Hill tucked in after him. Von Trips shook loose from the trailing pack and moved into third. It could have been a snapshot taken three months earlier in Monaco: the dark blue Lotus chased by a pair of red Ferraris.

Like Monaco, the Nürburgring was known as a driver's track—a circuit where handling counted as much as horsepower. It was the kind of track where Moss could use his artistry to build an incremental lead, and he did. By the end of the third lap he had a ten-second edge on Hill.

Now von Trips was agitating from behind. On the sixth lap he closed on Hill with a time of 9:08, the fastest ever recorded at the Nürburgring during a race. On the next lap he bettered his own record by four seconds. "Suddenly everything was wonderfully simple," von Trips said. "Inside of me was a sudden lift, a buoyancy. I was playing with what the course was giving me, not forcing anything. I was in a good spot."

Von Trips was driving faster and faster, but Moss and Hill kept pace in front of him. On the eighth lap Moss bested von Trips' record with a time of 9:02. Then von Trips overtook Hill and reduced the record to 9:01. A sea of fans waved German flags and white handkerchiefs from the hillsides.

Von Trips' record didn't last long. Hill stirred himself from third place to break the record once more, recording the first lap under nine minutes during a race. Hill and von Trips had by now scrapped their way up to within seven seconds of Moss. They could see him weaving in and out of the curves ahead, but could they pass him? "Seeing Moss is one thing," *Motor Sport* wrote in its recap, "and catching him is something quite different."

Then, in a heart-stopping moment, Hill seemed to disappear at Bergwerk, a tight right turn at the end of a long, fast section. Von Trips lost him in his rearview mirror and assumed that Hill had spun out, or worse. It would take at least thirty seconds to recover from a spin. Hill would never catch him, von Trips thought. Moments later Hill reappeared, as if by magic. He had been there all along, lost in the vibrating mirror.

On lap thirteen, with two laps left, a soft drizzle began to fall on the dark cedars. Moss, who had insisted on rain tyres, pulled away. He seemed robotic in his refusal to give in to exhaustion or make mistakes. He gradually padded his lead, leaving Hill and von Trips to fight their own private battle for second place and the six crucial points that came with it.

The two men raced side by side through meadows and dark tunnels formed by overhanging trees. On the penultimate lap Hill surged past von Trips. It was a critical manoevre. If Hill held on to second, he and von Trips would both have thirty-three points going into the Italian Grand Prix at Monza, where von Trips had a history of serious crashes. "When Hill went past me, I was very startled, appalled," von Trips said. "I tried to stay with it. It took a second for me to reconnect . . . Moss is forgotten. For me, it is now only about how to pass Hill."

Von Trips slid in behind Hill with the intention of slipstreaming in his wake before shooting forward for a final surge on the last lap. He was puzzled when Hill slowed slightly ahead of him; it looked suspiciously like an invitation to pass. It took von Trips a few moments to see the trap. Hill wanted to be the one to mount a carefully timed lunge from behind. So von Trips slowed with Hill. They exchanged sly smiles. "We drove for a little while side by side, playing a game: 'After you, sir,'" von Trips said.

They rode shoulder to shoulder uphill at 160 mph and into the final stretch when the Rhineland sky turned a dozen shades of grey. A wall of thunderstorms swept in, like special effects for the closing act. Both cars skated sideways on pools of water. "We were both all over the road," Hill said, "and damn lucky we didn't take each other out." Von Trips straightened himself out first and edged Hill by 1.1 seconds for second place under the eyes of his countrymen.

Von Trips sprang from his car in front of the pits. All around him people stood and clapped in the driving rain. Von Trips was handed a bouquet of red and white carnations, which he impulsively passed to Moss in the winner's circle. Moss in turn pulled von Trips under the winner's wreath with him. "It was a barrage of cameras," von Trips said. "I was full of happiness and thanks and fatigue and hunger. I put on my best Sunday smile."

"I felt close to Moss in this very special moment," he added, "but not as close as I had felt a few minutes before with Phil Hill."

Moss had pulled off his second major upset of the season against the Ferrari juggernaut, but it was von Trips who won the day. The driver who often seemed too impulsive for prime time had at last proved himself a model of steadiness and control. Hill, who had built his career on consistency, slipped four points behind in the season tally. His analysis: "I screwed it up."

Their duet would continue on the high-banked turns of Monza, the Death Circuit. Their summer of apprehension was almost over.

OVERDRIVE by Clyde Brolin, 2010

SPEED OF FLIGHT

So there you are, pushing like hell around the most famous twisty streets in the world. Battling for a crucial pole position, you are doing your frantic best to stop the 1,000 screaming horses behind your head from hurling you into the metal barriers when a realization starts to dawn. You're not driving your car. Someone else has taken over. You know that because you can see some guy clinging onto your steering wheel about three feet beneath you. You're just trying to work out why his yellow helmet looks familiar when it strikes you that you're no longer controlling your body either; he is. The good news is he's pretty handy. He's soon going nearly two seconds quicker per lap than anyone else. So you settle down to enjoy the view from this, the best seat in the house—until you notice you don't have a seat and you start falling . . .

Monaco, the most glittering icon of Grand Prix racing, is also its greatest anachronism. This cluttered hillside principality is not remotely suitable even for the Ferrari road cars stacked up outside the casino every evening—let alone the modern Formula 1 car. Three-time world champion Nelson Piquet Sr likened the experience to "flying a helicopter around your living room", while David Coulthard adapted that to the marginally preferable option of "riding a bicycle around your bathroom". It was in more standard equipment that a young Ayrton Senna announced his arrival to the F1 world at Monaco in 1984, nearly winning in the unfancied Toleman after a blistering drive in the rain.

Within three years he'd dragged his Lotus round to victory there, and the most challenging circuit of all would go on to host many of this Brazilian maestro's defining weekends. But his European base didn't just provide him with the chance to show off his talent; it changed his life. For it was in qualifying for the 1988 race, his first visit with McLaren, that Senna went into overdrive.

In an oft-quoted interview with Gerald Donaldson in his book

Grand Prix People, Senna revealed: "I was already on pole and I was going faster and faster. One lap after the other, quicker, and quicker, and quicker. I was at one stage just on pole, then by half a second, and then one second . . . and I kept going. Suddenly, I was nearly two seconds faster than anybody else, including my teammate with the same car. And I suddenly realized that I was no longer driving the car consciously.

"I was kind of driving it by instinct, only I was in a different dimension. It was like I was in a tunnel, not only the tunnel under the hotel, but the whole circuit for me was a tunnel. I was just going, going—more, and more, and more, and more. I was way over the limit, but still able to find even more. Then, suddenly, something just kicked me. I kind of woke up and I realized that I was in a different atmosphere than you normally are. Immediately my reaction was to back off, slow down. I drove back slowly to the pits and I didn't want to go out any more that day. It frightened me because I realized I was well beyond my conscious understanding."

It's not that unusual for an F1 driver to be on autopilot. One recent Grand Prix winner told me he often has spells when he is not consciously controlling the logistics of driving the car. Once he was leading a race by a minute when he noticed he wasn't even thinking about where he was braking; everything relating to his pedals, gears and steering wheel "just happened". It was the same when he was involved in a chase. His whole mind was taken up with the car in front, working out how to get close enough to mount an assault. At times like these, driving truly becomes second nature, and most of the drivers good enough to make it into Formula 1 are capable of switching off their fully conscious mind while racing. But Senna's experience was something else altogether. This was not easing off but going beyond concentration to the other side, a new "dimension".

Hollywood's modern classic *The Matrix* didn't come out until years after Senna's death, but fans of the film will recognize this kind of experience. In academic speak these special days are variously termed "peak experiences" or "flow". Sportsmen have a similarly wide range of phrases, from "hitting the sweet spot" or "the groove" to the most common term used by today's sport psychologists: "in the Zone". In Senna's case, "detached from anything else", he had unlocked an all-new level of hyper-ability and suddenly things became easy in these Earthly bounds of ours. He later described the sensation as "between two worlds". It sounds fanciful but this was no movie. It was very real and witnessed by millions around the world.

Of those, one had a clearer view of the magic than most. Switerland's Alain Menu, who went on to major success in touring cars, had been competing in the F3 race that weekend and his pit pass allowed him to stand on the inside of the chicane, looking back up the straight towards the exit of the tunnel. What came next remains etched into his memory as I discovered two decades later at a charity karting event at England's Rye House. We were talking about his own career when, unprompted, his mind travelled back to that Saturday afternoon in 1988: "There weren't many people around because it was a private area but I'm so glad I was there. Ayrton Senna was visibly braking eight metres later than anybody else, but it was his car that was amazing. All the other cars were a bit unbalanced and you could hear them banging around under braking. For him, nothing. As he braked the whole car just shook. You could hear nothing except for a noise that sounded like phphphphph."

Menu's exclamation is reminiscent of Hannibal Lecter recalling his favourite meal in *The Silence of the Lambs*, but breathing out rather than in. Spine tingling? You bet. "It made an immediate impression on me," he adds. "It gives me goose pimples to talk about it and I'm so glad I saw it. I'm sure some people who didn't see that lap would have heard what Ayrton said and thought, 'He's crazy.' If I hadn't seen it and I'd heard what he said, I would have said, 'Okay, whatever, it was just a fantastic lap, that's it.' But something definitely happened that Saturday and I believe it was special because I've never seen a racing car do this. Never, ever, ever. And I have no doubt it was the same the whole way round the lap.

"Ayrton was one-and-a-half seconds quicker than team-mate Alain Prost and two-and-a-half seconds clear of the next guy. Alain was a great driver but when he saw the lap times and the printouts he couldn't have believed it because it was so far ahead of what he was able to do. I'm very down-to-earth and I generally don't believe in this kind of thing. In my own career I've had some very good qualifying sessions but never anything like that. Sometimes when you are really at one with the car then things can look easy and feel easy and you do a fantastic lap because you never get out of shape or anything. So if that's what you call 'the Zone' then I've been there a few times. But I've never felt like that, where you almost forget what you've done. Later I heard Ayrton had to come into the pits because he could see he was looking down on himself from above the car. That was all so hard to believe but now I believe it because I saw it and I heard it."

After such an otherworldly experience it took only a day for Senna to come crashing back to earth. When the lights went out for the race he made the most of his hard-earned pole while archrival Prost lost a place to Gerhard Berger's Ferrari. It took the Frenchman fifty laps to get past, by which time the flying Senna was nearly a minute ahead with a quarter of the race left. Even so, when the Brazilian was told Prost had put in a couple of quick laps he unnecessarily set a new fastest lap in response. McLaren then ordered both drivers to ease off.

There are contrasting stories about how far Senna had gone into another trance, but eleven laps from the flag he clattered the barriers near the entrance to the tunnel—of the standard concrete variety. Prost could scarcely believe the sight of the stricken McLaren. He'd assumed his teammate might make a mistake on a quick lap, not after they'd slowed down. Senna never reported back to the pits but walked straight to his nearby apartment. When his McLaren colleague and confidant Jo Ramirez finally got hold of Senna late that evening, he was in tears. All because of a discomforting drop from the Zone to the comfort zone.

Senna later claimed to have had a "blackout" and resolved to find out how he'd let himself make such an error. He turned to his family for help and claims it took him two months to regain form—during which time this "underperforming" Brazilian took three poles and three wins from five race weekends. Sportingly for the lesser mortals on the grid he admitted people might not have noticed this "major part" of his development. But something must have worked as Ramirez confirms: "In 1988 he drove superbly—he just went in the car and he was inspired."

Of course Senna himself is sadly no longer available to expand on his cathartic weekend, so I turned to the man who prised the original account out of him. Canada's Gerald Donaldson is a veteran journalist who has written definitive books on racers including both Gilles and Jacques Villeneuve. He had never met Senna and he spent two days hanging around at an Imola test before the Brazilian beckoned him over during a rain break. This was a time when the partisan British press were making him public enemy number one for having the gall to go up against "Our Nige" Mansell. Senna was even booed on the Silverstone podium when he took third there in 1990.

"At first he thought I was British but when he found out I was Canadian he just mellowed," says Donaldson. "First he wanted to talk about Gilles Villeneuve, one of his heroes, and in the end I couldn't shut him up. He wanted to talk about everything, he went on and on and it filled two or three tapes. He was so eloquent and even people who didn't like him were mesmerized by his speaking style. Later in his life people were so spell-bound at his press conferences you could hear a pin drop. He was so deep, introspective and knowledgeable, and more than any other driver he could explain what went on in his head out there. That was the side of motor racing that fascinated him. For him racing was a metaphor for life. He used his experiences in a racing car as a means of self-discovery. He was fascinated by his temptation to go further and find new limits. He couldn't stop himself from pushing and he discovered that each time he reached the limit he was able to go further still. At the same time he knew he was exposing himself to more risk, but that fascinated him too.

"When the subject turned to qualifying at Monaco in 1988 he described the sensations he felt while he was driving. It sounded like it had happened before but never to such a degree. He said, 'Sometimes this happens to me and I can tell you this one example.' He said he sometimes felt like he was looking down on himself. He didn't use the term 'Out-of-Body Experience' but he said he could see himself from above shooting down this tunnel. I could tell that he was telling me something extraordinary because he was shaking, his voice was quavering and his eyes were misting over. That's the way Ayrton became when he got passionate—even in public press conferences. In this case the delivery was almost as important as what he was saying and it was clear he felt really deeply about it.

"Right after this I asked about God and he said it was after his crash at Portier the next day that he became more spiritual. He told me he had always believed in God but that experience helped accelerate or increase the depth of his faith. A spiritual side can either help or hinder in racing, depending on how you harness it. In some cases it might be an inhibitor to a racing driver if you think about it too much. Sportsmen are often told, 'Don't think about it, just do it.' If you try to intellectualize it, you can't. Ayrton did and he proved that you can. He was one of the most amazing characters who ever sat in a racing car. I believe Monaco was a lesson for him in many ways. When he became frightened he used his fear as a means of self-preservation, to prevent himself from going way over the top again. He may not have related another 'tunnel' experience but that didn't make him any slower. In fact he went faster than ever in qualifying after that and he never stopped taking pole positions. I spoke to him many times afterwards and I believe this didn't just happen to him at Monaco but everywhere."

Senna claimed not to have allowed himself back to quite the same extreme, even if he was enchanted by his glimpse of this other world. Yet, if his spirituality came to define the man, Senna's virtuosity over one lap still defines his on-track legend. The powers-that-be have regularly tinkered with qualifying, but in the mid-eighties cars were drained of fuel and sent on banzai runs with revved-up engines and ultra-sticky tyres designed to perform for just one lap—in short, the purest test of sheer speed motor racing has ever seen. This is where you really found "the limit", and Senna turned it into his private plaything, racking up sixty-five poles, a record surpassed by Michael Schumacher only after a much longer career. To see Senna on a Saturday was worthy even of F1's inflated admission prices.

"Qualifying was something special for Ayrton," recalls Jo Ramirez. "He was once half-a-second clear of the rest and no one could come near him. Everyone had used up their tyres but he had one set left. So he asked McLaren's Ron Dennis to let him go out again. Ron said, 'Why? No one's going to touch you. Why risk the car?' Ayrton said, 'No, I promise I won't go off. In my lap I could have gone quicker. I've just got to check for myself.' It was always his goal to do that. So Ron let him go out and he went even quicker. It was amazing how he could do that. He sought perfection every time—and, thanks to his approach, he often had this satisfaction.

"The way Ayrton could concentrate was incredible. Once in Monaco, he was holding the pole time. But early in one lap Gerhard Berger was a fraction of a second quicker. He made it as far as the chicane thinking, 'For once, God, I've got it!' It was fantastic. Then he got to the final corner, Rascasse, and screwed up. He forgot which gear he needed, made a big mistake and lost half a second. I love Gerhard—he's a super man and very honest. If he spun off he'd say, 'Forget about the car, it was my fault.' He didn't have to muck about. But after that he was so demoralized. Going up against Ayrton, especially in qualifying, was so frustrating to him. It got to the stage when he said, 'I never tried any more because I was just going to hurt myself. I'm never going to beat him.'"

Berger, paired with Senna at McLaren from 1990 to 1992, is no slouch, but perhaps his biggest mistake was to be part of the same Grand Prix team as a great at his peak. Another potential superstar, Berger took Benetton's first win in 1986 before landing two big-money deals at Ferrari in a fourteen-year career that included ten wins. He also made it to the Zone, even in Senna's domain on a Saturday afternoon.

"When you go out to start a qualifying lap it's a mixture between anxiety and pleasure and hating and hurting and every other emotion," says Berger. "Mentally it's tough because you know you have to squeeze more than 1,000 bhp around the circuit. In the turbo era the engineer would come in before the session and turn the boost to full. You'd had 900 bhp in practice but now you'd have 1,400 hp and no one in the world can say you didn't have respect for that. You'd have qualifying tyres so you could brake later and shift in different places. It was like being in a rocket. This was before 1994 when we still had walls next to the corners and no run-off areas. Round [Belgian Grand Prix track] Spa you felt you could touch the barriers and you'd think, 'Bloody hell . . .'

"But qualifying was when you could really find the Zone. On some days you are fighting the car, everything hurts and nothing seems to fit. Then you have no lap time and you know it. But at other times you could feel yourself reaching a higher level. Maybe the intensity could vary but you have some laps where you know you're on the ball. Everything would be just like in slow motion, everything becomes very smooth and very soft and you remember everything. When you're really on it, it's absolutely the best feeling in the world, one I still miss. Winning races is good but that's the feeling I liked the most. I liked the thrill of squeezing it around, taking this 1,400 bhp to the limit. When you made it through and you had a good lap time and you didn't break your legs, it felt like you were on the Moon . . ."

I guess that's where you're supposed to end up in a rocket—though as Berger explains, while he did make the leap into hyperpace he also had cause to ditch in the sea from time to time. Thankfully he had his own idiosyncratic way of getting going: "In qualifying you don't have time to build up so you need to be on it straightaway. It's very easy to go over the limit so you want to be just at the end of the surface without going over it. You have to get the balance exactly right so it's important to know your body very well and to know how you can activate it. It's not always possible but you train for it and you have certain things you do. It's like a switch. Some people do mental training and yoga, things like that. It never worked on me—when I tried it I fell asleep. So I used to drink an Espresso and a Ramazzotti or something. That was just to get a bit warm when you needed a small kick."

So there is another way to go over the limit in an F1 car. The grinning Berger wouldn't seriously advocate drinking and driving, but when you hear just what his Saturdays involved it would send anyone to the liquor cabinet. Still, with good reason racing drivers rarely deviate from their ubiquitous mineral water. The

Zone is such an elusive beast that a "clean mind" is a prerequisite for all sportsmen. Any outside worry lingering in your head will distract from the task at hand and pollute the pure ability to perform. That is clearly the prevalent attitude among sports psychologists, and one shared by the vast majority of drivers—but there are exceptions. Damon Hill took his finest win at Suzuka in 1994 despite a distinct belief that even his own team had lost faith in him. Then there's this tipsy rocket man.

Of all Berger's wins, it was his last one that stood out. Many paddock insiders thought he should never have travelled to the 1997 German Grand Prix. It is easy to understand such sentiments because his preparation had hardly been ideal. Berger had missed the three previous races while he had three operations for a sinus problem, and he was on antibiotics that affected his ability to train. He was also involved in a battle with Benetton team boss Flavio Briatore over his F1 future after a disappointing year for driver and team. Yet all of that was rendered insignificant when, three days before the Hockenheim race weekend, Berger lost his father in an air crash.

"That was a big time in my life and everybody—really, everybody—told me not to go to the race because I was just not ready," he recalls. "The team was against me, my health was against me and the emotion from the loss of my father was against me. When I arrived in Hockenheim everybody was trying to smile about it and saying, 'We're so happy you're back,' but I could read on their faces they were all thinking, 'What are you doing here?' That's what gave me the push. I just said, 'Sod it.' I went to one of the usual press conferences and, without warning the team, said I was going to stop driving for Benetton at the end of the year. Then I got into the car and said, 'Let's see what we can do.'"

What Berger could do was quite unlike anything he had ever done before. No aperitif needed, the dire circumstances leading up to the race paradoxically thrust him firmly into the right state of mind. He hadn't won a race for three years and Benetton hadn't topped the podium since Schumacher left two seasons earlier, but this weekend no one could touch this Austrian astronaut on a mission. Berger even dreamt he'd win and he duly

stormed Hockenheim, taking pole, race and fastest lap in a car
that was far from the best on the grid.

If I hadn't had all this shit, if my father hadn't had his accident,
if I hadn't been in hospital, if I hadn't come to Hockenheim and
had everybody say, 'Great that you're here,' perhaps it would have
just been another race," he admits. "Maybe I would have finished
third or fourth, I don't know. But I've always noticed in myself the
tendency that I've been stronger when I've been in the shit. If
contract time was coming or there was more pressure than
normal, I could find extra time within me. It seems the most difficult conditions made my body find more energy than ever before.
That's wrong, absolutely wrong. What you want is somebody
who can get this level of performance out of himself every day in
every kind of condition.

"My Hockenheim race was proof for me that so much comes
from your head. It was the most emotional Grand Prix of my life
and it showed me how much of racing is mental. In an F1 grid of
twenty-four cars, everyone can perform well—as a driver you
don't get there otherwise—but there's a range of performance.
Now I'd say one of the main things a driver needs, apart from
natural talent, is mental strength. If I have to judge a driver I'd
look into his mental situation and mental approach. In my case it
appears I could only produce my full natural ability in certain
situations. But you're looking for guys like Schumacher and
Senna who always exploit the maximum out of themselves no
matter what has happened. Ayrton always produced such strong
performances you could see one of his strengths was mental. Of
course we're all human so we all have ups and downs, but these
guys are consistent."

The self-criticism is modest and perhaps accurate, because
that Hockenheim weekend showed Berger could have had the
talent to dominate—if only he had been able to produce it every
time. Not everyone is built like that, though, and sport is the
better for it. It's what separates a Jimmy White from a Stephen
Hendry, and it makes for classic viewing when the guy overdue
his one day in the sun comes up against the guy with a perma-tan.
But the experience was enough to make Berger realize what separated Senna et al was an ability to set up home in the Zone.

McLaren boss Ron Dennis confirms as much: "Putting aside the talent, the performance and the results, Ayrton Senna was just the consummate professional. Senna the driver was total commitment, with a level of discipline you rarely find. He just enjoyed always being on the limit. Put him in a car and he was a complete driver, every bit as good at racing as qualifying. The more challenging the circuit, the more Ayrton enjoyed it. He would enjoy sharing all his pole laps, but especially Monaco.'

Incredibly, Senna's 1988 Portier crash was the last mistake he ever made in those streets. Nowhere exaggerates a driver's mental capacity more than this frankly ludicrous track round a built-up area on which you can't pass and even F1 cars have to dip to 30 mph at one point. But this genius at avoiding the goldfish bowl with his rotor blades took the chequered flag in each of his last five visits to the principality. Berger agrees: "Monaco was the most challenging circuit—such a tight line to walk on with no room to hit any white line on the tarmac or any small mistake. You need a special way of concentrating. That's why the same guys win most of the time. Look at Ayrton's six wins and Michael's five wins. They are very strong and known to have a special ability to concentrate. They can get the job done whether it's a normal circuit or on streets, in the rain or the dry, in a difficult car or an easy car.'

Difficult? Try this: the 1991 McLaren was only ready a week before the first race, leaving time for just a single test at Portugal's Estoril. Senna's new race engineer James Robinson reckoned the car "wasn't anything special" either there or when they arrived at the US Phoenix street circuit. Yet Senna shocked everyone, taking pole by a second. "The time he managed in qualifying bore no resemblance to his time from practice," says Robinson. "It was two seconds quicker, a phenomenal leap. He either had a very good way of defining exactly the level the car would handle grip or he could take it over the limit without actually doing so. He was always working the steering wheel, always on and off the throttle at a very high frequency. The feeling he got from all those little things would tell him the car was starting to creep away from him and put him on that level."

Senna went on to perform the unlikely feat of winning the first four races—then a record—before battling off Nigel Mansell's superior Williams for his third and final title. But that early-season charge nearly stalled as early as race two, at his beloved home circuit of Interlagos, but for an intervention from above. "Ayrton finished the last seven laps with only sixth gear," adds Robinson. "Riccardo Patrese's Williams was chasing us and they didn't know. Over the radio I heard a couple of comments in Portuguese and Ayrton was sitting in the car praying. That was a new experience for me but you soon understand that's him. His thought process led him to conclude somebody could help him through this and sort it out. That's what set him apart—his sheer belief in his own ability, that he could do what other people couldn't do. And, though he'd probably hate me for saying it, the belief that somebody else was looking after him. I've never experienced that with any other driver."

The pain of manhandling his car meant Senna couldn't even wave to his adoring crowd afterwards, let alone point to the heavens. But before Robinson is ejected from the union of rational engineers, we should stress Senna's interest in metaphysics was based on a firm grasp of physics. Only after he had exhausted all the avenues of his scientific knowledge would he call on his God for roadside recovery. Senna's pursuit of knowledge about the science of driving built up its own group of fans. He raised standards for being able to recall, precisely, every move his car made on a track. He used this mental database to describe tarmac with unerring accuracy, remembering details of bumps and how his car behaved—all while lapping at speeds his rivals couldn't match. But even in drawing on such an incredible store of information, Senna would say he was still using only a minimal portion of his brain. Few humans get even close to using their full memory capacity, and Senna was just doing his best to exploit an ability he would have considered endless.

Describing his pole lap for that Interlagos race, Senna said: "A billion things go through your mind and body. It all happens so amazingly fast, it is like a mystical feeling." A billion? Far-fetched? Hardly. Someone with a lot of time on their hands once calculated the total amount of connections in the brain amounts to the figure 1 followed by a string of zeros 10.5 km long. If Alain Prost earned the moniker "the Professor" for his ability to nurse cars home based on tyre wear and fuel consumption, this kind of knowledge means Senna has to run him close in the university hierarchy.

Such brainpower helped Senna produce the defining lap of his career, in the 1993 European Grand Prix at a drenched Donington Park, where he passed four cars on the first lap to lead. Legendary British commentator Murray Walker classes it as the best lap he ever saw and few disagree. Robinson recalls: "Everyone quotes Donington in the rain, but you only have to watch classic examples like that to see what Senna was made of. Even with such self-belief you just don't overtake four cars on the opening lap of a Grand Prix in the pouring rain at a track where there are guys who have driven more laps than you. But while the rest were thinking, 'If I go any faster I'm going to fall off because I haven't been round here for fifteen minutes and it's rained heavily since then,' Ayrton could judge the grip level on the outside. He would say: 'I've driven Donington in the wet so many times in earlier formulae I know where the grip is. It's not on the racing line when it's raining. In this corner it's here, in this corner they resurfaced it two years ago and it's here.' He'd just put that information together. That's what made him special."

Senna's explanation was simpler still. That Easter Sunday, hours before golfer Bernhard Langer declared his US Masters victory was extra special, coming as it did "on the day my Lord arose," Senna nipped in first. When Brazilian TV asked him how he'd managed to destroy the Donington field, he said: "God is great and powerful and when he wants nobody can say anything different." Einstein was right. God doesn't play dice; he drives fast cars and plays golf.

To uncover the changes Senna went through after Monaco in 1988, we have to look to this spiritual dimension. By pushing to the limit and beyond he found standard rational explanations no longer sufficed for what he could achieve. That weekend he found a way to tap into his subconscious, allowing his brain to work his body unhindered and his natural skill to flow. Only when he backed off did he revert to the standard conscious method with unfortunate results. But this gap in his knowledge left a vacuum over which there was only one suitable bridge. The difference was simply divine.

Years later, Senna declared this weekend his epiphany. This journey to "another dimension", coupled with his most human of errors the next day, led him to root his life in an unshakable faith. From then on he claimed everything he did was with God's help—though he didn't reveal that straightaway. A cursory trawl round YouTube reveals plenty of quotes about this God, starting from the day Senna won his first title months later at Japan after a storming drive from fourteenth at the second corner: "Now I can say without fear, and I never lost my faith, that from Monte Carlo to here it was God who gave me all the strength and power to win this title." He later revealed he'd felt God's presence with him during the Suzuka win, a "fusion" with the divine that was so powerful it again almost made him lose concentration on the job in hand.

This wait to be crowned before going public with his spiritual beliefs echoes Muhammad Ali, who changed his name from Cassius Clay and professed his Islamic faith only the day after he had become the king of the world. That approach is easy to understand because on a planet where so many are brought up to believe "winning is everything", it is harder to dismiss the words of those at the top. To a sports fan like me, those two giants alone were enough incentive to investigate the influence of this deity on such an improbable field.

Senna admitted his quest to learn about God had only just begun and that it would take "a long time". Western society encourages us to spend less than two decades of our allotted seven "learning" before paying our dues with a lifetime of "doing", but Senna did not settle for that. From his mystical Monaco moment at twenty-eight to his death at thirty-four, he used his accelerated journeys round the world's racetracks to accelerate his journey to spiritual understanding to a level rare in such a high profile Western celebrity.

"There is an area where logic applies and another where it does not," Senna told Joe Saward in 1993. "No matter how far down the road you are in understanding and experiencing religion there are certain things which we cannot logically explain. We tend always to understand only what we can see: the colours, the touch and the smell. If it is outside that, is it crazy? I had the great opportunity to experience something beyond that. Once you have experienced it, you know it is there and that is why you have to tell people."

Having been raised a Catholic, Senna began carrying around a battered Bible with a sticker on the front that read: "Cristo pole position". Gerhard Berger confirms the book formed a central part of his life in the bearpit, adding: "I saw Ayrton read the Bible a lot, especially before qualifying. It was always part of his thing, even around the paddock." Anyone who has visited said paddock will know such an approach is bold in a world where machismo mocks mumbo jumbo. Berger used to hide his teammate's reading material at regular intervals—but that was just in jest. Racing drivers, like boxers, are always searching for each other's psychological weaknesses. After the infamous clash with Alain Prost at the 1990 Japanese Grand Prix—where Senna proved he'd read the rather less than spiritual "eye for an eye" part of the Old Testament as he retaliated for Prost's similar move the previous year—the Frenchman memorably stated that Senna's faith gave him an immortality complex. This comment belies the fact that Prost was also a religious man who even visited the Pope after his first world title in 1984.

"Ayrton was very religious," says Jo Ramirez. "Before the race you'd always find him sitting in the motorhome, quiet, patient, alone with a chapter of the Bible. He used to calm his mind that way. He always used to seek twenty minutes to be on his own. Everyone respected that time and he just sat and did his own thing. Ayrton started off Catholic but to be honest I don't know whether he ended up a Catholic, a Protestant or whatever. He just believed in God. Of course people said he felt he could do things way above everyone else because nothing would ever happen to him. Even Alain said Ayrton felt God was in the car with him as his co-pilot. I don't know whether Ayrton did feel that way but it's possible. I'm a Catholic myself, but I don't really believe in that. I believe in God, but everybody is entitled to think their own way."

Senna's religious side didn't always pay off. McLaren may

have grown used to his sacred antics, but even they must have been taken aback by one incident recalled by Ramirez: "Once his sister brought a bottle of some kind of holy water. Ayrton put it on the back of the car around the engine and suspension. The mechanics just looked on. He didn't finish that race because, yes, the rear suspension broke. Afterwards he came in and said, 'Don't let this woman anywhere near the car . . .'" Moreover, while this dimension makes Senna's legend even more attractive to some, to a young Jacques Villeneuve, son of another idol who died behind the wheel, it was a drawback: "I was more of a fan of Prost than Senna—partly because of this religious side," he says. "I always had a hard time with that. It just depends where you were brought up. Brazil is a very religious country, that's all."

Faith is indeed not the taboo in Brazil that it is elsewhere, but not all the country's great drivers have been religious. Even a horror IndyCar crash that broke his legs and ended his career failed to stir a spiritual renaissance in Nelson Piquet Sr, though he admitted it was a blessing in disguise: "Motor racing is a dangerous sport and you have risks," he told me. "You might live, you might die. My accident didn't change my approach to life but it was good for me. It was a barrier to stop motor racing completely. I started a new business in Brazil and had new success."

Given the conflicting reports, I wanted to know whether Senna's beliefs were firmly based on the rituals of Brazil's national religion, Catholicism, or if he had a more direct line to God. Clearly only the man himself can answer that—and we've established he is no longer available for interview. But one evening at Interlagos his sister Viviane explained to me: "Ayrton believed in God and had a good relationship with God, and that was it. You can't narrow it down to any particular religion like Catholicism. It was much wider than that, better called spirituality than religion."

Materialists may not accept a difference between the two. One is mumbo, the other is jumbo. If so, spirituality is the jumbo, though even some faith communities greet the word with derision. Senna was simply a spiritual man, and McLaren boss Ron Dennis recognized that this was what mattered: "Ayrton's spiritual side only influenced his work positively," he says. "He was just at peace with the world. Save for all his frustrations about things that weren't right in the sport or his own country, he was a quiet individual who always looked in himself for solutions—or sometimes in the Bible. He looked to understand all the time. He wanted to try and determine why things were the way they sometimes are, even the negative things. I think he felt the Bible was a good place to look for guidance. That was an important book for him, but more in the spiritual sense than the religious."

Such an attitude ensured his legend was not hemmed within the borders of his own nation. Everywhere he went he was mobbed, and nowhere outside Brazil did his death hit harder than Japan, whose Honda engines powered Senna to much of his glory. The Japanese go beyond most when it comes to adulation of global superstars, but at Suzuka, the circuit of so many crucial title showdowns, people still talk of the Senna era as madness. These days, with patience, you can plough through the crowds at the funfair between the drivers' hotel and the circuit, but when Senna appeared the whole place ground to a standstill. Thousands would camp in the grandstand opposite the pit lane in the hope of glimpsing the great man on his way in or out.

When Senna was pronounced dead after the 1994 San Marino Grand Prix it was early evening in Italy, early morning in Japan. Immediately the Japanese began pilgrimages as best they could, and soon both Suzuka and Honda's Tokyo HQ were swamped. The giant motor manufacturer's base is not small, yet one senior employee told me they soon ran out of room for the flowers. As former Honda racing chief Shoichi Tanaka explained a decade later: "No one will ever overtake Senna in the eyes of the Japanese public. Senna is still much bigger than Schumacher. It was a mixture of everything—he was supremely talented, he raced for Honda, but also they were attracted to his personality and his spirituality."

If Senna had lived to tell of his Imola crash he would surely have attributed his good fortune to the same divine source. That he didn't means to many that his deity deserted him that day. Those who wonder why any God would prematurely deprive the world of one of His most committed and successful PR machines have a point. A mysterious way indeed, but the twentieth century regularly showed the power of an early departure to cement legend—think JFK, Martin Luther King, Marilyn Monroe, Elvis Presley, John Lennon, Princess Diana. If the name Ayrton Senna da Silva sounds incongruous in such company, consider that he was leading from pole when he went off. And it was all captured live on TV around the planet. Talk about going out at the top.

Senna continues to draw crowds today, even if all they see is a small bronze plaque in a green field in São Paulo. That's all there is. No elaborate tomb. No flashy headstone. Not much for a man honoured by three days of national mourning. Every visitor to Morumbi cemetery is greeted by the same phrase: "Nada pode me separar do amor de Deus." Nothing can separate me from God's love. So advanced was Senna's soul he would have considered that to apply equally to life, death and everything else. "Nothing." Not then, not now. Not before his death, not after. Not before the renewal of his faith, not after. Hardly original thought, but the pilgrims keep coming because Senna found a way to use that ancient wisdom to help reach the top in one of the most modern, hi-tech and ultra-competitive fields of human endeavour there is.

To the residents of São Paulo and a thousand other cities, Ayrton Senna was superhuman. He is admired the world over because he had so much natural ability, yet he gave it all and more, producing what can only be described as magic on the track. One of the most sublime talents to grace a motor race, here was a wizard who pushed his machinery to the very limit and often apparently beyond. Senna was convinced his conjuring was ultimately being channelled through him from a far greater power. But his faith doesn't mean he just shut his eyes and let his God take over. Without a work ethic to match his outrageous talent he would never have made it.

Senna's secret was that he did not waste time worrying about beating the rest. His primary mission was to beat himself. Perhaps that is why qualifying was so very special to him. Anyone who has played a video game that records your best lap as a shadow car will know the feeling. No matter how perfect your lap appears there is always a tenth to be found somewhere. Racing drivers don't have the luxury of a transparent version of their car giving such a direct comparison, but the addiction to finding those fractions is intense—and this Brazilian felt it more than anyone else. To return to Donaldson's *Grand Prix People*: "There are some moments that seem to be only the natural instinct that is in me. Whether I have been born with it or whether this feeling has grown in me more than other people I don't know. But it is inside me and it takes over with a great amount of space and intensity. I am intense about everything I do. I have an attitude about life that I go deeply into it and concentrate, and try to do everything properly. It's part of my personality."

Senna's nephew Bruno—who has gone on to become a racing driver himself—saw this intensity up close, adding: "There are many levels of concentration but my uncle was always able to reach a very deep level, particularly in qualifying. That is how he did his thing. It's all about reaching that level. I don't know if he sometimes went over it, like in Monaco, but I'm sure he went back to a similar level many times—perhaps even every time he got into the car for qualifying. Once when I was there he did a race with a heart monitor and his heart rate just before the race was 60 bpm. So while everybody was having a heart attack he was just chilling out. When a driver's at the very top automatic level, everything just comes to him. In a corner he knows what to do and how much throttle and brakes to use, but he's not even thinking about it."

After Bruno's early start in karts, Ayrton once declared his nephew an even bigger talent than him. But despite his pedigree, Bruno was a late starter in cars as his father Flavio Lalli also died in a motorbike crash and mother Viviane was understandably reluctant to see yet another loved one at risk. The gap of eight years would have finished off most careers, but not this one. Still, even on the day Bruno won his first GP2 race at Barcelona in 2007, he told me he didn't yet feel comfortable enough to reach this level in a car though he'd felt it in karting. Shortly before his Formula 1 bow three years later he'd still only flirted with this "automatic" Zone a couple of times in qualifying without nailing the perfect lap.

No surprise there. Bruno will face the hardest comparison every time he steps into a racing car, but it took Ayrton until he was twenty-eight before he started "flying" in Monaco—even after an uninterrupted life in the cockpit. What does Bruno make of the effect on his uncle of that sunny Saturday? "It's hard to say," he admits. "But at that time he was looking into his religious side. Perhaps by reaching a very deep concentration level with his outside senses heightened, it gave him a window into his soul."

As he flew round the harbour you'd think Ayrton Senna had truly located the limit. Instead he found himself tapping into a new force he had never learned about at school, one he'd never been told about in the news. This human, at the very edge of his endeavour, went beyond the five-sense, three-dimensional world that's supposed to be all there is. If this was Joe Bloggs talking we could legitimately dismiss his tales as the rantings of a madman. But it happened instead to one of the most supremely gifted, globally admired sporting figures of all—one whose ability was matched by a lucidity that could accurately convey the experience in words. In his finest moments Ayrton Senna found tiny glimpses of a heaven on Earth to which we can all aspire—with or without a pitpass, as we shall see . . .

About the Author

Clyde Brolin has spent over a decade following the Formula 1 circus around the world both as a journalist and a team member. But what still intrigues him most is the surreal, near-mystical mental state the top racers reach when they push to the limit.

Overdrive: Formula 1 in the Zone is based on more than 100 exclusive interviews with F1 greats—from legends such as Stirling Moss and Jackie Stewart through to Michael Schumacher and modern-Day stars Fernando Alonso, Sebastian Vettel and Lewis Hamilton.

The book delves into their minds and reveals Ayrton Senna is far from the only racing driver to reach such spectacular heights. From time slowing down to Out-of-Body Experiences, the racing cockpit is a treasure trove for these mind-bending moments. Yet the book also discovers you don't have to speed at 200 mph to find your own inner genius "in the Zone".

Overdrive was featured in the *Daily Telegraph*'s top ten sports books of 2010 while Brolin was shortlisted for Best New Writer at the British Sports Books Awards. He is still interviewing the planet's biggest sports stars about their finest hours and how they get there—and his next book is due to be published in 2016. To find out more please visit www.overdrivef1.com or @overdrivef1 on Twitter.

JOCHEN RINDT UNCROWNED KING
by David Tremayne, 2011

THE SUPERFAST LIFE OF F1'S ONLY POSTHUMOUS
WORLD CHAMPION

Chapter 15—Late Blossom At Lotus
August to December, 1969

"And we got to Watkins Glen and Jochen came to me and said,
'Colin wants me to drive for him for another year and has offered
quite a bit of money . . .' So I said, 'Look, if that's what you want
to do, that's fine.'"—*Jack Brabham*

Before he flew to Germany, Jochen told Heinz Pruller: "I feel my
bad luck is destroying me. And Chapman and I only seem to
argue with one another now." Things had got so bad, however,
with the four-sided conversations from Holland continuing, that
both men knew things could not continue as they were. They had
to do something about the toxic atmosphere that was threatening
to break Team Lotus apart.

Pruller remembered: "At the Nürburgring Jochen and Colin
sat together in the transporter for three hours—there were no
motorhomes in those days!—and even missed part of the first
practice session. Colin had a photo of Jochen that had been
published in England with the caption: 'Is this the perfect
racing driver?'

"'Don't you believe it, Jochen, you're far from it.' He told him.
'You're fine in the car, but you're a bastard outside.'"

Jochen considered that, then replied: "Perhaps I shouldn't talk
so much. Certain matters should be confined to the team. But
when one suffers so much misfortune, hard words come easily."

In the burgeoning mood of reconciliation, Chapman agreed
that selling the 49s had been daft, and that he worked the mechanics too hard, something Jochen had repeatedly been getting at
him about.

While refusing to accept that the two cars he provided for
Jochen and Graham were any different to those he had supplied

to, say, Jimmy and Trevor Taylor, he nevertheless offered Jochen unequivocal number one status for 1970, that Hill would be switched to Rob Walker's team though not kicked out completely after all the years of their relationship, and that he would get the full treatment and a clear number two driver. Chapman told him about an all-new two-wheel drive car that was under development.

"His character is very complicated to describe," Jochen said of Chapman. "Obviously he wants to win, but the way he wants to win and I want to win are different, that's probably the reason for arguments."

The tension was certainly eased after their lengthy conversation, but no magic wand had been waved. "He only argued with Colin about the way the cars were built," Ecclestone pointed out. "Their frailty. I know they spoke at length at the Nürburgring but I don't know whether they ever got things out in the open. But things quietened down a bit."

Pruller remembers going to a party that Team Lotus held at Bernkastel on the Moselle that evening, though it was a sombre affair for Jochen after the death that afternoon of the promising young German driver Gerhard Mitter, whose Formula Two BMW had crashed. Chapman was the perfect host, full of bonhomie, glad-handing everyone and helping to serve them.

"I turned to Jochen and said: 'In a year or two you'll both be saying we used to hate each other but now we're the best of friends,' but he almost shouted his reply. 'Me, stay at Lotus? I can't imagine that!'"

It was not just the idea of returning to Brabham that Jochen now had in mind. Behind the scenes he, Bernie Ecclestone and former McLaren designer Robin Herd had been dreaming up a grand new plan to start their own Formula One outfit. Herd, who had penned the unloved Cosworth four-wheel-drive car after leaving McLaren early in 1968 before his M7A design began winning races, would leave Cosworth to set up the new business. Wealthy Spanish racer Alex Soler-Roig, with whom Jochen had become friendly after his accident in Barcelona, was involved but would not be a financial contributor.

Later that weekend Jochen showed Heinz the initial brochures for the as-yet unnamed new venture, which would probably have born the same Jochen Rindt Racing moniker as the Formula Two team that he and Ecclestone had set up that year. At that stage, according to Heinz, it called for Herd to design an advanced two-wheel-drive car and a four-wheel-drive machine, with the idea of Jochen choosing whichever was most suitable for prevailing conditions at any given race in 1970.

Jochen and Bernie would finance the enterprise and look for backers later, and the plan was for the three to share the profits equally until Ecclestone, according to Herd, insisted that Jochen and Robin should get 45 per cent each and that ten per cent would be sufficient for him.

"At that time we were talking to Robin about doing that," Ecclestone said. "Robin's right about the shareholdings. I did say that. I think it was quite close to coming off. I don't know why it didn't happen in the end. I don't remember . . . It looked like it was going to happen, but to be honest it was one of those things where we didn't care if it did or it didn't, you know?"

Little by little Jochen had begun to develop reservations about the project, as more people became involved. The way Pruller tells it, Herd was mandated to appoint a team manager and chose Alan Rees. That was no problem, but then Rees brought in Max Mosley, a failed Formula Two driver best known back then as the son of infamous World War Two Fascist politician Sir Oswald Mosley. Mosley really wanted to start his own race car manufacturing business, and brought with him businessman Graham Coaker as production manager. Mosley wanted Jochen as their driver, but Jochen felt that the original plan was being hijacked.

Herd tells a slightly different version. "The Cosworth programme was falling apart, and after Max had said to me let's build our own car, Bernie came along and asked me to do a team with Jochen. He said, 'We'll form a company. I was very tempted. What I really wanted to do was the car that became known as the March 711, which we eventually ran in 1971, with ground effects. We did a little bit of that with what became the 701, but we couldn't do it properly with the rear suspension arrangement that we plumped for in a hurry. I wanted a long wheelbase, the radiators and the oil tank in the middle of the car . . .

"It was a difficult decision. I don't remember there being a name for the car or the company, but contracts were there to be signed. I was living in Northampton and Jochen came to the house one day and said: 'Robin, you have to do this.' He was an arrogant bugger, but he was very quick and for sure I liked him. You know how you come across some sportsmen for the first time, and somehow they just have something about them? Jochen was like that."

In the end, according to Herd, the Ecclestone project died, "Basically, because a) Max was very persuasive, and b) I knew Max and Alan Rees, and I didn't know Bernic."

Speaking of the deal around the time of his seventieth birth-day in February 2009, Robin let out a guffaw and gave a rueful shrug of his shoulders as he added: "Looking back, it was a mistake. I lost the chance to do the one thing I regret not doing, which was to design an F1 car properly. Bernie, the money, Jochen, the time . . ."

And, with hindsight, he added of Ecclestone: "You know, Bernie is the most able person I have ever met. People criticize him in certain cases, but my God, his ability is beyond compre-hension! He offered to buy Oxford United from me, and I turned him down. If I list my stupid mistakes in life, that was certainly one of them."

Thus March Engineering—Mosley, Rees, Coaker, Herd, with an added vowel—came into being instead for 1970 and a differ-ent chapter of F1 history was written. There were suggestions that the idea of a Herd/Rindt/Ecclestone project was nearly revived in 1970, but Robin denied that. "No, because I think by then Jochen had decided that everybody at March were a bunch of bastards! Beyond the pale!"

There was an amusing anecdote in all this. After Robin had opted to go the Mosley route, and was still trying to persuade Jochen to sign up, Jochen knew that the prototype Formula Three car that March was building was being put together in Coaker's garage because they had yet to find premises. And he let it be known that he had no interest in driving for a company whose sole product thus far was being put together in "Graham's shack". Except that with his heavy accent this came out as "Grem's shek",

and with typical ironic humour Herd and Mosley initially christ-
ened their nascent business Gremshek Engineering.

There was another little financial matter that Jochen had to
resolve at that time. "1969 was pretty much the same thing for
me as 1968," Michael Argetsinger said. "I got to most of the
Formula One races although my own racing was picking up a bit
so I missed a couple. I had Jochen's phone number at his home,
but I only remember having occasion to phone him a few times.
The time I remember most was in the summer of 1969. I was
living in Bremen. I had a call from Mal Currie, who was the Press
Director at Watkins Glen, and he wanted me to get Jochen to
agree to speak at something we called 'Inside Track Seminar' over
the US Grand Prix weekend. He said the fee was $500 [$2,950 at
today's rates]. I called Jochen and we had a nice chat. I told him
what the deal was and what we could pay. He said, 'Oh, Mike, it's
a problem. I'd like to do it but Bernie says I can't do anything
anymore for less than a thousand dollars.'

"I said I understood and was sorry, but that the budget
wouldn't go over five hundred. Without any argument Jochen
then said, 'Okay, for your family, I'll do it this time.' There was a
pause. And then he said. 'Just don't tell Bernie!'"

Told the story in 2010, Ecclestone chuckled and said: "Those
were good days . . ."

Later that summer, as his interest in the Herd project faded,
Jochen had brief talks with McLaren about joining as Denny's
partner, since Bruce was looking to retire from Formula One and
spend more time on the Canadian American Challenge Cup
series. Jochen discussed things with Bruce's business manager
Teddy Mayer, and his old friend Leo Mehl of Goodyear, but
McLaren's unwillingness to open up a slot for him in the lucrative Can-Am series ultimately stymied that avenue too.

There were brief approaches from Matra, which would return
with its own V12 engine and a works team run from France rather
than by Ken Tyrrell in 1970, and from Ferrari, but the possibility
that interested Jochen much more was an offer to return to
Brabham. Like Bruce McLaren, Jack, who had turned forty in
1966, wanted to retire from driving.

"When I was going to give it all up and go back to Australia,"

Brabham explained, "I thought that the best man to have in our bloody team would be Jochen."

Roy Winkelmann remembered having a coffee with Jochen at Heathrow that September, as Jochen prepared to fly to Monza for the Italian Grand Prix. "He said to me, 'Am I brave or stupid, driving a Lotus?'

"And I said, 'Colin always makes his cars light and they break. You understand that. And it's your choice. But you are also very brave.'"

Things had begun to reach fever pitch by then, but in the month-long break in the World Championship since Nürburgring, Jochen and Colin had started to get on better. And as other avenues closed to him, Jochen began to consider rather more seriously the possibility of staying with Lotus after all. It boiled down to a simple choice in the end: Lotus or Brabham.

One day at the Hotel de la Ville, where Team Lotus always stayed, Pruller realised how much things were changing when Chapman said to him: "Jochen made it very difficult for me to get to know him, it took me almost a year. But once I knew him I discovered that he had a heart of gold. He is so very, very blunt, but sincere and honest in all his opinions, and I have learned to respect this."

Part of the reconciliation had occurred at Oulton Park, when Jochen agreed to drive the hated Lotus 63 in the non-championship Gold Cup. It was still uncompetitive, but by spread-eagling himself on the ground at selected corners, Chapman was able to watch Jochen extracting everything the car had to give, and later declared that he had learned more about four-wheel-drive then than during thousands of miles of testing. Enough, it transpired, to steer well clear of it while working with designer Maurice Philippe on the new car for 1970 which would use two-wheel-drive only and be called the 72. The car that would take Jochen to the pinnacle of the World Championship one year hence . . .

Jochen dragged the 63 home second to Ickx's Brabham in Cheshire, and told Crombac afterwards: "Very soon after the beginning of practice I realised that the distribution of power between the front and the back had to be modified; a long-winded operation as one needs about a month to cut the necessary gears!"

Back on the championship trail for the Italian Grand Prix at Monza, his luck began to improve. Canny slipstreaming with his mate Piers Courage helped both until the McLaren drivers got in on the act too, which left the grid Jochen and Denny; Jackie and Piers; Bruce and Jean-Pierre Beltoise. And with the exception of Hulme, who dropped back with mechanical trouble, the other five would fight out a dramatic race which saw Stewart clinch his World Championship with a finely judged victory that yet again denied Jochen by the tiny margin of eight frustrating hundredths of a second.

Wings were still relatively new technology, and several drivers elected to race without them. Jochen, Stewart and Siffert were among them, but Beltoise, Hulme, McLaren and Courage elected to keep theirs.

Back then Monza had yet to be ruined by chicanes, and slip-streaming was the name of the game as gaggles of cars ran together. Jochen led Denny off the line, and the New Zealander buckled his left front wing against Jackie's Matra as they sped away and Jackie dived between the McLaren and the Lotus before they reached the first corner.

Stewart led Rindt, McLaren, Siffert, Courage, Beltoise, Hulme, Jack Brabham and Graham Hill at the end of the first lap and stayed in front until the seventh, when Jochen took a brief turn. Then it was Hulme for a lap, until Stewart went ahead from the ninth to the twenty-fifth, whereupon Jochen had another three laps out front. Jackie had the blue Matra in the lead again on laps twenty-eight to thirty, before Piers led a Grand Prix for the first and only time. Then it was Stewart again on lap thirty-three, Jochen on thirty-four, Stewart on thirty-five and thirty-six, Jochen on thirty-seven, then Stewart all the way from lap thirty-eight to sixty-eight.

It was not, however, as straightforward as mere words make it seem. Lap after lap positions behind the leader changed as the leading bunch—Stewart, Jochen, McLaren, Beltoise, Hill, Siffert, Hulme and Courage—drafted by one another. It was like Jackie and Jochen's battle at Silverstone all over again, with interlopers. Whenever the Scot led, the others were never far behind and were content to bide their time.

Hill had chased after and caught the bunch by lap twenty after losing ground initially, but as he closed up Hulme began to drift back with clutch and brake problems. Then Siffert began to lose ground as half distance approached, his Rob Walker Lotus showing the first signs of the engine problems that would end its race.

McLaren was content to ride along at the back of the six-car group that was left, but set a new lap record on the thirty-seventh tour. Subsequently Beltoise, who was also content not to be the leader, bettered that on the sixty-fourth, leaving it at 1m 25.2s or 150.96 mph. Nobody was hanging around.

Courage was the next to strike trouble. With twelve laps left his Cosworth DFV began to cut out at high revs, so gradually he began to lose the tow. Now there were five cars left, and Hill asserted himself in second place from the forty-sixth to the sixty-second lap as a mere two seconds covered them all. But then the reigning World Champion dropped a place to Jochen on lap sixty-three, and suffered a broken driveshaft on lap sixty-four, only four laps from home.

So now there were four, as Rindt, McLaren and Beltoise planned how they would unseat Stewart when it really mattered. What they didn't realise was that Jackie had A Plan . . .

Jochen led through the Lesmos on the final lap, but as they stormed towards Parabolica here came Jean-Pierre, leaving his braking impossibly late in the bewinged MS80. Jackie had also got past Jochen, preventing him from taking the optimal line, so momentarily it was a Matra one-two. But the Frenchman simply went in too deep and scythed across his teammate and the Lotus driver's bows, and as he began to understeer wide they nipped back inside him on the sprint to the finish line. And that was where the canniness of Jackie Stewart and Ken Tyrrell paid off.

They had invested a lot of time in practice on Saturday painstakingly choosing the right gear ratios. "It was an idea that Ken and I had," Jackie reiterated at Monza, forty years on, "and we spent endless laps getting it just right. I figured that if all I had to do was change from third to fourth coming out of Parabolica, rather than changing up to fifth gear as well, the saving of that gearchange could give me a crucial advantage. So I had quite a high third gear and a very long fourth."

But it went deeper than that. He and Tyrrell also carefully calculated the fuel weight the car would likely have at the end of the race, so that they got the ratios spot-on.

"It might have seemed a lot of work for such a tiny advantage," Jackie continued, "but back then every time you changed gear you risked making a mistake. At the sort of speeds we were doing, any momentary loss of momentum could translate to forty or fifty metres.

"I got through the Parabolica quite well, but I was hoping that I hadn't got through too well, because then you might carry a few too many revs and need to make that extra gearchange.

"In such a tight finish, that apparently insignificant attention to detail gave me those few extra inches, and that proved the difference between winning and losing because I won by a distance of just twelve inches."

Stewart made it to the line eight-hundredths of a second before Jochen, whose momentum swept the Lotus ahead of the Matra just after the finish. Beltoise was nine-hundredths adrift of Jochen, and two-hundredths ahead of McLaren. Bruce thought he had beaten Jean-Pierre; Jochen said he didn't realize it was the last lap. He had seen neither the official scoreboard nor Lotus's pit signal.

"That's what Jochen said," seventy-year-old Stewart scoffed good-naturedly in 2009. "But Colin was never very accurate . . ."

Jackie and Ken knew all too well that it was the last lap. They won more money as lap leaders than anyone else, and they won the race. But even more importantly, they clinched their first World Championship together. And they did it in style. Only nineteen-hundredths of a second covered the top four in their dramatic blanket finish, Jochen to Jackie's right, Jean-Pierre coming up on their left with Bruce on his tail.

But Jackie had to be convinced by Tyrrell that he had really done enough to win the title.

"There's no doubt. You're the World Champion," the hatch-et-faced entrant told him.

"Ken, listen to me. Are you absolutely sure?"

"Yes."

So what was that all about? "It was no good thinking I was

World Champion unless we were absolutely sure of it," Jackie subsequently revealed years later. "I wanted to be absolutely sure Ken had got it right, because there were dropped scores to be taken into account back then. I was so conscious of the fact that he might have thought we'd done it, when we hadn't quite."

Eventually, the Scot was convinced and finally allowed himself to celebrate his first World Championship. Later, he and his wife Helen had to escape the frantic Monza crowd via a window in the administration block's toilet. Momentarily they took refuge in the Dunlop transporter before they were spotted again—Stewart was already an icon with his long hair, sideburns and black Beatle cap—and were finally rescued by their friend Phillip Martyn who whisked them away to the Ville d'Este in his 6.9-litre Mercedes.

Jackie remembered that things really sunk in the following day when he overheard the concierge trying to reschedule his flights.

"This is important. It's for Jackie Stewart. He is the new World Champion, you know."

That was when he finally began to savour the moment, after what he described as "one of the most important races of my life".

Sadly, when Jochen's crown was confirmed a year hence, he would no longer be around to savour a similar moment.

In Canada Ickx won from the returned Jack Brabham, after the Belgian had inadvertently turfed Stewart off the road. The Lotus's Firestone tyres were no match for the Matra's Dunlops or the Brabhams' Goodyears round Mosport, and after leading initially Jochen dropped back to finish a nevertheless happy third, delighted that his car was finally reliable.

Motor Sport's Denis Jenkinson had never been a Jochen Rindt fan, as we have seen, and somewhat petulantly declared that he would never win a Grand Prix. He was so certain of himself that in Monaco he had bet that should Jochen actually win in 1969 he would shave off his famous beard.

He had done this once before, when he bet that Sydney Allard's dragster would not get below a certain elapsed time for the standing quarter mile. When it did, he saved his hirsute appendage by performing well in the car himself. But this time he would be held to account as *Motoring News* scribe Andrew Marriott was present when the comment about Jochen and victory was made, and challenged him once again to bet his beard. Commentator Robin Richards drafted a pledge on a table napkin and handed it to Brands Hatch supremo John Webb for safe keeping, once Jenks had signed it in front of witnesses Pat Mennem of the *Daily Mirror*, Anne Hope of the *Sun*, and John Langley of the *Daily Telegraph*.

"I was sitting at a street café having a meal," Jenkinson told writer Alan Henry. "To be honest, I was a little bored with the way it had suddenly become fashionable to be a Rindt fan and we got into a bit of an argument about it."

In the October 1969 issue of *Motor Sport* Jenkinson had even likened Jochen to Mercedes' second-rank inter-war driver Manfred von Brauchitsch, knowing the latter assuredly not to have been in the league of the great Rudolf Caracciola or upcoming Hermann Lang.

"Rindt would seem to be the re-incarnation of von Brauchitsch," he wrote. "A good, fast driver who seldom won GP races. If anything was to go wrong it seemed to happen to von Brauchitsch, such as tyre treads coming off while leading, his car catching fire at the pits when refuelling, spinning off and being disqualified for receiving outside assistance. He also drove in great opposite lock power slides when Caracciola and [Dick] Seaman drove just as fast without any tail sliding. Even if Rindt does not make excuses for being beaten, there are those who are only too ready to do so for him! Is he really the unlucky one, or is something missing from his physical and mental make-up? Does he lack that difficult-to-define 'something' that Stewart has got, and Clark and Moss had before him?"

There was no love lost between driver and writer. Jochen detested Jenks, and was oblivious to any embarrassment he might cause to third parties by steadfastly refusing to acknowledge the little writer's presence if they all happened to be in the same group. Jochen would talk with those he liked, and simply ignore those he didn't. Since Jenkinson would later revere Ayrton Senna, who demonstrated so many of Jochen's greatest characteristics, a cynic or somebody who knew Jenks' character well might assume that ranking him on a par with von Brauchitsch was a calculated insult born of the curmudgeonly writer's famed orneriness.

As he headed to Watkins Glen for what would be his fiftieth Grand Prix start, however, Jochen would finally have the perfect riposte. And unfortunately for Jenkinson, it would come while that issue of *Motor Sport* was still on the shelves.

It happened like this.

Jochen took pole position with 1m 03.62s despite a persistent misfire, which just beat Denny Hulme's 1m 03.65s for McLaren, with Jackie Stewart a tenth slower and Graham Hill another three-tenths further back. That pleased Jochen no end as there was a $1,000 ($5,900) prize for the polesitter, and because the Lotus mechanics gave him a fresh engine for the race since the cause of the irritating misfire could not be located.

After the drivers were taken round the circuit in open Chevrolet Sting Rays so that the 100,000 spectators could get a look at them, the grid formed in almost perfect Glen autumn weather. But the warm-up laps had effectively accounted for both McLarens, Bruce's blowing its engine, Denny's suffering top gear selection dramas. Then there were interminable delays, with cars held on the grid until some began to boil while the local cops followed up a reported spectator trespass on to the track. When finally the race got underway, Denny faltered immediately and nearly stalled as Jochen swept into the lead from Jackie and Graham. The McLaren was clearly history, and soon Jochen and Jackie left Graham and Jo Siffert to themselves as they resumed their Silverstone duel.

When Jackie took over the lead on the twelfth lap, courtesy of a small error from Jochen, they were already twenty seconds clear of Piers Courage who was into a terrific scrap with the works Brabhams. The Matra led for nine laps before Jochen pushed the Lotus back in front on the twenty-first lap as it was Stewart's turn to make a small mistake. Despite Jackie's best efforts, it was to the stay there until the end. As the Scot's engine note changed very slightly, Jochen edged gently away, so after thirty laps he was two seconds clear and in control. He was also about to lap seventh-placed Hill, who was ruing a different choice of Firestone tyre which had necessitated suspension changes that had upset his 49B's handling.

As he went into lap thirty-six a telltale plume of oil smoke emanated from Stewart's left-hand exhaust pipe, and the writing was on the wall for the Scot. He crept into the pits as the engine had blown an oil seal, and suddenly Jochen was holding a commanding lead of more than half a minute over his great friend Courage, with no prospect of anyone challenging him but the Gods who had so often tricked him. There were still seventy-two laps to go, and in some ways they were agony for him, as he pondered everything that was likely to go wrong. But nothing did. Lap after lap the red, white and gold Lotus sped past the pits, and interest centred on the increasing fraught battle between the three Brabhams for second spot, which Courage showed no inclination to surrender. Innes Ireland wrote in *Autocar*'s race report: "As Stirling Moss said, the only way to slow Rindt down would be to dig a ditch across the track!"

With eighteen laps left, Team Lotus's great moment was soured when ill fortune struck Hill. The veteran spun as he got on to oil and his almost treadless left rear tyre gave up the fight to grip. The starter motor failed to fire up the stalled engine, so Hill undid his belts, leapt out, and bump-started the car down a slight slope. Jumping back in he intended to head back to the pits to have his belts retightened. He never made it. The left rear lost its air suddenly at the end of the main straight on the ninety-first lap, and as the car spun and overturned several times the hapless Hill was thrown out and suffered a broken right leg and a dislocated left.

There were concerns for Jochen too, in the closing laps, as his engine sounded less crisp and mechanics waited on tenterhooks for him to dive in for fuel in a notoriously long race which played havoc with fuel consumption. He had broken the lap record three times—1m 04.40s on lap twenty-four, 1m 04.044s on lap fifty-four, and 1m 04.34s or 128.69 mph on lap sixty-nine—but had then eased back from 9,800 to 9,500 rpm, and he stayed out. He had judged it perfectly, having at one stage let Brabham unlap himself so he could slipstream Jack's delayed BT26A. There was to be no repeat of the Monza or Silverstone fuel pick-up dramas. Finally, Karl Jochen Rindt, from Mainz, crossed the finish line first to be greeted by the acrobatic Tex Hopkins. It was surely the sweetest sight he had encountered to that point in his career.

He had won a Grand Prix.

Actually, he had won the world's richest Grand Prix. He trousered $50,000 (the equivalent of $295,000 today) for winning and a further $2,000 ($11,800) and the Lentheric Trophy for setting the fastest lap, to go with $1,000 ($5,900) he'd won for pole position. And as Piers, himself $5,000 ($29,500) better off after being voted BOC Driver of the Day, led John Surtees home, he could be forgiven for feeling that the podium made it the happiest day of his racing life. Graham's accident apart, the only fly in the ointment was that Nina had stayed at home because of the expense of travelling to the US.

"I could not believe that I was going to win until I started my last lap," Jochen admitted, "and even then I was not sure what was going to break in the last few yards. I was happy, but it was odd because I had waited for this too long."

Subsequently, Jenkinson temporarily lost his beard. Half of the shavings remained for years, embalmed in plastic in a special position in the clubhouse at Brands Hatch. The other half was mounted and subsequently auctioned in the Doghouse Club (for racing drivers' wives).

Of all the people around Jochen, Jackie Stewart possibly best understood the frustration he had felt for so long in not being able to win a Grand Prix.

"I always thought he was gonna win one and I always told him that," Stewart said. "I won a Grand Prix in 1965, at Monza, and another in 1966, in Monte Carlo, and I had also won the *Daily Express* International Trophy race at Silverstone, so I had won three Formula One events early on. Jochen was winning regularly in Formula Two, and I was winning occasionally in Formula Two and so was Jimmy. So it was more than obvious that he was gonna win in Formula One if he could do that, but it was a question of going with the right team and the right people. And if there was anything lacking with his decision-making process, it was with whom you go and when you go with them.

"I went to BRM and it was the right move because the logic was that I would learn more from Graham Hill than I would from Jim Clark. I knew that Jim was linked very closely to Colin Chapman, without any fear of contradiction; ask Peter Arundell or Trevor Taylor, who were both very good racing drivers. But they were never going to do anything as number two drivers to Jimmy at Lotus because, albeit for different reasons, it was like racing at Ferrari when Michael Schumacher was there; it just wasn't gonna happen. So there wasn't a lot left.

"Jochen didn't have easy access to what I would call a top team. For whatever reasons Ferrari, I don't think, were in the loop for him somehow, despite the win at Le Mans in 1965. I had a big advantage, because for me to go with Ken Tyrrell in 1968 it was because I had done Formula Three and Formula Two with him and his preparation was faultless and his choice of mechanics and engineers was the best. He was very demanding, but he was incredibly good to people. Ken's people all had pensions. In those days that didn't happen. So I knew that the integrity of Ken Tyrrell for me would be so much better than anyone else. I was able to see that. And that's one of the missing links that some of the drivers didn't have who could easily have won as many Grands Prix as I did and as many World Championships and maybe they would have gone on a lot longer than I did.

"I'm not sure that Jochen had that thread early enough in his career. He got stuck at Cooper for a long time, and I think he could have done better. Everyone knew by 1964 that he was a hotshot about to happen.

"I have to say that I was delighted when he finally made that breakthrough, because he totally deserved it."

The final race of the F1 year was held in Mexico City on October 19. As in 1965 it was a Goodyear benefit, thanks to the latest G20 tyre, but this time it provided a runaway victory for Denny Hulme and McLaren, with Jacky Ickx and Jack Brabham endorsing the result with second and third places in their Brabhams. The new World Champion was fourth in his Dunlop-shod Matra. Jochen, having been narrowly outqualified for fifth place on the grid by Jo Siffert in Rob Walker's similar Lotus, ran fourth initially behind fast-starting Stewart, Ickx and Brabham, after Hulme's car wouldn't go into first gear. But Denny soon hit his stride, and as the orange car worked its way towards the front of the field, Jochen set about hounding Jackie after both Brabhams had passed the Matra by the ninth lap. The two friends were mere tenths apart until the Lotus's upper left-hand front suspension wishbone wilted on the twenty-first lap. Quite possibly the component had succumbed to pounding over the kerbs in the chicanes, in which case the Brabhams were markedly stronger as Ickx in particular used more kerb than anyone else, throughout the race.

So Jochen's run of points was over, but he ended the year in a much happier frame of mind than he had been in at the midpoint. The arguments with Colin Chapman were largely over, and they had patched up a lot of their differences. There was a lot more respect between them, and they could see a point to a future together. And Jochen, at long last, was a Grand Prix winner! If Jackie Stewart had stepped forward to inherit Jimmy Clark's mantle as the yardstick, Jochen had proved that on his day, when everything was right, he had the speed and the talent to take on the Scot and beat him. Stewart might be the best, but Rindt was acknowledged as the fastest man in F1. And as he had shown since Silverstone, he had learned how to keep it smooth.

And he was going to stick with Chapman and Lotus after all. In Canada Alan Rees had made a fantastic offer on behalf of March Engineering, to sign as its number one driver. If the deal with Ecclestone to take Jochen Rindt Racing into F1 had gone through, Jochen stood to make £35,000 (the equivalent of £405,000 today). Now Reesie offered him £100,000 (£1.15m), a huge sum at the time for an F1 driver and one which gave the lie to Jochen's comment to Pruller back at Indianapolis in 1967: "If there is one driver who can manage to become a dollar million-aire through the sport, then it's Jackie. I can't."

Pruller believed that if the money had actually been there, Jochen would have accepted. But of course it was conditional. That amount of money always had to be. Rees had yet to source the funding that would enable March to pay him so much. As things turned out, that would have bankrupted a team that would sail very close to the financial edge in its early years.

"All the hints Max dropped at the launch about big secret sponsors was all smoke and mirrors," Herd admitted years later. Jochen's £100,000 would never have materialized.

Meanwhile, Jack Brabham had become utterly focused on the idea of Jochen being his team's number one driver, in a new monocoque BT33 that would use the same Cosworth DFV that had transformed Brabham's fortunes after the appalling reliability of the four-cam Repco V8 in 1968. At one stage the deal was all done verbally, until Chapman pulled out a second trump card: he would help Jochen to run his own Formula Two team, which would be overseen by Bernie. Jochen loved F2, and found that aspect of Chapman's deal most attractive. Ecclestone flew to Canada to handle the negotiations on Jochen's behalf, on the understanding that if Brabham could make an offer that amounted to 75 per cent of what Chapman had put on the table, then he would sign for Jack. Ecclestone spoke with Goodyear, but ultimately Jack had to concede that he could not raise more than 50 per cent of the Lotus offer.

Leo Mehl recalled being so cowed by a brutal approach from Chapman at one race, where he told him with steely eyes that no matter how much Goodyear stumped up he would beat it, that he didn't bother pushing the issue any further with his management.

"As far as my personal observations, it was clear that Jochen was happier in the Brabham set-up than he was at Cooper," Michael Argetsinger remembered. "It's a shame really that Brabham was in a transitional year in 1968 because Jochen should have had great results there. It was an example of bad timing. Along with many others, I wish he had gone back there in 1970 but I have no unique insight on the matter. If it is true, as I have heard and read, that Chapman trumped Brabham's offer with a very substantial amount of money, then I would find it very believable. Jochen very much wanted to get out of racing with some real money. He was just pragmatic in that way. I don't mean to say that he was cold-blooded and calculating, because he wasn't. He was foremost a sportsman and loved the racing for itself. But he was also very practical and expressed the view to me—and to others—'There should be a way to make some real money at this.' He was always interested in finding a way to make some decent money in racing."

Just before the US Grand Prix, Jochen and Bernie agreed to

Chapman's terms. "Jochen knew that driving for Lotus, the chance of having a serious shunt was a lot more than they were with Jack," Ecclestone said, adding after a pause: "I . . . I suppose in a way I wanted to put him off, but he wanted to win the Championship, and he thought that was the best chance." Ironically, it was.

Now Jochen had to go and see Jack.

"I had a contract with him, he was going to drive for me and I'd retire out of it," Jack said. "And we got to Watkins Glen and Chapman offered Rindt a lot more money. So Jochen came to me and said, 'Colin wants me to drive for him for another year and has offered quite a bit of money . . .' So I said, 'Look, if that's what you want to do, that's fine. You go ahead and do that and I'll drive for another year with Ron,' which I did. We didn't have the money to do it with Jochen, and I thought bugger it, if he wants to go and drive for Colin, that's fine."

Many people misunderstood Jochen not just because of his curt and candid manner of speech, but because he had that Austrian pragmatism that we still see today in Helmut Marko and Niki Lauda. If some thought that Jochen's first words of English were "start money", a standing joke of the era, it was because he was totally pragmatic and honest about wanting to make money from practising such a dangerous profession. Who could possibly argue with that?

So Jochen signed a one-year extension to his Lotus contract, with the proviso that, whatever he might do if successful in his quest for the World Championship, he would undertake six months' worth of promotional and public relations activity on behalf of the team.

"I could've held him to it, our contract," Jack said in 2008. "And he said to me, 'If you really want me and you want to hold me to that contract, just say so.'" There was an air of regret still evident in the great triple champion's tone because he still felt slightly guilty not to have held Jochen to that contract, even though nobody could ever have foreseen the turn events would take. What he did was an honourable act between one friend and another. "It was his decision in the end," he said, and there you had it.

At the end of the season, *Autocourse* editor David Phipps rated Jochen second behind Jackie in his Top Ten, and ahead of Ickx, Brabham, Hulme and Amon. "With a little luck Jochen Rindt might have made Jackie Stewart work even harder for the World Championship this year," he summarized. "He was robbed of two races by wing failure, and of another by driveshaft trouble, and was always remarkably fast for someone in a two-year-old car running on one-year-old tyres. Next year, in a new Lotus, he must be one of the favourites for the Championship."

In November 1969 Jochen and Heinz went out for the evening to a small tavern in Vienna, where they talked all night. "In 1970 I want to be World Champion and the biggest name in motor racing," Jochen said, without braggadocio. "But racing will only be a part of my life. When I have the title, I am going to retire immediately. I just don't want to be finished and exhausted before I am thirty and just carry on racing because there is nothing else I want, can do—or am interested in. There are so many things which I would like to do. Time is the most valuable thing you can have. And listen: I will be living another, say, fifty years. But I have taken out from twenty-eight years more than you can usually gain. Isn't that simple to understand?"

"I asked him," Heinz recalled: 'All right, you get the title. But then, wouldn't you like the idea of becoming one of the all-time greats, maybe equalling Jimmy Clark's world record of twenty-five Grand Prix wins?'"

Jochen was always very Austrian in his views—forthright, devoid of romanticism. And Heinz always remembered that he had replied slowly, "Look where Jimmy is."

But equally his single victory was not enough. He still needed to prove something; that he was the best in the world. "Because today there are only two real top professionals. Jackie Stewart and myself. Because we drive with our heads."

Jochen was a supremely intelligent man, educated, well-read, who saw a bigger picture to his world than just motor racing. And the 1969 season had made him brutally aware of the risks.

"The possibility of getting killed is a big one," he said that evening. "It's just luck, good luck or bad luck, whether you can manage to survive. I have been racing for eight years now; no one can understand what it means. Eight years of racing! You know, my show alone can earn me my living."

That he was determined to quit if he won the title in 1970 was beyond question at that moment.

The decision to stay with Team Lotus would ultimately enable Jochen to pursue and, ultimately, realize his cherished dream. But it would also cost him his life.

PERFORMANCE AT THE LIMIT BY MARK JENKINS, KEN PASTERNAK AND RICHARD WEST, 2005

INNOVATING: THE DRIVE FOR CONTINUAL CHANGE
It's a mindset that demands that the company be structured to deliver results quickly, to constantly innovate and also not be afraid to give up on something if it's not working more or less immediately.
Scott Garrett, Head of Marketing, WilliamsF1

Formula 1 is a global motor-racing spectacle where each team relies on technology and the ability to continually innovate, in order to outpace the competition. Innovating is concerned with continuously enhancing performance. It is about creating new opportunities, whether these are related to a product, technology or process. The point of innovating is to create new sources of performance, to find new ways of doing things that may improve both the efficiencies and effectiveness of the organization.

Innovation in Formula 1 is usually thought of strictly in terms of technology, but the sport has in recent years witnessed remarkably creative ideas in the marketing arena. The impact of Red Bull's involvement in Formula 1, for example, has not just been related to an injection of cash, it has introduced a number of marketing innovations to Formula 1, as described by Team Principal, Christian Horner.

"Red Bull has brought a very refreshing appeal back to Formula 1 with the introduction of concepts such as the 'energy station', an open-house facility that is open to any Fl pass holder, which was unheard of previously. We also have some fun initiatives that we have partnered with—film promotions, Formula Una [a Formula 1-based beauty contest] or the Red Bulletin [a daily newsletter printed during a Grand Prix weekend], so there's so many initiatives that Red Bull has brought to Formula 1 which I think without which Formula 1 would be a much duller place."

The ability of any Formula 1 team to innovate is as fundamental as its ability to put a racing car on the grid. It has no option but to continually develop, both its cars and its ways of operating, in order to stay ahead. Consider the speed of the leading qualifying car (pole position) for the Monaco Grand Prix, which has always been held in the luxurious city of Monte Carlo.

From the Alfa Romeo of Juan Manuel Fangio, which took pole position in 1950 at a speed of 64.55 mph, to Jarno Trulli's Renault RS24, which was at the front of the grid in 2004, there is a difference of 36.5 mph. However, it is perhaps surprising to note that this represents a year-on-year improvement of just less than 1 per cent. Many companies today would not be able to survive at such a low rate of performance improvement. Nevertheless there is one factor that helps to explain this overall relatively low figure: regulation.

The regulators of Formula 1 have continually sought to reduce the speed of the cars for reasons of safety and to ensure more equal competition between teams. Speed is a direct factor in creating injuries, hence the regulators' continual efforts to reduce speeds for the sake of safer racing. In contrast to the regulators, the teams themselves are obsessively focused on getting ahead of the competition by building faster cars. They are continually inventive in coming up with ways to make the car go faster; based often on innovations which the regulators have not yet identified. All the design groups in Formula 1 operate on the principle that if the rules don't say you can't do it, then that clearly means you can!

For these reasons there is a "sawtooth" effect, where speeds are increased through innovation, and then reduced through regulation. For example, in the period from 1995 to 1997 innovations in car design meant that the average pole position speed at Monaco increased from 90.8 mph in 1995 to 96.3 mph in 1997. If this rate had been sustained until 2004 the average speed of pole would have been considerably faster.

A further example of the speed of development was provided at the 2006 Brazilian Grand Prix at Interlagos. This was Michael Schumacher's last Formula 1 race before retiring. He finished fourth, but he put in a superb drive through the field after a return to the pits due to a puncture. He also achieved one final record with the fastest lap of the race (at 1 minute, 12.162 seconds). This beat the previous year's record, achieved by Kimi Räikkönen in a McLarenMercedes, by just over 0.1 of a second. It says a great deal about the constant pace of improvement in Formula 1 that Schumacher's fastest lap in 2006 was achieved in a car with a 2.4-litre V8 engine, whereas Räikkönen's car in 2005 was equipped with a 3.5-litre V10 engine, which was a much more powerful unit.

So is innovation in Formula 1 different from that in any other kind of company? Williams F1's Engineering Director, Patrick Head, thinks not. "Today I don't think there's a lot of difference between a Formula 1 team and any other kind of business. They both have the same pressures of trying to bring product to market, generation of ideas, recognition and nurturing of the good ideas."

So what are some of the pressures and demands of innovating in the context of Formula 1? Two factors emerged from our research: speed to market and the challenge of innovating in organizations which are growing larger and larger.

Speed to market (or the track!)

Whereas there are similarities between Formula 1 and other businesses, there are real differences in terms of the pace and intensity of innovating. The pace of innovation is significantly faster in Formula 1 than other technology industries. Here innovating is a continuous process, with a constant array of design changes and new components being incorporated into the car. This is shown in the graphic on page 437, which was kindly provided by Ferrari. The top part of the figure shows how Ferrari develops its production engines which are used in its high-performance road-going cars. We can see that the total development period for a production engine is forty-two months, if we exclude the time for the concept study up to the start of production (SOP). In contrast, during the same period within Formula 1, there have been three new engines designed, built and raced each with three iterations or evolutions (EV01, 2, 3) of the design. It can also be seen that within Formula 1 the development process is continuous, meaning that the engine is being both raced and developed simultaneously. For Formula 1 we therefore see a total of nine stages of development compared to the single stage of development for the production engine.

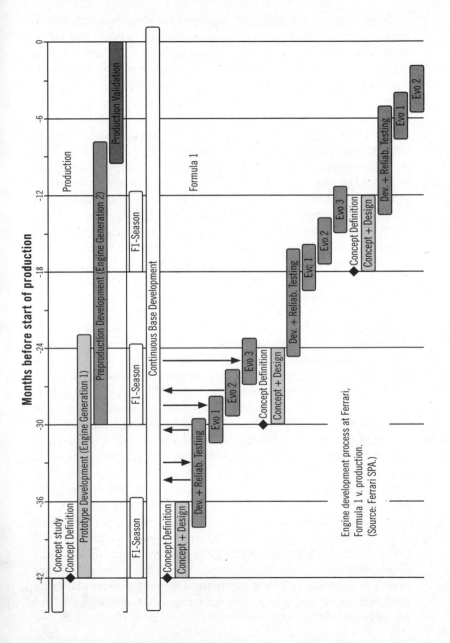

Months before start of production

Concept study
Concept Definition
Prototype Development (Engine Generation 1)
Preproduction Development (Engine Generation 2)
Production
Production Validation

Continuous Base Development
Formula 1

F1-Season

Concept Definition
Concept + Design
Dev. + Reliab. Testing
Evo 1
Evo 2
Evo 3
Concept Definition
Concept + Design
Dev. + Reliab. Testing
Evo 1
Evo 2
Evo 3
Concept Definition
Concept + Design
Dev. + Reliab. Testing
Evo 1
Evo 2

Engine development process at Ferrari,
Formula 1 v. production.
(Source: Ferrari SPA.)

One of the critical aspects of Formula 1 is that it requires all the teams to attend fixed race meetings around the world. Therefore, cars must always be ready to race at each event. Patrick Head, Engineering Director, WilliamsF1 said: "We are driven very much by specific dates and programmes. That means that with a new component the Chief Designer will know when the design needs to be issued, how long it's going to take to do the design, he'll know roughly how long it will take to manufacture and he can then target a test for the component to be evaluated."

All of these factors require a process in which individuals are familiar not just with the design process, but also the whole manufacture and development cycle; a fact which is underlined by top Formula 1 designer John Barnard: "The key to speed in innovation is being able to integrate the design and manufacture processes. A good designer will go and talk to the fabricators or machinist to find out how the part would be made; it may be that by making some small design changes at this point the part could be easier to manufacture and therefore both quicker to be released and potentially more reliable."

However, Barnard also believes that this means that more traditional management methods such as those used in aerospace are inappropriate for the flexible and responsive context of Formula 1. "The problem is that in a context like aerospace, people can almost determine their own production times, their own lead times. If you sit down and do the project timescale properly you just keep adding on and, 'That'll take this long and then I have to do that and once I've got that done I do this and so on.' You'll end up out here and that's your project lead time. It doesn't happen like that in racing because there's your lead time, there's your project time, that's the race, 'Now you make it fit', and you do whatever that takes."

The ethos of "doing whatever it takes" means that teams have to go beyond the bounds of their highly pressurised development programmes to stay competitive, as illustrated by WilliamsF1's COO Alex Burns: "In 2008 we were developing a new front wing aimed at Monaco, so it was planned to go to the Ricard test [the Paul Ricard circuit in the South of France] and then go on to Monaco; in testing we found a significant performance gain, so we actually rushed one through two weeks earlier to the Turkish Grand Prix, so Nico was racing with that at Istanbul. That call was made on the Sunday of the Spanish Grand Prix [two weeks before the Turkish Grand Prix], and we hadn't quite finished designing it at that point, so there's an opportunity to say, 'That's going to give us an advantage at Istanbul therefore can we push it through in nine days in order to get it out to Istanbul in time for the race.'"

A complicating factor is the variability of the different tracks used in Formula 1, which means that innovations are often track-specific. Monza and Monaco are very good examples of this. With the Monza circuit being the fastest circuit in the Championship year a major redesign of bodywork, wings and end plates is required to ensure the minimum of drag on the car while on Monza's super long straights. Likewise the brakes and their cooling require attention with the manufacture and fitting of different-sized cooling ducts and the like.

The exact opposite of this requirement is Monaco where due to the confines of this street-based circuit different ratio steering racks have to be fitted to cope with the very tight corners, and aerodynamics are almost the exact opposite of Monza to gain the maximum amount of downforce and therefore grip on the slow corners. Alex Burns bears this out in more detail: "A lot of the aero [aerodynamic] design work is track-specific. There is an underlying thread of upgrade that goes through the year; you are in a constant state of development but, we'll have a specific package for Monaco, a specific package for Montreal and for Monza. Then you'll have general upgrades that go through for the bulk of the races, so the high downforce races of Barcelona, Magny Cours, Silverstone, Hockenheim; all of these will be relatively similar."

Another key part of speed to track is the various simulation approaches that are used to predict performance prior to reaching the circuit and therefore it is in simulation technologies that Formula 1 has also made significant progress, as Alex Burns observes: "It is important that our simulation for brake cooling is accurate, so if we go to a circuit and we project the brakes will be at this temperature, we'll be using the same simulation package to predict it for the next races, even though the cooling will be a different package. But if our simulation package isn't correct, we need to correct it, otherwise we are going to have a problem. The brakes might run substantially cooler than we were expecting, that should flag that there's an issue with the simulation and we'd want to fix that before the next event, because you're losing performance by running them too cold and if you're running too hot, you're going to ruin the brakes during the race; you're not going to survive the race. So, you need to get these things right, you need to be accurate with simulations."

Innovating and regulating

There is a continuous battle between the innovation needed to increase speed and performance, and regulation to ensure that there is real competition and that racing is as safe as possible. While in the 1960s, 1970s and 1980s there were many radical innovations, in the 1990s and the 2000s growing levels of regulation have made it increasingly difficult for designers to achieve the "big breakthrough", as suggested by designer John Barnard: "Formula 1 now is much more a case of detailed development rather than innovation. The rules have boxed it in a lot more from what it used to be. Everybody has more or less found, because of wind tunnels, because of all the testing, all the test facilities that most of these Formula 1 teams have now, they've all ended up being focused down pretty much the same path in the same channel."

However, WilliamsF1's Patrick Head believes that the increasing intensity of regulation has meant that the nature of innovating in Formula 1 is changing rather than disappearing: "I think innovation in terms of bringing in radical new systems like active ride suspension [a hydraulic suspension innovation of the 1990s, now banned] is much more inhibited by regulation, but new ideas are still needed and the thinking tends to be on a more micro scale. We tend to look at adding a lot of micro-scale innovations together to give a larger overall effect."

Who gets the benefit?

The history of Formula 1 is littered with innovations where teams have created a step forward in performance, but the innovating team has not always been the one who has enjoyed the most race success as a result of its innovation.

As can be seen from the table below, often the innovator is not the prime-beneficiary of the innovation. While Cooper undoubtedly enjoyed success from its mid-engine layout, it was Colin Chapman of Lotus who took the design to a more refined stage and was able to overtake the Cooper concept on performance. Similarly, while it was Lotus who pioneered the development of the Ford Cosworth DFV engine, which required a different concept in chassis design (the engine formed a structural part of the car), it was Ford's decision to make the engine available to other teams. This meant the performance potential was dispersed across a range of constructors, effectively creating Grand Prix winners out of teams such as Matra, Tyrrell, McLaren and Brabham.

Who benefits from innovations?

Innovation	Innovator	Beneficiaries
Mid-enging layout	Cooper	Cooper/Lotus
Cosworth DFV	Lotus	All British constructors
Ground-effect	Lotus	Lotus/Williams
Fiat 12 engine	Ferrari	Ferrari
Turbo engine	Renault	Honda
Active suspension	Lotus/Williams	Williams
Six-wheel car	Tyrrell	Tyrrell
Composite monocoque	McLaren	McLaren
Semi-automatic gearbox	Ferrari	Ferrari
Tuned mass damper	Renault	Renault (+McLaren and Ferrari)

Renault entered Formula 1 in 1978 with a car using a light-weight turbo-charged engine. The regulations at the time stipulated that engines were either 3.0-litre normally aspirated or 1.5-litre turbo charged. It was generally believed that no-one would be able to build a 1.5-litre turbo which would be competitive against the 3.0-litre engines. Renault did so, but in the nine years it raced the turbo engine it failed to win a World Championship. In contrast, Honda, which entered with its own turbo-charged engine in 1983, was able to win three World Championships between its entry point and the end of the turbo era at the close of the 1988 season.

But there have also been instances where the innovating team was able to capture much of the value of its ground-breaking work. Ferrari's "Flat 12" engine was developed originally to be fitted into an aircraft wing, but found its true potential in the Ferrari 312T racing car. The powerful twelve-cylinder format with a low centre of gravity meant that it posed a significant threat to the well-established Ford Cosworth DFV. While there were attempts to copy the Ferrari format, most notably with Alfa Romeo supplying the Brabham team with a Flat 12 engine, these were uncompetitive and Ferrari enjoyed a prolonged period of success before the engine was rendered obsolete by new ground-effect aerodynamics.

These examples raise some important questions about how Formula 1 teams are able to protect their ideas. It did not take the significance of "Spygate" in 2007 to highlight that secrecy is a big issue in Formula 1, but it is something that Patrick Head believes can be taken too far.

"Some Formula 1 teams are so concerned about secrecy and the loss of IP that they literally build physical walls around departments to ensure that if someone leaves from the transmission department, they won't have an idea of what's going on in the suspension department; in contrast we have the view that providing we're progressing and developing it's more positive to have an open internal exchange of information than the risk of losing IP when somebody goes."

From Ferrari's point of view, its location outside of the UK's Motorsport Valley could be a benefit here. Ross Brawn, former

Technical Director, Ferrari, said: "If you've got an innovation you're lucky to keep it for three or four months, particularly once it goes out on the circuit. I guess we gain and lose from that because we don't have the grapevine feeding us, but generally I'm happier with the degree of isolation we have."

Another potential concern is the frequent movement of drivers around the teams, but in Patrick Head's view this is not a problem: "Most drivers are only aware of what we're doing on the surface, they know that if they press this button it does that, but they've got no idea of what goes on inside."

Of course, another possibility is that the potency of the innovation is masked by the fact that other aspects of the car perform poorly. Gordon Murray, former Technical Director at Brabham and McLaren, said: "Where we've had a massive innovation and we think we're going to walk it and the driver makes a mistake, the engine fails, you choose the wrong tyres or whatever and you have a series of races where other things go wrong. That happened to us a lot. It can be a bad thing if you cream the first race as everybody panics."

Innovating in public

WilliamsF1's Patrick Head identifies a further key point that, in Formula 1, the success or failure of innovating is a very public one: "You've got to do better and better each year, there is no hiding place if someone's not doing a good job. They can't tuck themselves away, it tends to become visible pretty quickly."

A similar sentiment is expressed by the FIA's Tony Purnell who also combines the point that it is both highly visible and immovable: "In Formula 1 you cannot disguise the truth about your 'product' because you are absolutely exposed to the reality of your situation every two weeks. And that's where Formula 1 is special, it really is."

Formula 1 therefore presents a very particular challenge to the process of innovating: because of the competitive pressures it has to be relentless, but it is also highly visible in terms of the success or failure of the process.

Balancing innovation with growth

Formula 1 teams enjoyed particularly high levels of growth in the period between 1993 and 2003. During this time, the typical number of employees in a Formula 1 team grew from around 100 to 500. This, however, created new problems for how they were going to maintain their flexibility and responsiveness, the essential ingredients of competitiveness.

The nature of technological growth in Formula 1 meant that there was a need for increased specialization, particularly around the areas of aerodynamics and electronics. This need for increased specialist expertise meant that the process by which a car was designed had changed from essentially a step-by-step linear process to one that now involves many activities occurring in parallel, as summarized here by WilliamsFl's Engineering Director Patrick Head.

"Probably fifteen years ago the design team would be working on the gearbox for one week and then the next week they'd be designing the rear suspension and then the next week they'd be in the wind tunnel sorting out the aerodynamics; you tended to be involved in every aspect of the car. Today we have specialist areas working in parallel, so we have to deal with the problem of how the transmission, for example, integrates with the rest of the car, how it satisfies the aerodynamic requirements of the diffuser [a structure which manages the airflow under the rear of the car], how it deals with the loadings coming from the rear suspension, etc."

WilliamsFl dealt with this problem by the senior management now focusing on the integration of these groups: "Fundamentally my job is to ensure that we are producing the quickest car, as op posed to the best transmission or the best rear wing mounting. It's a different way of working and it means that you have to have frequent contact between these groups. We have an open-plan design office and encourage people to liaise with the other departments who have an interest in their work."

The teams had to create structures which were able to bring together these specialists and the necessary equipment, but at the same time ensuring that these groupings did not become ghettos of specialists, detached from other parts of the team. Alex Burns relates some of the steps which have been taken to address this at

WilliamsFl: "We're trying to deal with this by creating smaller units and ensuring that we get the interaction between design and manufacturing and align this to the testing and racing operations. I think that once you get above 200 there's a real shift in the culture in a company. You ideally need fifty to sixty people to make things happen quickly. Within these groups you ideally need teams of no more than a dozen, and then ensure that they understand how they fit into the other groups.

In addition to creating appropriate structures, Burns notes: "It's also important that each group has something which is clearly related to the car, rather just, say, your delivery performance against your work's order due dates must be high. For it to work in Formula 1 everyone has to be able to relate their activities to the performance of the car."

While individuals such as Patrick Head have experience in all the component areas of the car, which they bring to bear when making trade-offs between different aspects of the car, many of those coming up have tended to have been specialists in one particular area, most notably aerodynamics. He said: "One of the problems created by this growth is that you see some people who were very capable in one particular area, such as aerodynamics, being head-hunted to be a chief designer or technical director in another team, completely wrong for the individual and the company."

Technology transfer

It has perhaps been one of the myths of Formula 1 that much of the technology from the sport finds its way into normal road cars. There are very few past examples of this, with perhaps the paddle shift gear system, which is now employed on a number of high-performance cars, being one of the more visible examples. In fact there are probably more examples of the technology moving the other way; some road cars had traction control systems well before they were used in Formula 1. Williams used a highly developed version of the DAF variomatic belt-based transmission system (known as Continuously Variable Transmission or CVT) in the Williams FW15C in 1993, but the system was banned before it was raced. Renault's Tuned Mass Damper system was also based on a concept which had been applied in the Citroen 2CV!

However, with the car manufacturers becoming more heavily involved in Formula 1there is a greater emphasis on creating broader value by seeking technology transfer that can be passed through to road-car manufacturing. One way that this is occurring is through the development of new technologies such as Kinetic Energy Recovery Systems (KERS), which is being introduced in 2009. For Honda's Ross Brawn, this provides a number of important benefits which are more to do with learning, ideas and processes than specific products: "For the KERS system the team at Honda that are developing our motor generator unit and control system for the race car have never had the budget, support and targets that we are giving them through the Formula 1 programme. In the normal economics of a road-car programme the balance between cost, performance and timing is different; it has different limits and objectives.

"For us, we can put in the support, we can put in the backing, we can put in the resources, and they're achieving things now that a year, eighteen months ago they didn't think they could achieve in terms of performance of the unit, in terms of weight of the unit, in terms of size of the unit. A road car cannot afford to have a Formula 1 system in it, that's a fact. We can justify different technologies, different materials and the economics for us are different, we are making a relatively small number of systems. But the people who are involved in that project will be the people that move into the road-car side with the 'can do' mentality that's come from the Formula 1 programme."

A similar situation is found at Toyota, as described by John Howett: "We have around twenty-five engineers based in our fundamental research company, which is outside of Toyota itself and they are working with our engineers on new materials technology. So even if we're not deploying it there are many discussions around how we use these technologies. We also have a machining activity in Japan with casting and machine shop. There's information transfer on the technology used, how to save money, how to make things quicker, and the alternative processes; so there is a constant interchange at all levels. Some of it's not exciting but it very often adds value to the business, either ours or theirs."

For BMW its involvement in Formula 1 is also a lot more than just a marketing programme; it is also looking to create value for the core business, through technology transfer. According to Mario Theissen, Team Principal, BMW Sauber: "It has to create technology transfer and that is why we set up the organization in a way which is different from any other Formula 1 team. We have a dedicated Formula 1 foundry, dedicated Formula 1 parts manufacturing plant and these two units are not run by BMW motorsport but by the respective departments who do the road car parts in Munich. So both the foundry and parts machining create new technologies and processes on their premises and we are the customer.

"What they learn working on Formula 1 they take immediately over to the road car. So we have real synergy, not in the way that you use any individual part of a Formula 1 engine or car but, you learn so much about technology and you speed up this learning process in a way that you couldn't do without Formula 1. Now it's becoming really interesting with the KERS system for next year. We are dealing with it on the power train side, which is in Munich, and we are now seeing that through the pressure, the time pressure and the weight constraints of Formula 1; we are achieving technology leaps within months. What we will have on the car on the grid next year in March [2009] is not available yet and the hybrid guys from road-car development are knocking at our door because they see the rapid progress we make in Formula 1. We can take higher risks, we need a shorter lifetime, we can take the risk of failure which is very different from a road-car project, so we are paving the way now and that will be very significant."

Technology transfer also works in a vertical, as well as a horizontal way in moving between Formula 1 and other motorsport series, as outlined by Peter Digby of transmission manufacturer Xtrac. It started out producing a new gearbox used in the demanding sport of rallycross which it found met a need in Formula 1: "We found that our rallying parts fitted straight into Formula 1 because although one has a lot more horsepower it's the torque demands that really matter, so in the late eighties our transverse Rally package at the time was able to go straight into

Formula 1. After this, Formula 1 began to advance faster and some of that technology has now trickled down to our other formulas. So, what was our standard Formula 1 gear in 1995 became our standard touring car gear in 2000. So, there was a lot of trickle-down of components, actual hardware as well as technology."

He draws an interesting contrast between motor racing and tennis; as the surplus from the All England Club, which runs Wimbledon, goes to the Lawn Tennis Association to support smaller tournaments and clubs within the UK: "Formula 1 does not take cheques and write them out to lower levels of the sport as I believe happens at Wimbledon, as much of the Wimbledon income goes into grass-roots tennis in the UK. That doesn't happen in motorsport, but what does occur, for example, is the F1 teams going off and developing expensive semi-automatic gearboxes or Kinetic Energy Recovery Systems that can later be installed on a normal car in racing or even road cars for a fraction of the price."

In Formula 1 there have certainly been a number of notable innovations over the last fifty or so years. Here we pick a number of particular examples to consider some of the general principles of innovating within Formula 1. The Ford DFV engine in 1968; the six-wheel Tyrrell of 1976; the "pit stop" Brabham of 1982; Ferrari's paddle gear change; the active suspension Williams of 1992; the all-conquering Ferrari F2004 of 2004 and the tuned mass damper of the Renault R25.

Changing the face of Formula 1: Ford DFV engine

The Ford DFV engine was a disruptive innovation in Formula 1. It changed the way in which Formula 1 cars were designed and effectively further shifted the basis of competitive advantage away from the engine to the chassis and aerodynamic aspects of the car. In many ways it was this engine that created the regional cluster of expertise in the UK, known as Motorsport Valley; now the core competence needed was focused on chassis and aerodynamics rather than engine design. Its contribution to Formula 1 and the motorsport industry more generally is highly significant.

The basic concept of the Ford DFV was that it replaced the

need to construct a full chassis along the entire length of the car. The DFV was part of the car and was attached to the chassis behind the driver with the rear suspension and gearbox attached to the back of the engine. This created a significant increase in the power-to-weight ratio of the racing car.

As an innovation the Ford DFV was a joint development between Cosworth Engineering which developed the engine, and Formula 1 constructor Lotus, which designed its type 49 car around the engine, allowing it to be attached to the rear of the chassis. The Ford Motor Company sponsored the project with a capital investment of £100,000, and after this the engine was also known as the Ford Cosworth with the famous Ford oval logo carried on its cam covers.

One of the main catalysts for the innovation was a change in regulation. In November 1963 the FIA announced that from 1 January 1966 Formula 1 engines would either be normally aspirated 3.0 litres or 1.5 litre turbo charged. Prior to this point the normally aspirated 1.5-litre engine had dominated, most notably with that produced by Coventry Climax and used by successful teams such as Cooper and Lotus. However, Coventry Climax decided that the development costs of a new 3.0-litre engine would be too high for it to bear; it announced its withdrawal from Formula 1 at the end of the 1965 season.

Colin Chapman of Lotus approached Keith Duckworth of Cosworth to see if he could design and build a new 3.0-litre engine. Chapman then sought support from Ford, which he received from Walter Hayes (the £100,000 referred to above), who was responsible for Ford's motorsport activities. Duckworth developed a novel lay out for the combustion chamber using four valves per cylinder. At the same time Ford had also commissioned a smaller four-cylinder Formula 2 engine using the same layout. The Formula 1 engine effectively doubled up two four-cylinder blocks into a VS formation. It was therefore given the name "DFV" for Double Four Valve. The car and engine were developed during 1966 and made their first appearance at the Dutch Grand Prix at Zandvoort on 4 June 1967. It won the first race and went on to dominate the rest of the season.

While Lotus and Cosworth were delighted with the situation

Ford's Walter Hayes was not so sure: "Almost at once I began to think that we might destroy the sport. I realized that we had to widen the market for the DFV engine, so that other teams could have access to it."

In 1968 the Ford DFV, which had been instigated by Colin Chapman of Lotus, became available to other teams for the sum of around £7,500 per unit. This started a tradition in Cosworth in building customer engines. In 2004 Cosworth was still supplying 'customer' engines to Minardi and Jordan, although the supply contract by then ran into millions of dollars per annum.

However, as has frequently been the case, while Chapman was the innovator he was not fully able to capture all the benefits of the innovation. Hayes' decision to make the innovative engine available to other teams ensured that while Ford dominated Formula 1 through the late 1960s and early 1970s, Lotus, although it enjoyed some success, did not.

Four isn't enough: the six-wheeled Tyrrell P34

Tyrrell Racing was one of the most successful Formula 1 constructors of the early 1970s (it was later sold to British American Tobacco as the formation "base" for the BAR Honda Fl team). However, its success, which had been based partly on the Ford DFV engine, had waned and Technical Director, Derek Gardner, was looking for a new way forward: "In about 1974 it was becoming apparent that the Ford engine had lost its edge, it was still producing the same horsepower that it always had, or a little more even, but with the success of the Ferrari, the possible success of engines like Matra or anybody else who came along with a Flat 12, V12 or 12 cylinder whatever, you're going to be hopelessly outclassed . . . I wanted to make a big breakthrough."

Gardner's idea was a radical one that had started in the late 1960s, when he had worked with Lotus on a series of cars for the Indianapolis 500: "So I thought about the six-wheel car and looked at it in a totally different light to the way I had as a potential Indianapolis car.

"I thought if I could reduce the front track and keep it behind this 150 cm [maximum body height stipulated by the Formula 1 regulations] then I'm going to take out all those wheels and their resistance, but above all I would take out the lift generated by a wheel revolving on a track."

Although Ken Tyrrell had his reservations he decided to give Gardner the opportunity to develop his ideas: "It was Derek's idea [the six-wheel car]; Derek had wanted to do it the year before [1974] but I didn't think that we were long enough established as manufacturers to go to something so radical. But he finally convinced me that we ought to try it, so we grafted four front wheels onto our existing car and created the six-wheeler. We decided to show that car [to the press], we explained this was an experimental car which we were going to test, and if it was any good we would race it."

A key aspect of the development of the six-wheel concept was the input of tyre manufacturer Goodyear, which at that time supplied all the Formula 1 teams with tyres. Gardner shared his ideas with Goodyear, which responded to the challenge by creating a tyre with a ten-inch width and a sixteen-inch diameter. The introduction of the six-wheel P34 temporarily restored the fortunes of Tyrrell racing, as can be seen below.

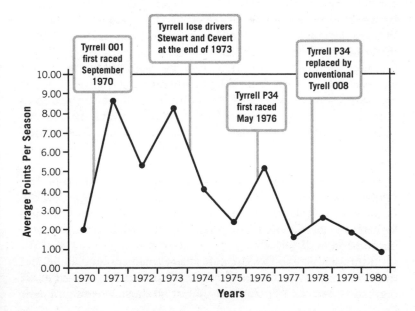

However, despite a promising performance in 1976, when Tyrrell finished in third place in the Constructors' Championship, 1977 proved to be a different story, with the P34 becoming uncompetitive relative to conventional cars. The reasons for this appear to have been due not to any fundamental aspect of the concept, but to the speed of development of specific components that were supplied by external suppliers to Tyrrell.

Ken Tyrrell said: "It became difficult to get big enough brakes to fit inside small front wheels. Because everyone else was using a standard front tyre, it became politically difficult for Goodyear to develop the small tyre for us. The car became too heavy with our attempts to put bigger brakes in it and at the end of the second year we had to abandon it."

Derek Gardner commented: "Where I think we went wrong was that Goodyear were supplying most of the teams with rubber, they were only supplying one team with very small front wheels. Therefore, the development of the tyres which are continually going on, it meant that almost with its first race the development of the front tyres went back—they just didn't develop as fast as everyone else. Whereas the rear tyres were being developed with the existing front tyres, so in effect you're having to de-tune the back of the car to stay with the front which was, really, not what it was all about."

As a result, Tyrrell returned to using a conventional chassis with the Ford DFV engine. Derek Gardner left Tyrrell to return to industry and it wasn't until he had retired and became involved with the Formula 1 Thoroughbred Racing Series, which race historic Formula 1 cars, that he was able to work with the six-wheel car once more.

Innovating the whole system: the Brabham pit-stop car

Pit stops have been a feature of Formula 1 for many years. But contrary to popular belief, they were not introduced by regulation to liven up the racing. Contemporary pit stops were created by the innovative Brabham BT50 or pit-stop car. The point of the pit-stop car was that this was not just a technical innovation, but an innovative race strategy which enabled a lighter, more nimble car to outpace the opposition to the extent that it would be able to enter the pits, stop, refuel, fit new wheels and tyres, return to the circuit and still be in front. It is a classic case of problem-solving and lateral thinking to win the race.

Gordon Murray, former Technical Director of Brabham and McLaren Formula 1 teams, said: "I was the one who introduced pit stops in Grand Prix racing by designing a half-tank car—to get the advantage of the lower weight, the lower centre of gravity. But it wasn't just pit stops; it was a plan that allowed you to achieve advantage through a faster, lighter car and a pit stop. It's just pure mathematics; you just draw a graph of the race—you draw the car's weight, the centre of gravity and the benefit per lap as a curve and then you take a chunk, a negative curve out of the middle bit where you lose say twenty-six seconds slowing down and coming into the pits and refuelling and if the total equation's better, then you do it."

But Gordon Murray's idea created other problems, such as tyre temperature. The performance of a Formula 1 car is very susceptible to changes in the condition of the tyres. When a race is started the tyres are relatively cool and performance is only optimised as they warm up to operating temperature. The challenge with the pit-stop car was that this problem was multiplied one, two or even three times in a race. So its success was also dependent on the tyres being able to get to their optimum temperature as soon as possible.

Gordon Murray said: "We developed these wooden ovens with gas heaters in them to heat the tyres up so the driver didn't lose the time and smacked those on at the last minute with the fuel."

One surprise for Murray was that the other teams were relatively slow to respond to Brabham's innovative approach: "In the first four or five races the turbo chargers kept failing, I said to Bernie [Ecclestone, then Brabham Team Principal], well that's it, we started in Austria, we only had four races left to end the year, everybody's going to turn up at Brazil with a pit-stop car. But no. Williams had it—but it was a kit that could be put on the car, it wasn't integrated into the design of the car."

Politics and innovating: the Ferrari paddle-shift

Perhaps of all the innovations in Formula 1 the paddle-shift gear change is the most recognizable innovation that found its way into road-going, high-performance cars. The paddle shift utilizes a semi automatic gearbox where the driver does not operate the clutch but selects gears by a pair of "paddles" located left and right on the steering wheel; pulling one side towards the driver changes up a gear and pulling the other towards the driver changes down.

It was developed by John Barnard, who was trying to find a way to improve the performance of the turbo-charged Ferrari Formula 1 car. However, the problem with this innovative idea was that it meant that the car had to be either designed as a paddle-shift car or a conventional gear-shift car; there was no possibility of producing a competitive car which combined these two features.

Barnard said: "There was a massive amount of politics around the whole paddleshift concept. It actually happened at the time when Enzo Ferrari died. Vittorio Ghiella, who was running Fiat Auto at the time, came into Ferrari to take over Enzo Ferrari's mantle. Towards the end of 1988 I was designing the 1989 car, which was a more developed version of the 1988 test car, but I was designing it such that it would not take a manual gear shift; you could only have a paddle-shift gearbox in it, which was a pretty big commitment to make.

"Ghidella was so nervous of the fact that it wouldn't work that he insisted that they built a manual version alongside it. I resisted heavily, because I knew that we didn't have the capacity to do that properly, but they did it and modified the car to put the manual version in and Mansell [driver Nigel Mansell] ran it for a few laps at Fiorano [Ferrari's dedicated test track] and said 'Forget it, give me the paddle shift again.' So that was a diversion caused literally by politics by the head guy at Fiat. I had to literally lay my contract on the line to be able to do it. My contract said that I had overall technical authority on all the cars and the race team and I used that. I put my contract on the line such that if it didn't work or there were unseen problems with it, then, effectively I go and commit hari-kari. So that was how it was done, which puts a lot of pressure on that you really don't need when you've got enough technical pressure as well."

Barnard went ahead with the paddle change on the elegant Ferrari 640, which won its maiden Grand Prix on 26 March 1989 at the Brazilian Grand Prix in Rio de Janeiro. However, the 640 suffered from reliability problems and despite also winning Grand Prix in Hungary and Portugal that year, Ferrari finished third in the Constructors' World Championship. The paddle-shift gearbox was quickly imitated by many other designers and is now a standard feature of a Formula 1 car. It is also used on some of the higher-performance Ferrari, Alfa Romeo and Subaru cars, as well being a common feature on many videogame steering wheels!

The gizmo car: Williams FW14B

The Williams FW14B won the first five races of 1992, with Nigel Mansell at the wheel; a record which even survived the dominance of Ferrari and Michael Schumacher in 2002 and 2004.

The FWl 4B was a highly innovative car in that it incorporated many of the leading-edge ideas of the day. Its designer, Patrick Head, had incorporated semi-automatic gearboxes, drive-by-wire technology and Williams' own active suspension system. The significance of these ideas was that many of them had been initially developed by other teams, such as the carbon-composite monocoque (McLaren and Lotus), semi-automatic gearbox (Ferrari) and active suspension (Lotus).

The source of advantage was therefore not one particular innovation, but the way in which they were all brought together, as summarized by David Williams, who was General Manager at WilliamsF1 at the time: "I think we actually were better able to exploit the technology that was available and that led to a technology revolution. We were better able to exploit it to the full, before the others caught up. It wasn't just one thing but a combination of ten things, each one giving you another 200/300th of a second; if you add them up you a get a couple of seconds of advantage."

The Williams car was so successful that many questioned whether this was a case of technology taking over Formula 1 and whether the skills of the driver were becoming replaced by the technology in the car. This led to further regulations to remove many of these so-called driver-aids from the cars.

The total package: Ferrari F2004

From 2000 to 2004, Ferrari dominated the Formula 1 Championships, with cars such as the Championship-winning F2004. So, what innovations were applied to make these cars so competitively outstanding? Ross Brawn, former Technical Director, Ferrari, said: "Ferrari doesn't have an individual feature, perhaps it never has had, but our innovation is an integration of the whole. Our efforts have always been to make everything as good as it can be, but to work together as a complete package."

Ferrari's innovation is in process and mindset rather than in the technology itself. Since the Ford DFV engine was first raced in 1967 it shifted the dominant design of a Formula 1 car to the chassis, with the engine simply being bolted into the rear of the car (instead of its conventional position in the front). This approach enabled many teams to be Grand Prix winners and developed into a situation where engines were invariably "outsourced" from engine partners; and, even where the engines are "in-house", these can easily be made at a different site, perhaps in a different country, as is the case with Renault Fl.

However, when he joined Ferrari, Technical Director Ross Brawn wanted to maximize the unique characteristics of Ferrari —having their chassis and engine design in one location, in the small town of Maranello near Modena, in Northern Italy. He said: "When I left Benetton we were using a Renault engine but so were Williams and there was always a conflict about what sort of engine they wanted and what sort of engine we wanted.

"I really felt that if we could get into a situation where the engine was completely integrated into the car then that must be the best situation. So one of the things that was very important to myself and Rory [Chief Designer Rory Byrne] was to have some-one here who understood that and luckily Paolo Martinelli [former Ferrari Engine Director] very quickly appreciated our ideas and was completely receptive to the idea of a fully inte-grated engine as part of the car package."

One of the key ways in which they achieved this was by

maximizing the integration between the engine and the other systems of the car, as outlined by their former Engine Director Paolo Martinelli: "I think the integration of the work [between chassis and engine] has been a continuous process and is ongoing, so I think year by year, we are continuing in this direction. I think it was very important that there was trust from the top management and the direction given from the top, from Mr Montezemolo [President] and from Jean Todt.

"We do have some cross-functional areas; for example, electronics. We do not have electronics for the chassis and a separate group for the engine and gearbox, they cover the whole car and they help us to integrate the designs between chassis and engine. It is the same for metallurgy, they cover the whole car. Within each area we have experts who also work together, for example, in the area of CFD [Computational Fluid Dynamics] where someone in the chassis group may be working on design of the airbox and someone in the engine group is working on the flow of gases in the engine, they may often share ideas and calculations."

The tuned mass damper: Renault R25

In 2005 Renault introduced the concept of the "Tuned Mass Damper" (TMD) to Formula 1. Interestingly, applications of the concept of the mass damper have been around for some time and can be found in objects such as the domestic washing machine and the Citroen 2CV. As with most great innovations the concept of the mass damper is ingeniously simple. Dampers absorb vibration and by locating a tunable damper—the damper is adjustable to absorb particular frequencies—in the body of the car, Renault was able to both create a performance advantage and develop an innovation which was difficult for the competition to imitate quickly. The value of the tuned mass damper is that it cancels out the natural vibration of the tyres, which is transmitted through the chassis of the car, thereby increasing the adhesion of the tyres and reducing wear.

Renault had exclusive use of the TMD during 2005, as the other teams had not been able to quickly recognize the system or develop a response. However, into 2006 it was believed that a number of other teams had developed their own interpretations of the Renault system. Controversially, the FIA made the decision to ban mass dampers halfway through 2006, at a time when Renault was leading the World Championship. Despite the ban, Renault was able to secure both the Drivers' and Constructors' titles in that year. Interestingly, following the ban both McLaren and Ferrari developed damping systems which were smaller and integrated with the suspension systems, and therefore, less likely to fall into the category of a moveable aerodynamic device— which was the basis for the ban by the FIA.

The 'J' damper (apparently a random letter used by McLaren) uses a spinning mass inside the device to absorb the vibrations from the tyres, thereby producing a similar beneficial effect as the tuned mass damper.

The drive for innovation

So what makes successful innovation in Formula 1 possible? One of the most gifted and influential designers over the last forty years is John Barnard. Many of his ideas form the basis of the conventional Formula 1 car today. He summarized some of his ideas around being innovative in design: "If it's a really innovative project then that means that I can't be 100 per cent sure that it's going to work. So the one thing I always try to do when I'm either sitting down to design something, or I've got an idea in my head, is to have a back-up solution. I would generally try and think as I'm doing it 'Okay if it doesn't work what do I do?' so that I'm ready for that catastrophic event that there is something that we haven't foreseen that is so bad there is no other way to go but dump it.

"I tend to approach things like that, because you're not going to get too many chances to be very innovative in any business and you have to recognize that everything is going to have some sort of problem. That problem is either fixable in a fairly short space of time, hopefully, or it's so big that you've got to think of another direction. Effectively, don't get caught, be ready for the unimaginable, that your brainwave idea doesn't work."

Underlining Barnard's approach is the fact that truly innovative thinking has to be methodical, structured and above all have the total commitment of those behind it. He continued, "Give it a bit of time. Get to understand more about what you're trying to bring this innovative idea into, what sort of field you're coming into and understand more of the problems, and strength of character really. Most times I would say eight out of ten people will rubbish an innovative idea. Carbon monocoques and all the rest of it all got rubbished by people in the business, paddle shift, all the rest of it, all got rubbished. 'Why, what's the point? It'll hit something, be a cloud of black dust!' Be ready for that and don't let it put you off because it's very easy to be steered away from it by someone you think should know what they're talking about."

THE MECHANIC'S TALE by Steve Matchett, 2000

Twenty years have rocketed past since I began writing *Life in the Fast Lane*, *The Mechanic's Tale* and *The Chariot Makers*: three parts of a simple trilogy telling of the hidden lives of those who dwell inside the pit lane garages of Formula 1. When I began to write its first page my team's recent signing, a promising young German, Michael Schumacher, had yet to win his *first* world championship. A vibrant golden age of supreme high-technology and fierce on-track competition was about to unfold before us.

Naturally, on inking that first page, I had absolutely no idea of how my insights into the clandestine world of Grand Prix racing would be received; would racing enthusiasts want to hear of the behind-the-scenes happenings of the industry's mechanics, engineers and designers, the builders of the F1 cars, or does their fascination remain solely focused on the derring-do antics of the drivers of our 200-mph chariots?

Over these last two decades the response of the trilogy's readership has been thoroughly overwhelming; and the books continue to reach fresh eyes. For example, *The Mechanic's Tale*, a chapter of which you are about to read, has to date been reprinted fifteen times; a house record for a motor racing title by Orion, my London publishers.

It's no exaggeration to suggest that these hastily scribbled pages conspired to rearrange the order of my life; in a very tangible way they were directly responsible for redirecting me away from the Formula 1 pit lanes to the Formula 1 television commentary booths; my having enjoyed the past sixteen years broadcasting the sport I so dearly love to its continually growing and wonderfully passionate American audience.

Along this most unexpected route I have encountered countless fascinating characters, have forged many lasting friendships, and have been blessed far beyond my worth; the whole adventure, from then to now has been *truly* humbling.

It hardly needs mentioning that I'm forever indebted to these three odd little books; each one is precious to me. I'm always thrilled to hear of new readers being introduced to them, too, with old dog-eared paperbacks passing along from hand to hand. To help with this I'm presently editing an e-book version of my first title. *Life in the Fast Lane—The Twentieth Anniversary Special Edition* will be available online in 2015, allowing me to share my loyal old friend with a whole new audience.

CHAPTER SIX—1993

The long hours, the allnighters, the hectic workloads that we are often forced to endure: all terrible facets of a race mechanic's daily life; sometimes it feels as if we are little more than glorified slave labour. That said, I've always considered the pay we receive in return for all this hard work to be reasonably fair; at first glance, depending on one's own circumstances, it may look better than fair but, if our annual salary is divided by the number of our working hours, expressed as an hourly rate, well, it soon begins to look like pretty paltry remuneration.

You've probably noticed that I've been jotting down a record of my earnings, onwards from my first pay packet as an indentured apprentice, and I'll continue to do this, year by year. I have always kept letters which refer to salary, and I thought it would be interesting to include them in these pages, merely to add a little extra colour to any future study of a trained craftsman's career; in my case a span of twenty years.

Who knows, in a thousand years from now someone might be interested to learn more of exactly what a mechanic used to do, his skills and functions, and how much he was paid for using those skills. There again, more likely no one will care a fig for what I did. Times they are a changing: it won't be long before any need of mechanics will cease to be. I strongly suspect that within the space of another fifty or sixty years our skills will be redundant; like the longbow maker, the knight's page, the roof thatcher, the barrel cooper (and Klondike Pete's Golden Nuggets) we will have faded into extinction, an all but forgotten footnote of our world's history.

I really don't think I'm overstating the situation, take the following as an example: I'm writing this book, *The Mechanic's Tale*, on a notebook computer (I still want to call it a laptop), powered by a 200MHz Pentium processor, with "MMX technology" (which, apparently, makes the whole machine much more fun). This computer is under twelve months old, and it cost a whopping two thousand pounds; an awful lot of money but a necessary investment, one I hope to recoup with my advance from Weidenfeld & Nicolson, my London publishers. In 2001, the start of the next century, a mere three years away, this machine will be thoroughly out-dated, worthless beyond its valve as a whimsical curiosity. Compared to what will then be the latest machines, this one will feel akin to scrawling on a slate tablet with a wedge of chalk. Within three weeks of my buying this machine "voice recognition" software was on the market, and shortly after that I noticed a 266 MHz model was available. The other day a 333 MHz machine was advertised in *The Times* for less than *half* what I paid only a year before. If that bizarre trend continues, sometime in the next century you'll be able to buy a 1,000,000 MHz (should that be 1000 GHz?) machine for less than the price of a cup of tea; that can't be right, can it?

Here we are in the very closing stages of the second millennium, with significant advancements in microchip electronics screaming past us on a daily basis: The world is changing around us at light speed; I'm sure that many of our traditional trades, such as jetfighter pilot, NASA shuttlecraft astronaut, or Formula 1 Grand Prix mechanic will soon be but quaint memories of a bygone era.

Hopefully, the prehistoric urge to blast one another with air-to-air missiles will soon become rather passé anyway, and all cargo aircraft will be flown by remote; as will the vast majority of our future spacecraft, with much of the colonisation of other planets achieved by embryo transportation. As for race cars, well, they'll likely be subject to a total worldwide ban, either that or else constructed from a single piece of composite material, built with a magnetic repulsion system which keeps them hovering precisely 10 mm above an electronic track. Though, I suppose, the engineers of the future would still want to lower the front float-height by "half a mil" to improve that niggling bit of midcorner understeer. Certainly, many hundreds of highly skilled professions will simply vanish. Arguably, the only safe job at the moment is that of head chef of a successful tandoori restaurant; both now and in another two thousand years, hungry earthlings are always going to need a good naan bread and a chicken tikka.

Actually, talking of converting our annual salary into an hourly rate, which we were, a little earlier, I once tried to do exactly that while enduring a long session of airport terminal boredom. My colleagues and I were in Heathrow waiting for a flight to Japan, the penultimate round of the 1992 World Championship. With nothing but time on my hands I fished out my rather ancient calculator (held together with several strips of ageing Sellotape) and totted up the total number of hours we had worked, both preparing the cars at the factory and servicing them at the circuits throughout the course of a single year. One of our chaps kept a meticulous log of all hours worked, so gaining this information was amazingly easy. Armed with these figures I then divided the sum by our annual salary, reducing our income to an hourly rate. I was rather expecting to see a figure smaller than the minimal legal wage, and was all ready to start complaining that we would be financially better off working the same number of hours in some ghastly fast food outlet, McGrizzler's or Cheeseburger Yourself Solid.

Actually, I was getting rather excited at the prospect of unveiling the injustice of it all, imagining a series of spontaneous, highly colourful protest marches breaking out with massed ranks of mechanics from every team in the pitlane banded together in solidarity, all walking arm-in-arm, daubed placards held aloft; all singing loud and rousing choruses (with just one or two chaps slightly out of tune with the majority), the air thick with the scent of rebel spirit; everyone vibrant and alive, bonded with fine esprit, proudly marching towards somewhere (I wasn't quite sure where), carrying a petition inscribed with the quilled ink flourishes of a thousand signatories; our rolled parchment, tied with a simple bow of crimson ribbon, would be ceremoniously handed to someone (I wasn't quite sure who) and, after a brief consultation between the recipients, an announcement would be made by three dignitaries, all three dressed in morning suits and sporting new top hats, the men themselves standing atop a flight of grand steps fronting some large and important public building. "A better deal and improved conditions for all!" they would gleefully shout, "Victory for *every-one!*" And holding pairs of delicate grey gloves in their right hands they would wave with great merriment and enthusiasm to the singing crowds assembled below. The mood would be jolly and there would be dancing in the streets, and grown-ups at hastily arranged long trestle tables would serve jelly and ice-cream to children who didn't really want it but who had never been consulted on the matter. A national holiday would be declared: Mechanics Day, and the people of the world would all wear chrome lapel badges forged in the shape of two crossed spanners . . .

Well, all right, perhaps that imagined scene is all a little too fanciful but there must have been a possibility that a similar scenario—even if somewhat tamer—might have unfolded as a result of my calculations but, when I pressed the "equals" button, I was more than a little mystified to see the figure 71077345 faintly flickering on the little screen. What was I supposed to make of that? Was that a poor hourly rate or a surprisingly good one? The thing about calculators is they never give an explanation for their findings, they merely publish, wait a few minutes, and then switch themselves off.

I looked at the calculator and noticed that if you held the thing upside down the numbers spelled the words SHELL OIL. That didn't throw any real light on the hourly rate issue so I tried working the sum again but, sadly, the calculator's days were numbered: the yellowing Sellotape finally gave up and several bits of aged plastic fell to the floor. A pity really, I'd had it for years. Nevertheless, the day wasn't a complete disaster by any means, the exercise was worth doing if only to see the "Shell oil" thing.

3 February 1993

Dear Steve,

I am able to advise you that with effect from 1 January 1993 your salary will be increased to £23,474 per annum.

Thank you.

1993 was a busy year.

At the end of 1992 we left for Japan from the old factory, in Witney, and returned from Australia straight into our brand new factory in Enstone. As far as the race team were concerned the whole move was seamless; a very neat, well organized idea. While we were out of the country for the final two races, the factory staff had been flatout, moving all their machinery from one place to the other: mills, lathes, drills, folders, guillotines, metal, carbon, everything.

On my first day inside the Enstone factory two things really struck me: our new home seemed enormous and it was also very, *very* white. There were great expanses of white wall, white work surfaces, white cabinets, white window surrounds, white litter bins (each sporting a Benetton emblem). And we were all issued with brand new, bright white lab coats (each sporting a Benetton emblem). The only thing that had managed to escape this minimalist colour scheme was the cushioned flooring, which was grey. Presumably we couldn't order it in white.

Back in our old Witney factory the spares department, the stores, had been a small, cramped affair with John and Alec, our two stores-men, gradually becoming buried under ever increasing piles of race team supplies and car components. As our team continued to grow, every spare inch of shelf space was eventually filled. The only option at this point was to store things in boxes stacked all over the floor, but the demand for more and more storage space continued unabated. Boxes were stored on boxes, then more boxes would arrive. Slowly the stores became engulfed, turning the task of finding anything one actually wanted into a sort of unearthly mystic art, requiring John and Alec to engage in a form of telepathy in order to locate something buried deep within the dark reaches of the storage pit.

"John, have you got a Schumacher brake pedal pad, please?"

"Brake pedal pad . . . for Schumacher . . . now, let's see . . . they came in from Inspection about two days ago . . . then they went to be plated . . . they were collected from there last night . . . so . . . they should be . . . in that bag over there, on top of that box to your left, just to the right of the stuff that's just come back from the cleaners. Unless, that is, Alec's already put them upstairs, in which case they'll be . . ."

The storage capacity in the new factory is gigantic by comparison, with more useable floor space than the total combined area used by the machinists, the fabricators *and* the stores when we were all housed in Witney. The machine shop is now four times bigger; the same is true for the fabrication shop and the composite department. The security house and its automatic gate (painted bright white) extend an initial warm welcome to our guests; there is even a helipad for those harassed executives who simply do not have the time to travel by road. Regardless of that utterly unnecessary folly it must be said that our new factory is an overwhelming improvement on what went before.

In 1993 this new building was said to be the world's most advanced facility for Grand Prix race car production. Now (I'm writing this in 1998) with McLaren's proposed new Paragon building, and the recent completion of British American Racing's new facility in Brackley, it's possible this claim could be challenged but, nonetheless, the Benetton factory, nestled amid the green fields of the tiny village of Enstone, Oxfordshire, is still extremely impressive.

The nearest town to the new factory is Chipping Norton, sitting approximately halfway between Oxford and Stratford-upon-Avon. Chippy (as the locals call it) is an old market town on the edge of the Cotswolds with a relatively small community, somewhere in the region of five thousand people. Actually, the Chambers's encyclopaedia of 1895 quotes the populace as being 4,222; reassuring (and a little curious) to think that in over a hundred years its population hasn't altered. Back then, at the close of the nineteenth century, Chipping Norton was a thriving centre for commerce, with its looming Bliss Mill turning out some of the Empire's finest tweed. The sturdy cloth was woven in the mill, loaded aboard steam trains and dispatched straight from the town's station to the world's most discerning clientele. How things have changed: The mill burned down, the rail station has closed, and the international tweed industry is now all but dead.

Despite the loss of Bliss Mill, Chipping Norton soldiered on until the arrival of the next great employer: Parker Knoll, the furniture manufacture. I always remember the name Parker Knoll because of their incredibly chic lounge chair, the Parker Knoll Recliner, which was an absolute "exec" musthave in the mid sixties and early seventies. The company has since been through a terribly lean period but they survived and they're still producing their range of quality furniture in Chipping Norton, though I suspect that demand for their once magnificent recliner has since dwindled.

To survive into the late nineteen-nineties, the town has needed to again diversify, the main industries now appear to be antiques and tourism. Shops selling such rarities as "an interesting turn-of-the-century cake tin" or "an amusing eighteenth-century wooden spoon" seem to spring up all the time. In the middle of the town, standing proudly at the edge of the market place is an impressive five-storey building dedicated solely to the furthcrance of the ever expanding antiques trade; they even have a tearoom where after browsing their veritable cornucopia of interesting and amusing kitchen utensils one can be refreshed with a welcoming cuppa, a nice slice of Dundee cake and a good sit down.

Whatever the century, whatever the business, the townsfolk of Chippy keep plodding on, keeping the place alive. I like the place, I like it very much. So much so, in fact, that I decided to buy a house here. As a place to live the little town has everything I need: a supermarket, a superb tandoori restaurant, two wine shops and an excellent second-hand bookshop. And a wonderful pub. *The Chequers*, surely one of the best pubs in the world. *The Chequers* is a real-life *Cheers*, a pub where Josh, the landlord, seems to know everyone. Go inside his pub but once and, the next time, he'll remember your name. Standing well over six feet tall, with a great mass of greying beard, Josh was born to be a publican. He possesses that rare, enviable gift, to be able to chat with *everyone*, possessing sufficient knowledge of any given trade, pastime or subject of study to sustain an absorbing conversation; be it on sports, farming, antiques, sixteenth-century philosophy, even motor racing, Josh will effortlessly converse in your language.

I appreciate that a pub shouldn't influence one of life's major purchases but, I have to confess, the fact that Chippy had *The*

Chequers was the deciding factor in my decision to buy into the town. That and its convenience for my work with Benetton.

There had been one other change within the Benetton workplace since Tom Walkinshaw's involvement with our team: We were now supposed to start work at 8.30 as opposed to 9.00 in the morning. Apparently, all the TWR companies started work at this earlier time and the alteration to our working schedule would bring us inline with them. The concept seemed universally unpopular. It also struck me as being rather unfair, for as our official finishing time was to remain set at 5.30 p.m. (for whatever that's worth when employed by an F1 team), we would be working for an additional two-and-a-half hours every week, all supplied without any compensation. Although I thoroughly pondered the merits of this proposed scheme, I eventually decided not to get involved, feeling it really wasn't something to which I could lend my full support. Consequently, I continued to arrive at the new factory at my usual time of 8.45; the only difference being that I now arrived fifteen minutes *late* instead of fifteen minutes *early*. Swings and roundabouts.

The unveiling of Benetton's new facility marked a wider reaching change too, a geographical one this time. The hub of the Formula 1 industry used to revolve around Woking, in Surrey, to the west of London, with McLaren International, Tyrrell, Ferrari (their Godalming drawing office), Benetton and Onyx all, at one time or another, having been based in the south of England. The desire to group close to one's competitors is certainly not unique to motor racing, one merely has to look to California's Silicon Valley or London's Fleet Street for similar examples.

Grouping brings many benefits, not least being an increased potential of successfully poaching staff from one another without need of expensive relocation deals. Also, any gathering of large like-minded companies invites an arrival of useful smaller businesses: the suppliers. In the case of Formula 1's geography this means that a selection of small specialists are all close at hand: electronic harness builders, precision machinists and metal finishers, decal manufactures, even quality dry-cleaners.

Much has shifted throughout the last ten years: Onyx folded in 1990 (and I was so relieved not to have been offered the job with them that I interviewed for). Also, as we discussed earlier in these pages, both the Surrey-based Barnard/Ferrari and the Barnard/Benetton projects folded. Eddie Jordan went completely against the grain and opened his team's headquarters next door to the Silverstone circuit, in Northamptonshire, to the north of Oxford. At the close of 1998 Tyrrell ceased to exist, with their new incarnation, British American Racing, choosing to build their new factory in Brackley (very close to Jordan's factory). Stewart Grand Prix have established their base of operations in Milton Keynes, and TWR-Arrows have built their headquarters in Leafield, just outside Witney, also to the north of Oxford.

Unwittingly, through a series of incremental moves, the sport has gradually relocated itself further north, resulting in a complete shift to the industry's nucleus. McLaren are now the most southerly team, with Williams, TWRArrows, BAR, Jordan, and Stewart all encamped in a circle around Benetton. Following the close of their once-proud tweed mill, I suspect the good townsfolk of sleepy Chipping Norton had no idea that their little Cotswold market town would one day become the beating heart of the FIA's Formula 1 World Championship.

The Italian driver, Riccardo Patrese, joined us from Williams for 1993, replacing Martin Brundle to become Michael Schumacher's third Benetton teammate since 1991.

From a point of view of improving our new car, signing Patrese was a perfectly sound move. He had been with Williams throughout the entire evolution of their phenomenal FW14B (in fact, he had been with them for five years) and, as the B193 was to be an "active" car, having such an experienced driver onboard, one thoroughly proficient with the most successful active suspension system in the world, this could only be seen as being a major advantage. The wealth of experience that Riccardo had gained with Williams, all of which he now brought to Benetton, undoubtedly saved us untold weeks in terms of testing and developing our brand new active suspension and semi-automatic transmission.

Riccardo's contribution and worth to our 1993 campaign would prove invaluable.

That was the good news. On the downside, however, was the fact that he didn't really want to be with us. It was a sad state of affairs really. Riccardo had been happy at Williams, he had won races at Williams, he had been a runner-up in the Drivers' World Championship at Williams (a title he rather poignantly referred to as Vice Champion). He knew the team and the team knew him but, when Frank Williams announced that Alain Prost would be joining his team in 1993 (following a year's sabbatical after his dismissal from Ferrari in 1991) Riccardo assumed this would put him out of a drive. It was a fair assumption, too, Nigel Mansell had just stormed the 1992 Drivers' Championship, and the chances of Frank getting rid of Mansell in favour of Patrese must have seemed close to a billion-to-one against.

It had been announced early in 1992 that Brundle would not be driving for Benetton the following year and, knowing that Flavio Briatore had not yet made a decision on Martin's replacement, Riccardo moved quickly in order to consolidate his position: move fast and try to secure the next best seat (and if I'd been in Patrese's shoes I would have done exactly the same thing). With 240 Grand Prix starts, six wins, and 261 points to his credit, my boss Flavio and his new oppo, Walkinshaw, must have been thrilled at the prospect of having Riccardo in the team: here was a seasoned professional, a skilled and winning driver (one that just happened to be carrying all of that invaluable active suspension experience) asking to join the Benetton ranks: Why, yes, of course you may drive for us, Mr Patrese, please, sign here, and allow me to loan you my new Montblanc pen.

Then, shortly after Riccardo had duly signed for Benetton, Mansell's own negotiations with Williams collapsed and Nigel swiftly departed for America, taking his Drivers' Championship with him. Nigel's transatlantic move meant that Patrese's drive at Williams was still there for him, he would simply have partnered Alain Prost instead of Mansell. Riccardo, however, had already inked his Benetton deal . . . Damn! Damn! Damn!

Now, I don't know this for certain but I suspect, as far as Frank was concerned, Riccardo could have rescinded his newly signed

Benetton contract and stayed with Williams. In fact, Frank would likely have been more than happy for their working relationship to continue for another few years (securing Riccardo's knowledge of the Williams active project would have been justification enough for some sort of contract extension) but I don't imagine Benetton's management were remotely keen to release such a valuable new employee. All right, perhaps an agreement *could* have been reached whereby Patrese was reunited with Williams, but Riccardo decided to honour his Benetton agreement, a decision that speaks volumes of his character; he is a real gentleman, a man of ethics, one genuinely true to his word.

With Patrese prepared to honour his commitment to us, it was now up to Frank Williams to find a replacement for his team. Enter: Damon Hill, the Williams test driver, a man who knew their team very well; one who had also driven thousands of miles in a Williams active car. And who would now replace Damon Hill as the team's new test driver? Enter: David Coulthard, promoted to the heights of Formula 1 from the junior ranks of F3000. I must say, with names such as Patrese, Prost, Mansell, Hill and Coulthard, *all* mentioned in the previous few lines, *all* associated with Williams Grand Prix Engineering, Frank certainly knows how to attract quality drivers.

Personally, there was only one aspect of all of this that did not sit well with me, and that was the idea of Mansell turning his back on F1, deciding to drive in the US IndyCar series the year after clinching his Drivers' Championship. I didn't agree with his decision back in 1993, and I still don't agree with it today.

To walk away from the sport without defending his crown in the following year's competition seems somehow wrong to me. Perhaps it's only me that feels this way but I believe if you win something then you should compete against one's rivals to defend the right to retain it or, in the face of superior talent, compete and then graciously pass the prize to the new winner. We ought to pass the torch from generation to generation. This is not a situation unique to Mansell, of course, there are many similar examples: Prost did exactly the same at the close of 1993. I don't know, I just cannot agree with it.

I must stress, however, my criticism of Nigel's swift departure is in no way a criticism of the man himself. Mansell is a wonderful personality, a true showman, one full of real *character*, something that many of the new breed of drivers seem to lack. In his prime years he was the Bulldog Drummond of Formula 1, a driver capable of taking an average car and wringing every last drop of performance from it; either that or the car would simply explode under his strenuous demands. I remember watching him drive out of the Ferrari pits one year with such fury that he tore both drive-shafts clean out of the gearbox, *both of them*! You can't write that stuff.

Another of Mansell's great appeals is his ability to grow a most astonishingly splendid moustache, a real classic, huge and bushy and with a personality all its own. The media loved Mansell too, not merely because of his wonderful moustache (though this must have been an added attraction for the cartoonists) but mainly because he always had a story to tell about how tough his weekend had been, or how much his foot was hurting or that he had a headache like there was no tomorrow . . . Things like that just don't seem to happen anymore. The sport has changed. The drivers have changed. Anyway, there we are, at the end of the day Nigel Mansell *did* take his F1 crown and leave for the States, so, well, that was that, really. Onward and upward.

Annoyed at himself, perhaps, but determined to stick to his word, Riccardo Patrese joined Benetton, and he and I started working together as teammates in the winter of 1992. It was a long, hard winter too. For every one of us encamped in Enstone, producing our hightech B193 car resulted in a fiercely steep learning curve.

To try to make life a little simpler, the initial idea had been to retain last year's chassis, the B192, but with some severe upgrades, removing its old passive suspension and installing our own (Benetton designed) active suspension. The 1992 manual gear-box was also scrapped; for the new season we would build and develop our own semi-automatic transmission. Because of these major mechanical modifications, as well as additional changes to the aerodynamics, the chassis was deemed sufficiently altered from its original 1992 specification to warrant being renamed: the revamped B192 was now known as the B193.

With all of the upcoming test sessions we had planned, we would have more than enough work to keep all of us fully occupied throughout the winter. The FIA then stipulated the compulsory use of narrower tyres for the 1993 season. In testing them, however, regardless of the improved ride of our new active suspension, the 1992 chassis *hated* these new narrower tyres. The only genuine solution to our revamped car's handling problems would be to design and build a brand new chassis, one with its weight distribution and suspension geometry perfectly matched to the characteristics of the new, regulated tyres. But, at this late stage of the year, coupled with all the development work required of our new active suspension system *and* the brand new semi-auto transmission, to then embark on such a mammoth project as producing a completely new chassis would be utterly crazy, wouldn't it? Indeed, merely to suggest that such an undertaking was even *possible* would surely be seen as being no more than an act of sheer optimism, wouldn't it?

The decision was taken: we would build a new chassis.

The bonding glue had barely dried on our B193 chassis plates before they were pulled back off and changed again. This time the hybrid 1992 chassis was badged B193A; when it came into production the brand new 1993 chassis would be known as B193B.

It would have been impossible to have started the new racing season with this true 1993 car, there simply weren't enough hours in the day for that to happen, so the policy we had used in previous years was deployed once again: we would use the old car (B193A) for the opening two rounds, the intercontinental races of South Africa and Brazil; then the new car (B193B) would be introduced as its replacement once we had returned to Europe.

I mentioned earlier that when an active car is performing well it's a dream to work with. In the pit garages the mechanics are still having to adjust the wing angles, check for leaks, circuit damage, keep an eye on the brake wear etc. but the vast majority of suspension alterations are merely *software* changes, not *physical* changes:

connect car to computer, tap, tap, tap, unplug computer from car, send car back to the circuit. How long did that take, about six seconds? Okay, well it possibly took a little longer in reality, but not much. All well and good. But the amount of work it took Benetton to get the car to this stage of efficiency was mind-numbingly *immense*.

In Estoril, during the winter, where we were testing our new active suspension system shortly after its introduction, we worked on the car for three days and two nights (with no more than four hours sleep throughout the entire test) and it was only on the third day that we finally managed to coax the car into doing one, *extremely slow* lap past the pits. For two frustrating days and two freezing nights the car sat in the garage doing no more than being a complete stubborn bastard. We had the thing sitting on corner-weight scales (four electronic pads, one placed under each wheel, which are used to check for correct cross-weight distribution), with the hydraulic flushing rig maintaining fluid pressure to the car while Ross Brawn and his intrepid R&D engineers tried to coax the computerized electronics to communicate with some slight semblance of harmony with the system's high-pressure hydraulics. The car did not want to play nicely.

Software was installed into the car's onboard controllers and system hydraulic pressure was applied but, rather than react remotely logically to these basic commands, the car would instead violently shudder and shake, as if in the throes of some hyperactive, convulsive fit.

Hydraulic pressure was cut, data was scrutinized, teeth were thoughtfully sucked.

We tried again (several times, in fact), only to achieve slightly more disconcerting results. More sucking of teeth and then the Moog valves were blamed as being the most likely culprits. Moog valves are the electronically controlled flow valves which regulate fluid movement around the car's hydraulics. The mechanics dutifully changed the array of suspect Moog valves for another, identical set of valves. This did absolutely nothing whatsoever to improve the car's vile attitude towards us; if anything it only hated us even more.

* * *

Working on such complex highpressure hydraulics was a new experience for me, it was new to us all, and knowing that the system's operating pressure was in the region of 2,500 psi— certainly more than enough to blind a mechanic or cause an *instant* fireball in the event of a pressure rupture—I was always careful to ask the R&D engineers, the people who had designed and pioneered the concept, for advice of exactly when it was safe to disconnect a particular component. Once, one of these learned chaps told me that it was perfectly fine to disconnect a Moog valve from the car's fluid manifold: ". . . the system's dead, there's no pressure in the line now," he said. I looked at my colleague standing next to me, another mechanic, and he seemed distinctly unconvinced by this assurance. I asked this same R&D engineer again, just to confirm that we were all happy before I started unscrewing the Moog valve's four mounting bolts, as working on a high-pressure active suspension system is not unlikc defusing a bomb. This time, however, even the R&D engineer himself seemed unsure of the validity of what he had said: "Yes, it *should* be okay to remove it . . . but just be very, *very*, careful when you do so." And leaving us with that he swiftly disappeared to the very back of the garage. Very reassuring.

The problems with our "active" car continued: chins were stroked, heads were scratched and numerous telephone calls were exchanged between Portugal and England. New software was written and installed, more Moog valves were changed; then the original Moog valves were substituted for the ones we had just fitted. More software changes, followed by more coffee drinking. Moog valve changes were followed by ever more bouts of hydraulic flushing. Long and tedious hours were followed by longer, even more tedious hours.

For two full days the car never turned a wheel, we just continued to exchange one (possibly) faulty component for another (possibly) faulty component. We sipped coffee, we became increasingly tired, then exhausted, and then we became thoroughly sick to death of one another's company.

It must be remembered that these monumentally excessive hours are par for the course during the start of any Grand Prix year and, regardless of whether a team is designing an "active" car or not, it makes little difference to the inevitable massive workloads. Preseason testing has always been grim and the arrival of any new design is a potential allnighter in a box. A redesigned chassis, a new transmission, a different engine, new suspension, whatever, it makes little difference; until the freshly starched creases of the new design have been thoroughly ironed out, the chances of a good night's sleep between the months of January and April are extremely limited.

Exhaustion aside, what was really depressing to me, right throughout the period of our active suspension troubles was the fact that, as a mechanic, there was very little that I or any of the mechanics could do to speed things along. We couldn't *see* any fault. The parts that we removed from the car (because they were apparently faulty) looked as good as the parts we refitted. As mechanics, we are used to looking at any particular component and checking it for serviceability: too much play in a bearing or not enough lubrication on a gear; stretched threads, damaged wiring looms etc., etc., etc. Locate the fault, see or feel the problem, think of a solution, fix it and move on. But if the potential fault is caused by one wrong digit in the writing of a piece of software, well, what can we do but feel helpless, our role has gone.

I'm all for the advancement of Grand Prix technology but my formal training in *mechanics* never allowed for such speed of advancement in *electronics*—not even my BMW training included lessons in troubleshooting and correcting computer programming errors. Throughout the Estoril test the mechanics could (and did) ensure their machines were in pristine *mechanical* condition but until the electricians and the software specialists had worked their "virtual world" magic, no more could be done. The mechanics were completely side-lined. "Beautifully prepared car you've got there, mate, shame it doesn't actually work but, yeah, *very* nice, *very* shiny." All terribly frustrating.

That Estoril winter test was a career low point for me, not the lowest, not by any means, but I hold no fond memories of it whatsoever. Well, perhaps just one: another of our mechanics, Kenny Handkammer, created some wonderful imagery at some point during those three days when, tired and ashen faced, he described our "active" car as being like a moose: "Steve, you've got to help me," he said, "I can't go on anymore! I'm so knackered I can hardly walk, yet it feels like I'm having to carry a huge burden around with me, it's like I've got a massive moose flopped on my back, its front legs draped over my shoulders, its great head lolling to one side, and I'm having to lug it around with me, it's really starting to drag me down . . . The Benetton Active Moose, and I've got it! It's jumped on my back!" I *think* I know exactly what Kenny meant, too.

I thought I'd explain a little about semi-automatic transmissions at this point, plus I have to admit that I quite like them and, speaking as a mechanic, I think they are one of the most interesting things to come out of Formula One design.

At the close of the 1992 season, in Adelaide, I was standing below the rostrum with my teammates, watching Michael Schumacher and Martin Brundle receive their awards for finishing in second and third place. Both drivers looked delighted, of course, and when Martin saw us applauding he raised his right hand aloft, his thumb held upright in the classic gesture, thanking us for the reliability of his car. I remember noticing some strips of black tessatape which I'd used to bandage his right thumb and a couple of his fingers, prior to the race. I remember, too, how odd his bandaged hand looked as he stood there smiling from the podium, and thinking that millions of people around the world would be wondering what had happened to his hand, how had he injured himself.

This bandaging was something I used to do for him before the start of every race, it was protection against friction blisters, for the fingers and thumb of his right hand, caused by the knob of the car's gearlever. Regardless of the added protection of his leather racing gloves, the constant gear changes of the manual gearbox used to wound his fingers. As he stood on the podium, smiling and waving to the Australian crowd, I knew he must have been in considerable discomfort, yet his happy demeanour betrayed nothing of his pain, the adrenaline of competition keeping the soreness at bay, probably for another hour or two.

The Australian Grand Prix of 1992 was the last time Benetton Formula Ltd. used a manual transmission; those old romantic days of stick-shifting the gearbox have now long gone. Gone forever, I assume.

Back then, in the old days of yore before contemporary semi-automatic systems, the driver had to be brave enough to take his hand from the steering wheel in order to manually select the desired gear ratio. He was also expected to remember which gear he was presently using *and* be able to study the rev counter to coincide his next gear change with the engine's peak power output. Not only that, while downshifting through the gears he had to blip the throttle pedal (using the engine revs to synchronise the speeds of the gearbox shafts) *and* be careful not to change down prematurely, as this would overrev the engine, causing pistons and valves to over react and get all panicky: BOOM!

All of these gear changes were achieved via a hand-operated control knob connected to a titanium lever, connected to a joint, connected to a carbon rod, connected to another joint, connected to the selector mechanism, mounted on the gearbox casing. Not forgetting reverse gear, of course, which at one time was selected by pulling a cable mounted behind the driver's shoulder. It was all *wonderfully* quaint.

Well, "quaint" it may have been but it was also far from *efficient*. It was slow and cumbersome, certainly so when compared to the current state-of-the-art, computer-controlled, hydraulically operated systems: transmissions capable of changing gears within thirty milliseconds (0.03 seconds!) and doing so with *perfect* synchronised timing and with *zero* risk of over revving the engine. A brave new world.

In order to make these gear changes so rapid the selector mechanism utilizes the same high-pressure hydraulics used to control the active suspension. The higher the fluid pressure, the faster the fluid reacts to commands, and the faster the selector mechanism moves. Because of this extreme operating pressure (didn't we say the active pumps worked at 2,500 psi?) the hydraulic fluid travels around the car through equally high-pressure resistant hard-pipes and flexible hoses; anything less than jet aircraft specification equipment would likely result in a most dramatic fire.

Each time these hydraulic lines are disconnected from the car, to allow the engine or gearbox to be changed, the fluid in the system needs to be purged of air in exactly the same way as the active suspension components. In order to reduce the frequency of these necessary purges (remembering that track time is a highly precious commodity) the industry's top teams utilize a series of quick-release couplings known as a "dry-break". This zero-loss connector is a piece of pure aeronautical engineering magic: a push-and-click-on, click-and-pull-off device. When a dry-break is uncoupled from the car's hydraulic circuit it does so without losing a single drip of fluid, allowing the various mechanical components of the semi-auto transmission to be removed/ repaired/refitted without the need to purge the system. In terms of saved time, dry-breaks are worth their weight in gold; they are also reassuringly expensive.

With semi-auto transmissions (instead of the driver using a manually operated lever to change from gear to gear) the gears are "requested" via two micro-switches mounted on the car's steering wheel, each switch operated by paddle levers. Pulling the right lever changes *up* through the gears; pulling the left, changes *down*. I think it's interesting to note that all teams have their up-shifts and down-shifts co-ordinated like this. For some reason it would appear illogical—almost contrary to human conditioning—to reverse this, right-for-up, left-for-down design. Given the choice how would you prefer the paddle layout?

When the driver pulls the up-lever all manner of exciting things happen: the microswitch sends a signal to the computer processor located in the gearbox controller, asking if it would be possible for the selector mechanism to shift up a gear. The controller hears the micro-switch and ponders the situation. The controller checks with the potentiometer mounted on the rotary selector (which is similar to a motorcycle system) as to which gear the car is currently using. Next, the controller checks with the rear wheel-speed sensors to see how fast the car is travelling, and then checks the same information with the engine revs, courtesy of the crank-speed trigger. Now the controller looks at the ratio data of the desired gear—and because the ratios are constantly being altered, from circuit to circuit or (due to a change in wind direction) from session to session—this specific ratio information is pre-programmed into the controller every time the gearbox undergoes a ratio rebuild. The controller reads the information concerning the desired ratio, looks at the engine speed, looks at the wheel speed and does some (rather swift) calculations, working out the anticipated rpm that the engine will achieve should it allow the desired gear to be engaged. Armed with all of this information, the gearbox controller is now finally ready to go to work.

The controller tells the clutch-actuator to engage the clutch slave-cylinder and disconnect the transmission from the engine. The clutch-actuator dutifully obliges. Next the controller orders the actuator working the throttle butterflies to blip the engine rpm to a figure *slightly* higher than it anticipates the engine to be running at once the new gear is engaged. The throttle-actuator also does as it is told. Now, to bring the rpm to its correct figure, the gearbox controller asks the engine management controller if it could sparkcut the ignition system until the desired figure is reached. The engine management controller checks the operating condition of its engine and, on finding that all is well, it willingly agrees to assist the gearbox controller, and the ignition system is momentarily deprived of its voltage. So far so good. Now the gearbox controller orders the actuator operating the gear-selector mechanism to pull the original gear out of mesh, to then rotate the selector mechanism and to engage the newly desired gear. This is all duly done. The gearbox controller seems pleased with its work but it still has a final check to see that everything is as it should be, that nothing in the chain of events has become lost or confused with what has been asked of it. On finding only good news the gearbox controller then orders the clutch-actuator to reengage the transmission with the engine. And *voilà*: one gear change satisfactorily completed, all accomplished in less than a third of the time it takes to click one's fingers. Oh, and just to make life as easy as possible for our derring-do driver, the gear-box controller even instructs the dashboard display to illuminate a number from one to six, indicating which gear has been selected. And all of that for one single gear change. A Monaco race distance requires 4,300 of 'em.

*   *   *

Anyway, after that little digression, back to the teething troubles of our B193A chassis. Finally, after several weeks of tediously long and laborious tests, the car eventually began to calm down and even to respond favourably to treatment. More filtration and purification of its software, more delicate alterations to its hydraulics, and the 1993 Benetton active suspension car gradually matured, learning and growing through the season to become a very quick and nimble Grand Prix race car.

Throughout its relatively brief life we never once stopped refining the B193B; for the final two races of the year there was a big push from the design and engineering offices to equip the car with a four-wheel-steer system. I have no idea why this was considered necessary, certainly the car had no need of this additional complexity, it was as though some within our team felt ego-driven to produce the industry's most hydraulically sophisticated F1 Grand Prix car before the end of the year, at which point this exciting technology would be banned by the FIA.

The whole thing struck me as rather bizarre. We had thoroughly tested the four-wheel-steer system in England before we shipped our cars to Japan but, when I talked with our two race drivers regarding the potential benefits of this rear steering, both Patrese and Schumacher could find no advantage whatsoever in using it. Riccardo told me that it certainly made the car feel *different* (which it would, of course, the rear trackrods were moving . . .) but it was also a feeling he didn't like. And I didn't notice any improvement in lap times either; as far as I could tell the pre-Japan/Australia test proved the system was not needed.

True, the rear steer system was designed to be able to be switched on and off, so the drivers could elect to use it or leave it switched off (in which case the hydraulically operated rear steering-rack would merely work as a solid link) but it still had fluid lines connected to it, which meant that it remained a reliability risk.

The theory is this: any individual component of a race car carries a *potential* risk of failure. Consequently, the philosophy for theoretical ultimate reliability is not to have *any* components: if it ain't there it can't go wrong. Now, obviously, it's not possible to do away with *every* component . . . but the next best "real world" philosophy is to remove anything that doesn't *need* to be on the car. The fewer components we have, the less chance we have of encountering unreliability. Simple, common sense.

In the event of a hydraulic fluid loss, the rear steering rack was designed to failsafe in the straight-ahead position (again, replicating a solid link)—so there was no danger to the safety aspect of the car—but a loss of hydraulic fluid would, nonetheless, put the car's *entire* hydraulics out of operation (active suspension, clutch, gearbox, throttle butterflies included) effectively forcing the car into a race retirement. Personally, I was dead set against the four-wheel-steering system being incorporated into our race cars but, in the end, the decision was made to include it in the Japan/Australia race car build specification. In my opinion that decision was misguided.

I seem to remember that for these last two races of the year (perhaps in recognition of this additional steering system) the cars were redesignated: B193C but I don't think anybody took any notice of this new suffix. *Autocourse*, the Formula 1 end-of-season review continued to refer to the Japan and Australia specification Benetton cars as B193B, so either nobody told the book's editors of the name change or they simply chose to ignore it.

In the practice sessions (both at Suzuka and Adelaide) the drivers conducted a number of practice laps using the four-wheel-steering but both of them elected *not* to have the system activated during the races. And, as things transpired, we did indeed fail to get either of our two cars to the finish line of these final Grands Prix but, I must stress, the rear steering system performed trouble-free throughout both the Japanese and Australian race weekends.

> *Beware the Jubjub bird, and shun,*
> *The frumious Bandersnatch!*

One of the most memorable events of 1993 occurred in Spa. Unlike 1992 we didn't win the Belgian race, although we did finish second and sixth, which was a far better result than we could ever have hoped for, considering the quite atrocious start we made.

Half an hour prior to the start, as the cars left the pits to form up on the grid, both drivers were asked to try their cars' startcontrol at some point on the circuit, a systems check before arriving at their respective grid positions. Michael confirmed that all was well, but Riccardo said it didn't feel right to him, the engine seemed to bog-down; he couldn't define what the fault was exactly, but he advised against its use. It would be safer, he suggested, to switch the device off, to disarm the system and instead make a conventional start, selecting the gears manually. The automatic startcontrol software is a system allowing the clutch and the gears to be worked without driver input, matching gear changes with peak engine revs for optimum performance; the system is interlinked with the car's traction-control software, so avoiding any wheel spin as the car shot from the grid.

While we, the car's three mechanics fussed over its engine, gearbox and suspension, making our final pre-race checks that all was well, the team's engineers checked the streams of data stored in the onboard controllers. Finding no obvious fault with their startcontrol electronics the decision was taken to overrule Riccardo's advice and to use the system as planned . . .

If you ever have the chance to watch the start of that race on video you'll notice that as the rest of the field scream off towards *La Source*, Spa's tight first corner, both Benettons make the most *appalling* starts, chugging and spluttering off the grid as our rivals dash and weave their way around our two mobile chicanes. We had qualified third and sixth on the grid but by the first lap we had dropped to ninth and seventeenth. In light of our thoroughly awful start, getting Michael Schumacher to the second step of the podium was a pretty impressive result.

All of that opening lap caper aside, for me it was neither the start of the race nor its result which made the 1993 Spa Grand Prix so memorable, it was one of our pitstops, Patrese's stop on lap seventeen.

I'd now changed my role within the structure of our pitstop crew. In 1991 our then chief mechanic, Nigel Stepney, had set me to work on the left-rear corner of the car, but since the start of the 1992 season I had been in charge of operating the rear jack. I've already noted in these pages that I don't particularly enjoy this facet of the job, but working the pitstops is an intrinsic role of a race mechanic's lot, you either accept that or you leave; there seems no third alternative.

There is, however, one small advantage in using the rear jack as opposed to working in any other position on the team: to allow the car to drive into position I have to stand well clear as it speeds down the pitlane towards the other mechanics. Conversely, the distinct *disadvantage* of controlling the *front* jack is that the driver is aiming his searing hot, 200 mph race car directly at you.

The split second the car shoots past, I'd dash in from behind, throwing the jack forward and engaging its liftbar onto the two hooks, known as jacking plates, located at the base of the rear wing. For increased speed and efficiency, it is best to have this operation completed while the car is *still moving forward*; the airguns should also be in position, attached to the wheel nuts at this stage. As soon as the car comes to a complete stop I would heave down on the jack, lifting the rear and allowing the mechanics to finish removing the wheels. The same thing is happening at the front of the car, too, of course. I would then watch and wait until the gunmen had raised their arms, signifying that they had finished fitting the new wheels and tyres, and then glance towards Kenny Handkammer, operating the front jack, just to confirm that we were both satisfied that all was well. The car would then be dropped back to the ground.

Throughout all of this the mechanic with the BRAKES ON board—known as the lollipop—is in total control of the car; it is a position of great accountability, and one usually undertaken by the team's chief mechanic. When (and only when) he lifts that board away will the driver be allowed to speed off. The man holding the lollipop has sole responsibility to ensure that all work has finished on the car, that his mechanics are clear of danger, and to check that the pitlane is clear of traffic before giving control of the car back to the driver.

This season was to be the last before the reintroduction of mid-race refuelling, something that would inevitably slow pitstops by over seven seconds. This saga would all unfold the following season, however, in 1994, but for the present, in Spa, in 1993, with no refuelling permitted all we had to worry about was successfully changing the four wheels. The pit-stop performances of all teams were getting faster and faster throughout the season but, when Riccardo Patrese pitted for fresh tyres, we managed to service the car in a scintillating blur of just 3.2 seconds. A world record, the fastest pitstop in F1 history. I wonder for how many seasons it will stand?

For me, that world record pit-stop time set in Spa was the season's highpoint, while that horrid winter test of discontent in Estoril was, undoubtedly, the season's *extreme* low point. That said, I also remember Estoril for a much happier reason that year: Benetton and Michael Schumacher won the 1993 Portuguese Grand Prix; Michael's second career victory, and my fifth since joining the team.

Because of a glitch with the software settings on the race car's active suspension system—a problem caused by "finger trouble" when the electrician was keying information into the car's onboard microprocessors—Michael drove the spare car in that Portuguese race. The major significance of this race victory, achieved in the team's spare car, went largely unnoticed by the majority of the world.

One of our test team mechanics, Carlos Nunes, had temporarily joined the race team, just for this one race, standing in for Paul Howard, the car's "number one" mechanic who had been unable to attend. Carlos himself had worked in Formula 1 for many years, mainly with the struggling March team, and had only joined Benetton after March had finally ceased their F1 programme.

Here was Carlos, a Portuguese (an F1 veteran who had never had the good fortune to win a single race) running the Benetton spare car, in Portugal. Now, a Portuguese F1 mechanic is an extreme rarity, and the local media were already demanding interviews and photographs of him working in the Benetton garage. After all, merely for Carlos to be a member of a Grand Prix racing team had made him a minor celebrity. But now, for Carlos to be working with a *leading* Grand Prix team, a team that was a genuine contender for the race win, now that had made him *extremely* newsworthy. Immediately before the start of the grand prix, however, when word broke that Michael

Schumacher was to be racing the *spare car*—with Carlos Nunes himself in charge of that car—well, the local media became *very* excited. This story was now a guaranteed sports page headline. But then, when Michael achieved the near unthinkable, when the Benetton ace crossed the line to *win* the race, *driving the spare car*, then the Portuguese media went into a complete jubilant frenzy! "Hold the *front* page! Carlos has won! Carlos Nunes has won the Grand Prix!"

Local Boy Makes Good

Schumacher & Nunes Storm Estoril

King Carlos—The Hero of the Portuguese Grand Prix

Joan Villadelprat, our team manager, and the very top of the Benetton management were equally thrilled. Before long Carlos found himself promoted to the position of chief mechanic, and soon after that he was promoted again, this time to become the team's test team manager.

Carlos is a steady, easy-going chap, a man who enjoys the responsibility of his position, and one who will always look after the people serving under him. When Joan asks him to work he willingly does so, but when there is a chance to go home, to spend time with his wife, his family, his prize-winning Koi carp, then he will ensure that everyone gets home as soon as possible. Simple man-management skills and common courtesy, some people have them, some people will never have them. Carlos does and I'm sure he'll continue to prosper; he enjoys the work and that makes an enormous difference to his outlook, to his staff's perception of him *and* Benetton's genuine appreciation of him.

The Estoril Grand Prix weekend also saw Alain Prost secure his fourth Drivers' World Championship, making him the second most successful driver in F1 history, just one championship behind Juan Manuel Fangio. It's reasonable to assume that if Prost had stayed with Williams for 1994 he would have equalled Fangio's record. In my opinion he should have remained there, too, if only to defend his world title; certainly it would have made for an exciting season . . . but he didn't.

Perhaps Nigel Mansell quit at the end of 1992 because he didn't want to race against Alain Prost in equal equipment? Perhaps Prost quit at the end of 1993 because he didn't want to face Ayrton Senna on those same terms? I don't know their reasons for walking away but I do know the sport was poorer for it happening.

Once again Benetton decided not to retain Michael Schumacher's teammate for the following year: Riccardo Patrese's contract was not renewed. After a total of 256 Grand Prix starts (another world record which looks set to stand for many years) my thoroughly likable driver for this season decided to call it a day and he, too, retired from Formula 1. Like the years I spent working with Nelson Piquet and Martin Brundle before him, I really enjoyed my time working with Riccardo. I'd miss him.

Once again Benetton didn't win any championships, with seventy-two points we finished the year in third place. Williams continued to dominate the sport, winning ten races and yet another Constructors' Championship. With 168 points in the bag they scored exactly double that of second-placed McLaren; all bar eleven of those eighty-four points were scored by Ayrton Senna. Talking of Ayrton Senna, 1993 was his last season with McLaren International, he managed to win five races throughout the course of the year, and, in Adelaide, on 7 November, he won his last race for them. It was a fine finish to the season and a most admirable way for two such great names to part company.

And, as fate would have it, that win in Australia was also destined to be Ayrton's final Formula 1 victory: in Imola, just six months later, he would be killed.

What more is there to say?

THE PIRANHA CLUB BY TIM COLLINGS, 2004

WELCOME TO THE PIRANHA CLUB

Money, money, money

No one outside those who were present knows what happened at the Villa d'Este on the night of 5 September 1991. Few of those who were there, as the drama surrounding a complex "transfer deal", which saw Michael Schumacher leave Jordan for Benetton and Roberto Moreno move in the opposite direction, albeit reluctantly, and despite winning a court action against the Anglo-Italian team, have been prepared to reveal the whole truth. But it was there, in the imperial foyer of one of the most spectacular and famous hotels in northern Italy, on the edge of Lake Como, that their clandestine wheeling and dealing earned Formula One's business operations a new and sinister-sounding soubriquet: the Piranha Club.

That night, said many long-standing observers and critics of the sport, something about Formula One changed forever. The once-Corinthian world of Grand Prix racing, a world of pleasure and danger fuelled by adventure and funded by private means, crossed an invisible and barely-perceptible line, shifting irretrievably from sport to business, from competitive rivalry to something more Machiavellian. It was the end of the last vestiges of the golden age, of gladiators in cockpits, of mortality, parties and fun, and the start of a new era of money, politics and intrigue.

Schumacher had driven in just one Grand Prix meeting for the Jordan team after being hired as an emergency replacement for Bertrand Gachot, a temperamental "pilote", who was sent to jail for assaulting a taxi driver by using a gas canister in London after a traffic altercation. Moreno was a journeyman, but, he had a contract; a source of scant protection, it turned out, once the lawyers and power-brokers set to work on that hot night late in the Italian summer. It was a night that left many unwritten codes trampled in the dust and dew of a magnificent September morning and a night that confirmed Formula One's future was in the hands of ruthless businessmen and ambitious competitors.

The story of Schumacher's switch from green to blue, from the blarney boys to the knitwear family, stunned the paddock the following day at Monza where Milan's beautiful people, including many models, several racing enthusiasts and soccer stars, had gathered with the regular circus of drivers, owners, engineers, mechanics, reporters, friends, cooks and motor home staff to prepare for the annual Italian Grand Prix at the Autodromo Nazionale. The late Ayrton Senna, heading towards his third Drivers' World Championship in four years, was so aggrieved he spoke out with passion at the way in which his friend Moreno, a fellow Brazilian, was treated—like a pawn in a grand masters' game of high-octane chess with stakes to match.

But throughout the night before, as persuasive talking and financial offers were filling the hours downstairs, Senna slept above it all in his bedroom. Amid a flow of smoked sandwiches and fine wine, men like Flavio Briatore, Tom Walkinshaw, Eddie Jordan and Bernie Ecclestone, not to mention the Schumacher entourage including lawyers from International Management Group (IMG), argued with other legal representatives and agents, including those working for Moreno, as Schumacher, then just a new tyro from Germany, barely ready to accept his mantle as the "wunderkind", still growing accustomed to the place settings of his cutlery, and its purpose, at formal dinners, was moved from one team to another, after just one weekend as a Grand Prix driver.

It was cloak and dagger stuff, sparked by his sensational form at the previous meeting at the Spa-Francorchamps circuit, in Belgium, amplified by two applications for court injunctions, one in London and one in Milan, accompanied by many dire threats and wild reports of extraordinary behaviour and completed by a late-night agreement, manipulated by many deft hands, that ensured the outcome was that which the sport's controlling interests had sought. The Piranha Club was born and won the day.

Despite every effort they could muster for the fight to retain Schumacher, who had been nothing less than sensational in practice and qualifying at Spa-Francorchamps, Eddie Jordan and his men were unable to hang on to the best young driver of his generation. Their application for an injunction, in London, preventing Schumacher from driving for Benetton, had been heard in the High Court the previous day, but had failed. Yet they fought on, supporting Moreno in his case to retain his seat by taking protective action in court in Milan. Moreno won. He had, it seemed, and thanks to the Italian legal system, kept his seat.

Benetton were thus informed that they had to keep him in their car, alongside his compatriot and friend Nelson Piquet, for the Italian Grand Prix and, furthermore, were warned that they faced possible expulsion from the Championship if they failed to do so. Indeed, on Thursday afternoon, Italian bailiffs visited their garage in order that the cars were impounded and law, and with it Moreno's right to a chassis, was upheld. But, by power of attrition, the weight of pressure and the lure of money, Moreno was weakened and overcome and Jordan's scrap for Schumacher was defeated. "To begin with, when Bernie advised me to see Flavio, we were not moving," said Jordan. "At that stage our position was strengthened by Moreno. He said he was driving and Piquet said he was driving. He had a legitimate case. We were happy to discuss Schumacher driving for us for the rest of the season. But then the whole thing became unreasonable and fell apart. We wouldn't move. We are not a team which is going to be bought. We were not going to prostitute ourselves on an issue we felt so strongly about. In the end, though, we went . . ."

Benetton, with Briatore, Ecclestone and Walkinshaw working for a mutual interest, negotiated their way into creating the vacancy they needed in order to hire Schumacher. By three in the morning, after a week of legal attrition, days of tension and hours of argument, Jordan's powers were so weakened that he was unable to stop a deal taking place which put Schumacher into a Benetton and ensured that the Italians, with all their multi-various international marketing interests, were able to exploit the young German's burgeoning popularity in his fatherland. The television broadcast rights for coverage of Formula One in central Europe also leapt in value overnight. Jordan was outmanoeuvred, Moreno was forced to leave the job he loved, his contract was dissolved, Schumacher had switched to a richer and better-supported team, Benetton had a driver for the future and an entry into the German market and Ecclestone had seen his television rights' values soar as another part of his masterplan was executed. And the Schumacher era had begun.

It was a decisive move in the creation of the so-called Piranha Club's reputation, one act in a drama that was part of a far grander play that many felt was a scheme to develop Formula One into a global television spectacular. Significantly, for Jordan, a man of natural intelligence, warmth and sharp instincts, it was also a bitter learning process as he discovered that, in his first season as a team owner, it was wise to move with alacrity and, to borrow a phrase from professional football, get your retaliation in first for the sake of self-protection. "There was a lot of shouting and roaring," said Jordan. "There was a problem between the lawyers representing Moreno, and Benetton. Something silly was said that was interpreted by the lawyer as something quite unethical . . . The meeting went on for hours."

The old days of friendly deals done with a handshake, honoured by gentlemen and confirmed with a happy toast over dinner in the evening were gone. Long gone. This was the night when the lawyers and moneymen took utter control of the sport. The rich men won. The men with helicopters to convey them above the morning mist and dew the following day to Monza were the winners. Those, like Eddie Jordan and his right-hand man Ian Phillips, who stood and watched as the helicopters clattered and rose into the sky, who were to drive down the valley in a hired road car, in their case a Fiat 126, were the losers. It was a lesson they heeded. Like the rest, they knew the Piranha Club would be running the sport from that morning onwards. And they knew, as in the old phrase, if you can't beat 'em, join 'em.

Eddie Jordan is one of the most infectious and likeable characters in Formula One, if not British sport. He has played drums on stage all over the world, he has been featured in the popular British television show *This is Your Life* and he has dodged, ducked and dived his way through a chequered career from early days as a bank employee in Dublin to becoming one of the richest men in the land. He has done it all with a smile and a joke, never losing his sense of humour or his ability to talk himself in and out of trouble with consummate skill.

As a Formula One team owner, he has stamped his brand on the paddock, the sport and the public imagination. As a survivor, he has demonstrated that anyone can do it—climb, that is to say, from the army of those who live outside the paddock fences, peering in, to become one of the chosen and lucky few whose professional lives carry them from continent to continent as part of the Formula One road show. Jordan rubs shoulders with rock stars, actors, film stars, sportsmen, models and statesmen in his life as a popular public figure, but he has never forgotten what happened to him at the Villa d'Este. Nor has he ever lost touch with his humble beginnings.

Born in Dublin, on 30 March 1948, Jordan was expected, by his family, to become a dentist when, in his late teens, he was enjoying life in his native city. His wit and gift of the gab made him a popular man and the promised comforts of life as a professional fixer of teeth seemed, in some ways, suited to his generous nature. But it was not to be. Jordan wanted something else out of life. So, in 1967, instead of sticking to the path towards oral hygiene, he took a different turn and steered instead towards the acquisition of knowledge and wealth. He joined the Bank of Ireland as a clerk. It was a wise decision, an inspired change of direction and it left him well prepared for what lay ahead later in his life as a mogul of the pit-lane and paddock.

For three years, he worked happily in the bank, gaining more and more insights into how the world of money operates and how businesses succeed. He was a quick learner, especially where making a deal was concerned. It is, of course, a family trait that he had inherited and one that he has passed on to his own children. Self-confidence, after all, has never been lacking in such talkative Irishmen.

In 1970, however, luck played a part in the life of young Jordan. Due to a strike in the capital, he was posted from Dublin to Jersey, the famous tax-haven island off the French coast. There he came into contact with motor racing for the first time and it left him stunned as he watched racers in karts fighting it out on the Bouley Bay hillclimb. Inspired, Jordan returned to Dublin and bought himself a kart. He taught himself, went on to win the Irish championship and, in 1974, moved into single-seater racing in Formula Ford 1600 cars. That proved satisfactory enough and, with the racing bug biting hard, he decided to uproot from Ireland and move to Britain to compete.

Alas, a major accident at the end of the year left him with two broken legs. He was ruled out of the 1976 season, a blow of almost equal severity to his personal morale, as the fractured legs were to him physically. But he remained determined to race and, in 1977, after much wheeling and dealing, he was in Formula Atlantic racing. The next year, he won the Irish title, a success that prompted him to move on and partner Stefan Johansson in the British Formula Three championship, racing under a banner of Team Ireland. Johansson, of Sweden, was to remain close to Jordan, as, significantly, did most of his early colleagues, drivers and friends. In 1979, Jordan tried racing a Formula Two car, at Donington, and also tested a McLaren Formula One car, recognizing, by the time he reached this pinnacle in his personal racing career, that he was unlikely to go much further. At the end of that year, he created Eddie Jordan Racing.

This was the start of the real thing for this ebullient, funny Irishman. His true personality, his business acumen and his sheer enthusiasm for life in general, and racing in particular, came to the fore in a blend of qualities perfectly suited to the world of motor sport. By 1983, he had established himself and he ran Martin Brundle in the British Formula Three championship. Brundle finished second to Ayrton Senna. Competitive and driven by his ambitions, Jordan pressed on. Finally, in 1987, he won the British Formula Three title, with Johnny Herbert driving. Successes followed in the European Formula 3000 series the next year with Herbert and Martin Donnelly paving the way before Jean Alesi lifted the title in 1989 en route to his own long and spectacular Formula One career. The Frenchman's arrival at the top was the final stimulus to Jordan himself and, in 1990, to a chorus of warnings and some disapproval from those who regarded themselves as more prudent, he established Jordan Grand Prix.

The headquarters were a lock-up unit on the industrial estate at Silverstone, within the circuit and close to the track. It was a small beginning. When his first car was built and ready for a public testing, Jordan asked the former British Grand Prix winner John Watson, an Ulsterman, to perform the ceremonies on a cold, damp and foggy day at the old Northamptonshire circuit. A small group of reporters were invited to this inaugural occasion and were served tea, instant coffee in polystyrene cups and sandwiches to mark the day. The car ran in black, unpainted and unsponsored. It was a far cry from the glittering splendour of a hot day by the Mediterranean at the Monaco Grand Prix, but for Jordan it was a first important step towards fulfilling his dream and the relative modesty of his hospitality was in stark contrast to the warmth and sincerity of his welcome as he revealed himself to be an enthusiast fired by a perfectly simple ambition.

In 1991, he entered the FIA Formula One World Championship with Gachot and Andrea de Cesaris as his drivers, one the Luxembourg-born son of a Brussels-based Eurocrat, the other a veteran Italian with a reputation for unpredictability and extraordinary eyes. At the same time, in Formula 3000, he ran Damon Hill, having had Eddie Irvine, Heinz-Harald Frentzen and Emanuele Naspetti the previous year. The names are important because, as time went by, they would recur in the Jordan story, usually to argue the toss after being traded from one team to another, but more often to chew the fat and enjoy his friendship. Jordan, of course, was not only their employer, but for many of them also their personal manager and so, by virtue of controlling their careers, he was able to dictate what they did, who they raced for and how much of a cut he could take from their contracts. His penchant for management spread far and wide, his flair for making money equally. Like Flavio Briatore, at Benetton, Jordan could see that running a Formula One team was not the only avenue of business open to him through which he could make a decent profit in the early 1990s. Formula One, after all, as he was discovering, was a very, very expensive business and it needed very expansive funding.

But Jordan was no quitter. Nor did he lack personal resources in terms of energy, determination and imagination. As a younger man, as he would frequently recount, he had sold woollen carpets and out-of-date smoked salmon in the Dandelion Market in Dublin. He could sell snow to an Eskimo. When his son, Zac, was at preparatory school, several years ago, Jordan received a telephone call from the headmaster. He was told that boys had been caught smoking. Jordan, a fitness enthusiast, was very upset. But he was less distraught when he heard what the headmaster had to say. "Mr Jordan," he told him. "Zac was not smoking. He was selling Benson and Hedges cigarettes to the boys at £3.50 a packet. Do you have any idea where he may have got them?" As his team was sponsored by Benson and Hedges, at the time, it was not difficult to solve the puzzle. He was delighted. Zac, like his father, was a natural wheeler-dealer, a chip off the old block. And his improvisational business skills were a further inspiration to his father to survive as he began hacking his way through the intensifying financial jungles towards secure membership of the "Piranha Club".

In 1991, his first full year as a team owner in the competitive and ruthless world of Formula One, Jordan experienced a wide range of blows and surprises. His car, the nimble, pretty, sleek, green Cosworth-powered and Seven-Up-sponsored Jordan 191 was the envy of many in the pit lane and frequently out-qualified and out-performed Cosworth's chosen factory-supplied team, Benetton. His drivers, also, performed with an élan unexpected by many seasoned observers. It seemed they reacted well to Jordan's unique form of man-management, a personal approach that included generous rations of fun and games mixed among the serious preparations. The team often, for example, wore shorts—including Eddie Jordan himself—and treated its sponsors with a refreshing sense of candour and mischief. Jokes and japes, a shortage of cash and a team spirit unmatched by any other team in the championship were the strongest features as Jordan's happy voyage survived one financial knock after another and just about stayed afloat. When they finished fourth and fifth, thus scoring their first points, in the Canadian Grand Prix in June, the team celebrated raucously and Jordan let it slip that, using a financial arrangement he had discovered in Belgium, he had "bet' against his own success, through a form of insurance, in order to enjoy a payout which would allow him to cover the bonuses for success. The financial house in Brussels was not delighted to read this news. It had agreed a confidentiality clause with Jordan, in advance.

By the time of the late days of the English summer, when Gachot was arrested for assaulting a London taxi driver by using a gas canister as his personal weapon, the bills were piling up and the budget was exhausted. Financially, Jordan knew his team was on its knees. Debts were all around him, including a mounting and worrying amount owed to Cosworth for their engines. If ever he needed to wheel and deal, to duck and to dive, to borrow and beg and find some cash, this was the time. Cosworth were pressing for their money and threatening legal action. Without Cosworth, Jordan could not continue and, even if they managed to struggle through, he needed engines for 1992. It was a grim scenario, made worse by Gachot's idiotic arrest and the attendant bad publicity.

The likelihood of this Irishman, with his hard-working, fun-loving team of men in green, developing an outfit to challenge the likes of McLaren-Honda, for whom Ayrton Senna was on his way to the Drivers' Title, or Williams-Renault, for whom Nigel Mansell and Riccardo Patrese were heading towards second and third places, seemed remote. But, remarkably, he did. And he survived not only the threats of his creditors and the wiles of the Piranha Club to do so.

"When we arrived at Spa, the cars were sealed away," he explained, referring to the Belgian Grand Prix at Francorchamps in 1991, when Schumacher made his debut. "The local bailiff had taken action. Well, we had just had a torrid two weeks. We had Gachot go to jail. We needed to find another driver and when Michael Schumacher and Ian [Phillips] were trying to get the contract finished, on the Thursday night, we went to the track to find that local Belgian bailiffs had locked up the trucks and the cars. So, we couldn't get them out. They claimed we owed some money . . ."

Jordan's memory of the sequence of events, he admitted, was hazy. But, according to Phillips, money was owed to a Belgian driver, Philippe Adams, and his claim resulted in the local courts approving an order to prevent the team from using their cars or trucks for fear that they would be leaving the country without clearing the debt. Action was taken by the bailiffs, said Phillips, on Saturday afternoon after qualifying. "We found a way of clearing it up, so we could race," he added.

By this time, both Jordan and Phillips were deeply entangled in a desperate bid to keep the team going, to hang on to the dazzling talent of Schumacher—who qualified seventh at Spa-Francorchamps before destroying his clutch at the start of the race—and to try and find an engine supplier for 1992. It was a juggling act, in terms of energy and resources available, every day for each one of them.

It had started when Schumacher tested for the first time, immediately before the Belgian race, and it continued through the following two weeks to the Italian weekend at Monza. Phillips was busy trying to tie down Schumacher while Jordan was seeking engines and Cosworth, fed up with waiting for their bills to be paid, issued a winding-up order on Jordan Grand Prix. To make matters worse, Schumacher's management team were suspicious of Jordan's plans for the future. They were being fed information, suggesting the team was in a perilous state, which was designed, deliberately, to persuade them to avoid signing any long-term contract.

Schumacher, happy enough to have stayed in a breezy holiday chalet that he shared with his new team chiefs in Belgium, was also reminded that other teams had higher standards of hotel accommodation on offer. Few leading outfits, after all, would have considered asking one of their Grand Prix drivers to share quarters, including the bathroom, with commercial or marketing staff. To his credit, Schumacher was never troubled by this and always said, later, how much he had enjoyed his brief time with the Jordan outfit.

But Schumacher, like Moreno, was a mere pawn in the moves made by the major players at the time. His value was not only in his talent, but also in the fact he was German. Formula One needed a good German driver, but in a successful team with real prospects for the following season, not a team facing a wind-ing-up order from its engine suppliers with little prospect for real sustained success. Furthermore, Benetton, owned by the Italian fashion chain, were looking for ways of using Formula One for marketing their products, their image and the services of the other companies in their group. It was obvious, in hindsight, that Benetton had wanted Schumacher immediately they saw how fast he was at Spa-Francorchamps on the opening day of practice in the valleys of the Ardennes, if not long before.

Walkinshaw, who was then director of engineering at Benetton, where he worked in alliance with Briatore, knew all about Schumacher. Having run Jaguar in the group C world sports-car championship, he had enjoyed close quarters experience of him. He had seen him racing for Mercedes-Benz in the same series and knew he was an exceptional talent. He was, also, in close contact with Jochen Neerpasch, an agent working on behalf of Mercedes-Benz, as their competitions chief, as well as for International Management Group (IMG), and their drivers, particularly Schumacher, who remained under contract to Mercedes-Benz, where he had been developed as a junior driver.

"Neerpasch had been canvassing several teams to see if they were interested in Michael and, of course, I had been impressed with his performances in the Mercedes sports-car and I was keeping an eye on him," he admitted. "I said I was interested in running him (at Benetton), but understood he had a prior commitment to Jordan. I was only interested if the Mercedes lawyers could give me clear legal advice that he was not committed elsewhere. I would want my head examined if I didn't go after a driver of his obvious calibre."

It was the beginning of the end for Jordan's interest in retaining Schumacher long-term. Walkinshaw, a resolute Scot with a reputation for winning at all costs, was not a man who wasted words. He was well-connected within the paddock and throughout the global automotive industry, entirely aware of who was approaching who for drivers and engines for the following year and, together with Briatore, he was building a team to win at Benetton. He had recognized in Schumacher a driver he needed to help achieve his ambitions. Like Jordan, he was a proven entrepreneur, but a man also of widespread success throughout many parts of the motor industry. He had enjoyed success in many automotive programmes ranging from touring cars to sports cars to road cars. But success in Formula One had proved elusive until he caught sight of the potential of Michael Schumacher. With him, in a car packaged to provide the power and the performance he needed, Walkinshaw knew that he could direct Benetton to the World Championship.

He saw all he needed of Schumacher, to confirm he could perform in Formula One, in just two days at the Belgian Grand Prix. Walkinshaw was unhappy with Moreno and he wanted to replace him. "I wanted someone in the second car who would liven things up a bit," he explained. With that in mind, he contacted Neerpasch, regarded by most of the players in this particular game as the man in control of Schumacher's future, alongside his manager Willi Weber, of course. When he telephoned Neerpasch, he interrupted the German while he was soaking in his bath. Walkinshaw laughed. But they talked.

"Jochen agreed to come over to London to discuss the possibility of Michael coming to Benetton for the next race at Monza," he said. "They came over with their lawyers and a contract and we examined it and satisfied ourselves that indeed, there was no agreement with Jordan. The only thing was that there was a block in the contract that if Mercedes-Benz came back into F1 racing, in the following three years, then they would have the right to take him back. I thought that it was worth taking that risk and we signed him for Monza."

Briatore, like Walkinshaw, saw Schumacher as a key figure in Benetton's future. "The first time I saw Michael was in Spa," he said. "The first time I spoke to him was in London, a few days afterwards, in my house. He was with me, and Neerpasch, and we discussed Michael's position with Jordan and it was confirmed to me that there was no contract. It was only a one-race deal. I told Michael I was ready to put him in the car and that we didn't need any money from personal sponsorship and that was how we did the deal. It was very important for him to get into the car immediately. For me, that was no problem. I felt it was important to find someone for the future of the team. We all felt very strongly that he was our driver for the future.

"My only worry was that he would not have enough laps, but we did not expect instant miracles. We knew too that we did not have the best car, but we were working on it. We wanted a driver for the future. I knew my situation at the time was not a winning one, but we were looking ahead and Michael was the first really important step. Nelson [Piquet] and [Roberto] Moreno were our other drivers at the time. I had already decided what to do and two weeks before I met with Michael, I had told Moreno what I wanted to do in the future; that he would not be driving with us any more and that it was not our intention to renew his contract."

Briatore and Walkinshaw's matter-of-fact approach to the signing of Schumacher made it seem obvious that Jordan had no right to believe he had any hold on him. In this age of harsh legal realities, any notion that Jordan, having been the man to have run him first in a Formula One car, would have a moral first call on his services was treated as an irrelevance. A naïve irrelevance. "I had no problem with Eddie applying to the court [for an injunction to stop Schumacher signing for Benetton]," said Walkinshaw. "He tried on several counts and the judge dismissed every one of them. I think there's been a lot of nonsense on this. The fact is that Schumacher, for whatever reason, had no contract with Jordan. He was a free agent. How anyone can allow a talent like that to be walking around the paddock, I don't know. That's their business. When we were informed of that we went about the proper way of securing him."

To a tough, experienced operator like Walkinshaw, this manoeuvring was all in a day's work. To the struggling new boys at Jordan, fighting on all fronts, every hour was a scrap for survival. Having "found" Schumacher, thanks to some help here and there, and having survived the legal and financial difficulties of racing at Spa-Francorchamps, they had become engulfed in other problems which distracted them from the more urgent need to sign Schumacher promptly, beyond agreeing a deal with heads of agreement in principle, and prevent him being charmed away in the night. And these distractions, together with the information exchanges that are all part and parcel of the dealers' market that is the paddock in Formula One, undermined them.

"I had a call from Weber on Friday afternoon," explained Phillips. "Michael was due to come for a seat fitting on the Monday [eight days after the Belgian race and three before the opening of the Italian Grand Prix]. Then, instead of Michael, Neerpasch and Julian Jakobi turned up! We wouldn't let Jakobi near the place. He was working for IMG then."

As Phillips explained the story, running back through his mind to recall clearly these days now a distant decade away, Jordan interrupted to add his own recollections. "I refused to let him in. I'd just come back, from Japan, with a deal for Yamaha engines and, at that stage, we were skint. Absolutely skint. And . . ."

"Neerpasch presented a contract to us," Phillips continued.

"Yes," interjected Jordan. "And Flavio [Briatore] was restricting us a little, too, because we were a bit embarrassing to Flavio's Benettons with our Jordan, with a customer engine, so he made sure that we got stung . . . He made sure we wouldn't get the [Ford Cosworth] engines the following year. But, it was worse than that. Cosworth, at that time, put a winding-up order on Jordan, so I had to run like hell to find an engine!

"I had heard—from Herbie [Blash] and Bernie [Ecclestone]— that Walkinshaw, and a few others, were desperate to get this Yamaha engine and that if I didn't move quickly, I would never get it. So, I'm gone to Japan and Ian [Phillips] is looking after things—and all this is flying around the place when I come back on the Monday to find out that Weber had actually had the good manners to ring us and to say, 'Look, please, be careful.'

"And, by that stage, Bernie and Flavio and Tom had, I believe, concocted how they were going to get a German into Formula One their way. There had not been one for such a long time.

"It was a very big market that was sports-car orientated and this was their big opportunity. And, it was no secret. Bernie will tell you. He helped to cement the deal between Flavio and Benetton to get Michael Schumacher to join them. At that stage I could, possibly, understand because the chances of Jordan surviving were, in our view, quite good, but in anyone else's view, quite limited.

"He was probably aware that there was a winding-up order about. Flavio would have told him about that and about Cosworth's move, their petition. But we had a Yamaha engine, at that stage, untried and untested . . ."

The significance of the engines was one of the critical issues in the story. Phillips explained. "There was something else," he said.

"Neerpasch, on Sunday morning in Spa, came to us and he said we had to run Ford the following year. We were all sworn to secrecy on Yamaha and so we said 'Oh, yeah, we will be . . .' But what we didn't know was that Tom Walkinshaw, who was the engineering director at Benetton, was so involved by then. He had run the Jaguar sports-cars against the Mercedes. He was the only other person in the world who knew how good, really, Michael was . . .

"And, at this time, we didn't know of Bernie's involvement in all of this. After Spa, Bernie disappeared to Sardinia, I think, where he was buying some land. He knew we had got the Yamaha deal because he had been instrumental . . ."

Ecclestone has helped Jordan by using old contacts and advising each to help the other to make the deal work.

Phillips continued. "We had been lying about Yamaha, because we had to! So they [Neerpasch and Weber] were using the big Ford angle and when they went to see Bernie and said 'What shall we do?', he told them Jordan had got a deal with Yamaha . . . He squared that away. No problem."

Jordan joined in again, adding his own personal recollection. "So, then the fun starts! We said '**** this!' I knew it was happening. I had spies at Benetton. And they must have had spies at Jordan. So everyone seemed to know exactly what was going on. So, it went straight down to court injunctions. They fought it like crazy. We went in London. We knew what was happening. So, we collared Moreno. We paid Moreno to use the same injunction to save his arse, but in the Italian courts. He won. We lost."

As Jordan and Phillips recalled it, Moreno had the opportunity to stop the entire affair from happening at all. Having won his case, in Monza, to save his seat with Benetton, he was, it was claimed, offered half a million dollars to walk away. The court action in Italy was concluded on Thursday, a day of simmering drama and comings-and-goings in the Monza paddock as the affair unfolded gradually.

"At that stage, they were reconciled to losing him [Schumacher]," said Jordan. "The financial penalty for losing Piquet was so huge that they would have had to retain him. And they had to give a seat to Moreno. And, then, we were all invited up to the Villa d'Este."

"No, we were summoned. Summoned," said Phillips.

"I'd never seen so many lawyers coming in and out like that," said Jordan.

"And we didn't find our own hotel until Friday night," said Phillips.

"Ian and I shared a single bed," Jordan added. "And Flavio paid for it! He sent sandwiches to the room and they were mouldy old things! And he—Ian—he is not easy to sleep with!" At this point Jordan added descriptive reasons, relating to odours and habits, for this assertion. They are too colourful for publication. "To do this in a small, single bed in one of the staff out-houses at somewhere like the Villa d'Este is not very pleasant. On top of losing your driver."

Phillips continued his version of the story. "We got summoned to the Villa d'Este. When I got to the circuit, I saw Bernie and I told him about the courts and what had been happening. He had got the plans for his house in Sardinia out and he said, 'This looks great,' and he pretended he knew nothing about it. Then, he said, 'Leave it to me. I'll sort it out.' Then, at seven o'clock that night, he said 'Get up to the Villa d'Este' and we drove up to Lake Como in our little Fiat 126 rent-a-car . . . Bernie was there and we saw Michael."

Phillips was struck by the change in circumstances since the last race. "In Spa," he said, "we stayed in a holiday camp. Five-pounds-a-night. I shared a bathroom with Michael and Willi. It was a bit like a dormitory! And the next time I see Michael, he is in the Villa D'Este! And they are all sitting there, having dinner, and we were outside . . . and Bernie said, 'We might as well let you go, there's no money, Moreno's not going to get it and he's not going to get paid' and it went on and on and on.

"Eventually, Bernie came out. I'll never forget it. Eddie, Bernie and I, we went round the side of the staircase and there was this lovely glass cabinet and I turned to Bernie. We had told Moreno, 'Don't settle, don't settle' and Bernie told us he had to settle. He said there was no money. I told him that 'If nobody sits in that car by the time official qualifying comes on Saturday, at one o'clock, then Benetton will be excluded from the Championship. They cannot take *force majeure*. Moreno is going to sit it out. He isn't going to give in . . .' And, half an hour later, he came back and said they had offered Moreno half a million dollars."

"No," interjected Jordan. "They offered it to us first. And we said we wouldn't take it."

"And we told Moreno not to take it," said Phillips.

"And, to be fair, the lawyer believed he wouldn't take it," Jordan continued. "And Moreno needed to keep his seat, at the time, and we needed money. We weren't prepared to take the money. And we were so strong about not taking the money they then offered it to Moreno, who did take it."

Phillips said: "Yeah, at two-thirty in the morning, Moreno took the money . . ."

"That is correct," said Jordan. "And that is how the Piranha Club was formed. Word just got around the paddock."

So, it unfolded, Moreno accepted an offer to give up his rightful seat with Benetton for Schumacher. Jordan, without a second driver, then accepted an offer of $125,000 from Moreno to fill the vacancy.

"And we did the deal that night," Phillips said. "We had no idea where we were staying. Tom [Walkinshaw] had been wandering around like the wine waiter, spilling it on the carpet! And we still did not know where we were staying. No idea."

"And," said Jordan, "you just couldn't believe how many lawyers there were present there that night. There were waves of them coming in to that place. There were Benetton lawyers, Flav's people and Bernie's people. But they were all so shocked that the family was upset that, in Milan, we had won a case against Benetton!"

"Everyone had been looking for Moreno the whole of Thursday," Phillips added. "But he was there, actually, all the time, holed up in our motor home. Nobody, but us, knew where he was. And, of course, the other thing was that we had Marlboro begging us to take [Alessandro] Zanardi. But Walkinshaw had told us, 'You can't have Zanardi because I've got him under an option.' He had given Zanardi a seat fitting . . ."

"Yes, but at three in the morning, we couldn't afford to take another risk," said Jordan.

"So, when we left at six-thirty, having had just three hours' sleep in that single bed, we had to go and tell Trevor [Foster, team manager of Jordan] what was happening and to get Moreno fixed up for the car and all the rest of it," said Phillips. "Then, we pull up at this set of traffic lights, on the way to Monza, and there's the old Marlboro man there. And he pulled up and said, 'Why don't you take Zanardi?'.

"We told him he had an option with Walkinshaw. And he said, 'No, he doesn't' and we go on then, the first time we meet Zanardi, he turns up at our motor home at half-past-eight in tears. 'I really want to drive this car,' he said. But it was too late then. Walkinshaw had told us what he did to stop us having Zanardi."

"And then Ron came out with this thing about the Piranha Club, just as we were walking across the paddock at Monza," recalled Jordan again. "He said, 'Welcome to the Piranha Club.' I can remember the moment."

Phillips went on. "Everybody was agog. Bernie pretended he knew nothing about it, when he had orchestrated the whole thing. He was able to say we had Yamaha."

And Ecclestone, it appeared, also had the vision to see how valuable Schumacher was to become to the show as a whole. His talent, in a successful team, would open up a vast new German market. According to one German source, Schumacher, at the time, was under contract to RTL, the big German broadcaster, whose arrangement with Formula One dictated that if a German driver entered the sport it would double the fee to be paid and if a second entered, it would increase again.

In short, Ecclestone knew that his television business would profit from the arrival of Schumacher and Germany in the Formula One paddock. Of course, the same would apply, in principle, for Jordan as for Benetton, but the latter were considered a more powerful and successful operation with a stronger brand image.

This manoeuvring, furthermore, affected the careers of other men. The Swedish driver, Stefan Johansson, for example, was in Monza during the Italian Grand Prix build-up waiting for a call from Jordan. As a Marlboro-backed driver, he hoped that the fact Jordan were a Marlboro-sponsored team would help assist him in taking the chance to slip into the vacant seat created by Schumacher's departure to Benetton. He never received the call.

Zanardi, the runaway leader of the International F3000

championship, was also waiting for a chance to step into a Formula One car and had pinned his hopes on Jordan. But, on Thursday afternoon, Benetton having decided to dismiss Moreno anyway, they gave Zanardi a seat-fitting. Clearly, Zanardi was being lined up as first reserve if the team failed to land Schumacher cleanly.

Having received £150,000 from Mercedes-Benz to give Schumacher a drive in Belgium, knowing that they were prepared to pay £3 million the next year to keep him there, Jordan had built up high hopes for the future. In theory, they had secured, he thought, a driver of huge potential and the backing of a dependable supporter. But even before this arrangement had begun to form in his mind, he confronted problems. Prompted by various sources, Neerpasch had started to "move the goalposts" and make life difficult, as Phillips explained.

"When Neerpasch turned up at the factory on the Monday before Monza, he presented a contract and he said 'These are the conditions for Michael' and he had, basically, taken the space on the whole car! So, we worked all night on doing a version of the contract that we COULD sign. And then, in the morning, we were sitting around waiting for them to turn up when a fax came through from the lawyers with two lines, from Michael. 'Dear Eddie, I am sorry I am unable to take up your offer of a drive. Yours sincerely, Michael.' And, at that very time, Michael was at Benetton, having a seat-fitting. We knew because someone phoned somebody at Benetton."

By the end of Tuesday, on the basis that they could prove he was having a seat-fitting at Benetton, after having agreed heads of agreement with Jordan, Jordan had requested his lawyers to apply for an injunction against Schumacher driving for Benetton. The case was heard in London the following day, Wednesday, twenty-four hours before they were due to arrive in Monza.

"We had heads of agreement, which under normal law would have been good enough," said Jordan. "But what happened was that the judge said he could not restrict someone's right to work unless we had signed that other contract, which was unsignable. 'On the basis that that contract is flawed, I cannot give you the injunction to stop him driving for another team until this contract is settled and signed,' the judge said. But they had no intention of signing that contract. So, it was a mistake I will always remember. But, in the heat of the moment, no one was at fault."

Neerpasch recalled it all, at the time, with some detachment. "Michael Schumcher signed an agreement with Eddie Jordan on the Thursday before Spa," he said. "It was an agreement to talk about an agreement. What he signed was a letter of intent. Eddie Jordan offered him the drive, but he needed money. Mercedes-Benz agreed that money and asked for sponsors' space. We talked with Eddie about the rest of the season and also the future, but only on the condition that our money would guarantee a certain space on the car.

"I went to see Eddie Jordan on the Monday and we could not agree. A number of teams were interested in Michael and we went to Benetton. They wanted him and it was a straightforward deal. He is paid as a driver. I think the Jordan was a very good car for this year. There was no need to change. Michael wanted to stay with Jordan, but Eddie would not agree with our requirements for sponsor space and he wasn't prepared to discuss our contract. He wanted Michael to sign before Monza. Michael was still a Mercedes-Benz driver, but we released him for F1. At the end of it all, I think it is very important for Germany to have a competitive driver in F1."

Neerpasch said also that Mercedes-Benz wanted to have Schumacher back for their planned re-entry to Formula One in 1993 with Sauber. "It would have been against our strategy to release him," he said. "We built up the drivers and we wanted them for our own team. It was sensational, of course, for the F1 people to see Michael at Spa. Everyone was interested in him. We discussed a lot of things. In the end though, Eddie said he was using Yamaha engines and we discussed this and we decided that Yamaha were not going to be reliable or victorious. We wanted Michael to have a season in F1 and we wanted to finance the season for him. We wanted him ready for the following year, for the new Mercedes team for 1993.

"At that time, we saw 1992 as a preparation year. We wanted both Michael and Karl [Wendlinger] to get F1 experience. We discussed this with Michael and he decided to stay at Jordan. They were a very nice team and made him feel welcome. He did not want to change. But we discussed it for a long time and, finally, decided to change."

The whole episode, known at the time as the "Schumacher Affair", threw a shaft of stark light into one of the dark corners of Formula One and contributed towards the creation of the Contract Recognitions Board, an organization to formalize the legitimacy of drivers' contracts with teams. But that came later, a long time after Moreno was manipulated and paid off amid rumours of threats and retribution among the various lawyers and participants at the Villa d'Este.

Senna, the great and glorious champion of the time, was disgusted at the events. At the hotel, when he learned of what had happened, he told Briatore what he felt of his treatment of his compatriot and friend Moreno. On Saturday afternoon, he claimed pole position. At the subsequent news conference, looking serious and thoughtful, he took the opportunity to give vent to his feelings. He spoke with controlled passion.

"It's difficult to comment in a clean and fair manner, without knowing all the clauses in the contracts, but, as you know, even the best contract in the world, drawn up by the best lawyers, is only worth anything if both sides are really working for it . . . What has happened was not correct. It's always the people in the top teams who are written about the most. So, I feel, that unless one of us speaks about it, something like this just goes by and people get away with it.

"Moreno is a good driver. He's dedicated. He's a professional. And he had a contract for the whole season. But people just push others who are maybe not in a strong position and they threaten, use their apparently strong position to get a driver to change his mind and to accept things. As a principle, I don't think this was a good move. There were commercial interests involved and future prospects which made certain people do these things . . ."

Moreno, the fall guy, the man pushed out of Benetton to make way for the new wunderkind's arrival, deserved a word. Upset, threatened, cajoled to take money and give up the seat he had proved was his through legal action, he turned to God for support.

"I think everyone in the paddock was surprised by what happened," he said. "Unluckily, for me, I was alone at the time I was told, as my wife and my daughter were in Brazil. I didn't even tell my wife afterwards because I thought it would hurt her too much. It is very difficult for a person to go through all that alone. Fortunately, I am a religious person. I believe in God. I opened the Bible and I asked God to put me in the right direction and it opened at a good page. That gave me self-confidence and I kept myself together. It was very stressful.

"My only problem was that I caught a virus before the race in France, in July. I went to the doctor and took some penicillin. It upset my stomach and I was not recovered for the race. I think it is the only problem I had this year. I took legal action because I just wanted to defend myself. I had to defend my rights on the contract I had for this year. On the Thursday night, I slept for only two hours and had my seat fitting at seven in the morning. I got in the car, I concentrated and I tried to do my best . . ."

TALES FROM THE TOOLBOX
by Michael Oliver, 2013

Tales from the Toolbox Introduction

With today's curfews preventing Formula 1 mechanics from routinely working late on their cars, things are very different to what used to go on thirty or forty years ago. In that era, teams did a lot of their development at race meetings (they did not have simulators or CFD to help with this) and so often a huge list of modifications would be given to the mechanics for them to make overnight. The "all-nighter'—working throughout the night on the cars—was therefore commonplace up and down the pit lane in those days.

About *Tales from the Toolbox*

Tales from the Toolbox is a unique collection of behind-the-scenes stories and anecdotes from the world of Formula 1, focusing on the era between 1960 and 1980. It is available from all good book retailers in print and in e-book form; 40 per cent of all royalties from the sale of the book go to the Grand Prix Mechanics Charitable Trust.

The Grand Prix Mechanics Charitable Trust, which was founded in 1987 by Chairman Sir Jackie Stewart, exists to help out current and former Formula 1 mechanics and their families financially in times of need. For more information about its work, go to www.gpmechanicstrust.com

About Michael Oliver

Michael Oliver is a journalist and author who attended his first motor race when he was just a few weeks old. He learned to count from the numbers on the side of racing cars and continues to enjoy going to motor sport events to this day. His books include definitive works on the Lotus 49 and Lotus 72, as well as *Tales from the Toolbox* and he continues to specialize in researching the Lotus marque's history.

Formula One team owners had two distinct philosophies concerning the practice of their mechanics working through the night. For some, like Team Lotus boss Colin Chapman, it was *de rigeur*, an accepted part of the job that gave extra hours in which to prepare your cars or make last-minute modifications that other teams wouldn't be able to do. For others, all-nighters were to be avoided at all costs, since they resulted in tired mechanics, and tired mechanics made mistakes. There were a number of reasons why the all-nighter was so prevalent in Grand Prix racing during the 1950s, sixties and seventies. A major factor was the small number of people employed by the big teams, as Tony Robinson points out. "All-nighters were fairly regular; that was accepted. Things were different then. Part of the reason was that, with Stirling and his 250F, and even in the 1960s when we were racing the BRP cars, it was invariably just two mechanics per car." Another reason, as Robinson remembered, was the timing of practice sessions at Grand Prix meetings. "We used to have practice and qualifying somewhere round about five or six o'clock in the evening on a Saturday. I often thought it would have been a good idea to have the last practice at midday on a Saturday, to give a full twenty-four hours to prepare the car."

Denis Daviss agrees that practice session timing had a lot to do with it; this was a particular issue at Monaco due to the fact that the circuit was on public roads, so sessions were held first thing in the morning and last thing at night. "I used to be quite good at all-nighters; I'd had a lot of practice. Most race team guys worked all night because there was always a late practice. At a place like Monaco, they had an early morning practice, which was fine, then whatever went wrong we spent all day fixing it; evening practice it was the same sort of thing, doing whatever needed doing, meaning working over the next night ready for the next morning. Quite often we didn't even bother to go to the hotel, even though it was just across the road. Fortunately, the garages had a shower . . ."

Tony Cleverley notes that another factor contributing to all-nighters in the early years was the poor reliability and build quality of the cars. "The all-nighters came up because cars just weren't made as well as they were later on. The difference between the later cars and an early Cooper or Lotus was huge. The engines wouldn't last as long, Colotti gearboxes would let you down, and you'd be doing an all-nighter to put a clutch and a gearbox in a car. The reliability of the cars was pretty poor, that's where the difference comes in. We used to do a lot of all-nighters, I must admit. You wouldn't think about going to Monte Carlo without knowing you were going to do at least two all-nighters, whatever happened."

Alan Challis recalls that, at BRM, all-nighters were also an integral part of life as a Grand Prix mechanic. "At the time, it was part of the job. You accepted that and, as long as you got the cars finished in the morning, you'd done your job, what you were paid for. The first race I ever did, at Reims, I ended up fast asleep. The first time we'd ever filled this bloody car up was on the Saturday night, and we were about six or seven gallons short to be able to get through the race. Fortunately, we had fabricators with us, and we set to and made loads of little aluminium fuel tanks, which we fitted all around the car, and Bendix fuel pumps to pump it from the little tin tanks into the main tank. Apparently, I was sitting at the front of the car waiting for this bloke to finish off this thing and ended up fast asleep, laid against the radiator. Behind me they were banging a big sheet of aluminium with a hammer but I still didn't wake up until they gave me a kick. I think that weekend we had gone to work Thursday morning and the first time we actually stopped was when we loaded the truck after the race."

Before the Formula 1 teams became sufficiently organized to present themselves as a coherent package, race promoters used the carrot of start money or appearance money to attract entries for their races. This also reflected the fact that there were literally dozens of relatively minor non-Championship races littering the calendar, and sometimes there would be a clash between two far-apart events or races on the same weekend.

Many car owners and drivers were able to survive an entire season by moving between races and living off the start money they earned, regardless of whether they managed to finish the races and win some prize money. However, that system meant that it was absolutely essential, at all costs, to make the start of a race. A crash in practice, particularly at some of the more far-flung circuits such as Enna or Syracuse in Sicily—which took the best part of a week to reach—could be financially disastrous, and mechanics would go to any lengths to get a "start money special" to the grid.

Dick Scammell remembers one such occasion. "You always had to start the car. At Porto [the 1960 Portuguese Grand Prix], with Jim Clark driving, he went off and wrecked the car and we just set to and welded it up. The only trouble was, the only welding wire we could find was in a fence outside. It probably wasn't the best . . . I can remember Colin saying 'I think we'll just do a couple of laps shall we, gently?' In the end Jimmy finished third. In those days, of course, mechanics got 10 per cent of the prize money shared out between them. It didn't come to much but it all helped."

Ralph Gilbert, who worked as a mechanic for Bob Gerard Racing during the 1950s and sixties, recalls another similar occasion. "Our most memorable all-nighter was in Sicily. One year we went down with two cars for John Rhodes and John Taylor, a Cooper-Climax and a Cooper-Ford with a 1500-cc twin-cam engine. Taylor drove the "twink" and Rhodes was in the Climax-powered car. Rhodes went off and bent it severely. We spent all night repairing it and got it all going. I remember Bob came up to Rhodes on the grid after the warm-up lap and said 'How is it?' John looked at him and replied, 'Well, it oversteers a bit.' 'That's normal isn't it?' said Bob, and John said 'Yes, but it oversteers on the straights.' The car looked more like a forty-five-gallon drum. I think it did ten laps before it ran out of water. The chassis tubes used to hold the oil and water at that time, and as soon as it got hot, the tubes expanded and opened up the cracks."

Stan Collier also recalls a desperate patching up operation in the same country. "When we were down in Sicily with Formula 2, our cars were going all right but John Campbell-Jones had a shunt and broke his rear upright. Nobody had a spare rear upright with them in those days, we didn't carry them, so he patched it all together, strapped it up with bits of metal and bolts and all that, and put it back on the car, just to get started in the race. He had to do that or else he wouldn't have got any money. To go all the way to Sicily and get nothing, was a long way to go."

Another venue where it was vital to make sure the car started was Watkins Glen, home of the US Grand Prix, and the most lucrative race of the year. Although they didn't actually pay start money, in effect they did because prize money was awarded even to last place. It was therefore essential to at least take the start, as Collier remembers: "Olivier Gendebien was driving one of the BRP cars. We were standing in the pits at Watkins Glen waiting for him, and the first thing we saw was the car coming around the corner, end-over-end. He wasn't hurt, just had a few bruises, but we had a couple of all-nighters getting that car ready for the race. If you didn't start, you didn't get your money. We got it all patched up and he went out.

"I had another one at Watkins Glen with Piers Courage, he had a big shunt in practice and I worked all night there, welding up the back frame and putting everything together." Ex-BRM mechanic Pat Carvath, who was seconded to the Parnell team that was running Courage's BRM at the time, remembers the occasion well. "Parnell said to Stan and myself 'Get that to the starting line and you'll be on a good bonus;' he came eighth in the race. Later, Parnell gave us about $100. I said 'Are you sure you can afford that, Tim?' He'd promised us about a thousand dollars I think but, anyway . . ." In the race, Courage stopped first in the pits to change an ignition transistor box and again to replace a missing bolt in the rear suspension, but finally ran out of fuel just before the chequered flag. Being classified eighth was more than enough compensation, however—for Parnell, at least.

That mechanics received a share of the winnings goes a long way toward explaining their determination to place cars on the grid—even if it meant all-nighters—and also why they were so keen to ensure their cars reached the finish through addition of the so-called "mechanic's gallon".

As Arthur Birchall points out, at Lotus, there was another financial reason as working all-nighters was a way to top up your wages, albeit not by a great deal. "We would do an ordinary working day, which finished at something like six o'clock, then, if you worked until ten o'clock you got something like one and sixpence, which is probably about seven-and-a-half/eight pence, for fish and chips. If you worked from ten until two you got another one and sixpence, and, if you worked every night, most of the night, you could make up probably a quid at the end of your week's work in expenses. It was not a lot, it would just about buy a packet of chips, the one and sixpence. But then, you didn't work there for the money . . ."

Sometimes it wasn't just a case of money; pride came into it as well, particularly when a car had made a long journey to get to a race. Bob Sparshott recalls one such occasion when he was running a March 761 for Brett Lunger in the early part of the 1977 season. "We went to the tyre testing in South Africa about two weeks before the race. What the teams used to do was take that opportunity to do some hot weather running, and then stay out there and take part in the race. We were nearly at the end of the testing when Brett went off and did a mighty lot of damage to this car.

"When it came in on the back of the wrecker, I saw that the left-hand side was all stoved in. I said to the lads 'Well, we don't have a spare car, we'll be going home, that's it.' But then, when I went round the other side, it was completely intact. I began thinking 'Hang on, let's have a better look at it.' It didn't look like there was too much trouble with the bulkheads, but all the outer skin had gone, and the left-hand front and left-hand rear suspension, the wing, everything. I said 'If they've got some spares in England and they shipped them out, together with what we've got, I reckon we could make this ready in time for the race.' My mechanics thought I was nuts. I said 'Well, it's up to you guys. If you don't all want to have a go at it, then we are not going to be able to do it.' So they said 'Yeah, let's give it a go.'

"I did it in the full knowledge that Ken Gillibrand had a work-shop about a mile down the road from the circuit. He'd worked in England and he had a European-type shop with proper welding equipment, a flat-bed for chassis, all the kit. I immediately went and saw him and said, 'Look, I can only do this with your help.' Of course, he was rubbing his hands together because he was going to earn some money. I worked with Ken on the chassis and left the rest of the team up at the circuit building up bits and pieces, so that when we got the chassis back they'd just be able to bolt it up.

"We just did a ten-day ball-breaker, basically, with no or very little sleep. It got painted on a Sunday night by a local paintshop guy and then taken up to the track and they built it up and we were there for first practice. Even Bernie [Ecclestone] came up to me and said, 'That was a bloody fantastic job, I didn't think you could do it,' and I replied 'Nor did anybody else.' And it finished fourteenth in the race. Not very high but it finished . . ."

Mechanics also rose to the challenge when it was a sponsor's home race and a car was damaged, as clearly it was one of the most important races of the year for their backers. Roy Topp remembers one particular instance in Montreal, Canada, when working for Walter Wolf's team. "In 1978, we had Bobby Rahal driving for us and he damaged his car when he went off in a big way. The previous year's WR1 was in Canada somewhere as a show car on display in one of the hotels, and they decided that they would get it out; this was after last practice, with the race the next day.

"It was in show condition, so didn't have a proper engine or gearbox. We got it back to the circuit and worked all night to rebuild this thing into a race car and Rahal drove it. Actually, he was going really well. Unfortunately, it didn't finish. I can remember Jody [Scheckter] saying, 'Don't work on that, work on my car,' because I was chief mechanic. But I had to do what was best for the team."

A downside to working constant all-nighters was that mechanics became tired, crotchety and rebellious. Mike Barney recalls this happening when he was working at Coopers. "When we did the IndyCar in 1961, we'd been there for two all-nighters on the trot and were just wiped out. We weren't doing very much by then, we were just too tired. It was silly, really. When you look at it now, it was a total waste of time. Anyway, we'd staggered on and it was a Friday; it had to be a Friday because that was the day *Autosport* came out then, and Noddy [chief mechanic Michael Grohmann] had gone down the road to buy a copy. He came back and we were stood there looking at the pictures when Charlie Cooper came in. He walked up to Noddy, who was already a bit ratty by this time, and snatched the magazine out of his hand.

There was then an almighty row, dear oh dear. The language, you've never heard anything like it. The pair of them were effing and blinding at each other, over the chassis of the IndyCar. Charlie Cooper went out of the place in a huff and sent the works manager, Major Owens, up. He said 'We can't have the tail wagging the dog, you'll have to go.' But he didn't go because they were just about to go off to Indianapolis."

As former Cooper mechanic Denis Daviss explains, sometimes, if a number of teams were doing an all-nighter at the same time, it was important to gain a psychological advantage over the competition. "With the Lotus 49, when it won its first race out, we were all working in a garage in Zandvoort. Our hotel was across the road, a place called the Myerscough. All the Lotus mechanics were there as well as Chapman and Keith Duckworth, they had been having wheel bearing problems with the 49s. At midnight, I did a little deal with one of the waitresses across there and she tripped along in all her uniform with a big silvery-coloured tray, a great big coffee pot on it and all the cups round, just for our crew, not for them. Because that's part of the game, isn't it? I had to grease her palm a little bit but that was all right, it was only money."

Just occasionally during all-nighters, inquisitiveness about their working environment would get the better of mechanics, as Roger Barsby explains. "We were at Monza with the old BRM P139—that was a sod. Surtees was driving for us, and we always used to have the garages in the Parco Hotel Monza just on the entrance to the park. Surtees got in the car and said 'That's well down on power,' so we put the spare engine in. He didn't like that, so we put another one in and he didn't like that either, so we went back to the original. That was four engine changes in three days, and, bearing in mind it took about six hours to change an engine, we got some all-nighters. One night one of the blokes said, 'I wonder what's over that wall?' So we stacked up some tyres and looked over the top: it was a wine cellar. We weren't very popular the next morning, but there we go . . ."

Probably the only thing worse than having to stay up all night and work, is having to stay up all night and sit around not working. Ben Casey remembers when this happened with the BRM team in 1971. "We went to Ontario Motor Speedway in California. It was Formula 1s versus their Formula As, and we had our P153s. At the back, in-between the gearbox and the engine, was a butterfly plate that the suspension used to hang on, and we found that these were all falling apart, cracking. We hadn't had a good weekend; we were working until midnight or more the first two nights, and then this occurred. So we had to strip out all the cars. Aubrey [Woods, BRM engine designer] knew somebody who worked for Boeing or Lockheed, and he went off into Los Angeles with them, got these plates riveted and repaired, and then we had to refit them ready for the race in the morning. That was a long night because, once we'd got everything stripped out, we were sitting about waiting. If you are working, the time goes quicker. We couldn't go anywhere, so just had to sit there and wait for him to come back. It was a bit of a shame really because we were ten miles from Los Angeles but I never saw the bloody place."

Another cause of greater-than-average team workloads came when two drivers of very different heights were used and they needed to swap cars. Mike Lowman, who worked as a mechanic for Dan Gurney's Anglo American Racers, vividly remembers one such occasion. "At Monte Carlo in 1967, we had a driver who was six foot three [Gurney] and one about five foot eight [Richie Ginther]. After the first qualifying, they decided that the engine in Dan's car was better than the one in Richie's. To do an engine change on it was an all-nighter anyway, so we thought, why don't we change the rest of the car, leaving the engines alone. Put Dan's stuff in what was Richie's and Richie's stuff in Dan's car?

"It turned out that this was no quicker, and we ended up being up all that night and the next doing the same thing in reverse. Because that car was a complete monocoque, to get down into the footwell area was an 'upside-down job,' you used to get into the cockpit upside down and just work above your head inside it, which was not the most comfortable place in the world. There was just so much to change on it but they wanted to feel comfortable in their cars. It was a nightmare changing that. And then Richie didn't qualify . . ."

Lowman discovered another, altogether different pitfall of working all-nighters later in his career when he was with Don Nichols' Shadow Formula 1 team at the 1973 Belgian Grand Prix. "The first year we went to Zolder, we were working late and [Tecno Formula 1 car designer and constructor] Allan McCall and his team were still at it on the Tecno. We were ready to go and they were struggling, with just tiny little flashlights working on the car. I decided I'd lend them some of our neon lights, so I took them over.

"I went to get out later and the gate to the paddock was padlocked. I went up the control tower, which was about eight floors high, and there was not a soul about. I said to the lads, 'Okay, get back to the garage and get me my hacksaw.' I cut off the padlock, out we got and just pushed the gate shut. I didn't think any more of it. The next day, we were working on the cars when this police officer came into the garage and said, 'I understand you cut the padlock' and I said 'Yeah, we couldn't get out last night, there was no one there, we were locked in and we needed to go back to get some sleep and some food.' He said, 'Ah, you cannot do that, I must have your passport.' Fine; I gave him my bloody passport and said to him, 'I'll tell you what, if there is any problem, I'll just go back over there and weld it up for you, I don't have the slightest problem in doing that.' So we wheeled the old welder over to the gate and welded the thing for him, which defused the situation a little bit."

It was very rare to find a team that actively avoided all-nighters if it could, though Team Tyrrell was one such outfit, as former chief mechanic Roger Hill explains. "Yes, we had a few all-nighters, everybody does but we tried not to get involved in that. We would work a long day, and you can get a lot done in a day if you really want to. If you think you've got all night as well, that's all right but it's not very good the next day or the day after. We managed amongst ourselves, got stuck in and tried to do the job so it didn't eat into the time when you should be having a rest. We tried to get organized and Ken was the instigator of that. He knew damn well that, if you'd worked all night for a couple of nights or a couple of days, whatever, then you weren't doing such a good job."

The main exception to this rule within the team was with the 1000-cc Formula 2 cars, where the highly-strung motors needed a lot of attention, as ex-Tyrrell mechanic Neil Davis explains. "We used BRM and Cosworth engines, and just about every night you had to take the head off the Cosworth engine and do all the valves and springs and God knows what, because they'd been over-revved and the clearances were closed up. It was quite stressful, and it was two or three in the morning before you got between the sheets, and then more or less got out the other side and start work again. But I wouldn't have changed those days because they were an experience, and we were the sort of age that we could cope with it."

A passionate morning in Monza

Tired and hungry mechanics did not make happy people. However, they were still usually able to maintain their sense of humour even under extreme provocation, as ex-Cooper man Denis Daviss illustrates. "In 1966, as soon as we got back from the German Grand Prix, I went rushing off with an engine-less car [a Cooper-Maserati] on a trailer. I arrived in Modena and the engineer, Alfieri, said, 'We need to go testing tomorrow.' Just then Surtees arrived 'Yes, we need to go testing tomorrow.' That meant I had to put an engine in. I wasn't alone, I had an assistant this time: Maserati had a roving Alsatian that guarded the place at night. He liked me, for some strange reason, possibly because when I'd been there at other times I'd pinched some stuff out of the restaurant kitchen for him. Anyway, I was laying down on the floor, doing the underneath stuff. To start with the dog would come and lay on me and put his head on me. He got a bit pissed off with doing that, so when I put down a spanner, he'd pick it up and take it away.

"In the morning, when the engineer, Alfieri, and Surtees arrived, I was just about to go and have a shower and something to eat because I hadn't eaten since I'd left Byfleet. They said: 'We need to go testing' and I replied, 'I really need to go and get some food.' 'You can have some when we get to Monza,' they said, so we loaded up this thing and off we went to Monza, which was about four hours' drive away because it is on the edge of Milan.

We got there and they were all fit and ready to go. John got into the car and did a little run to make sure nothing was leaking, because it was getting very close to lunchtime. He did his lap and he's sitting in the car; he said 'You've got some springs and roll-bars with you, haven't you?' and I replied, 'Yeah, I've got some stuff, not a lot.' He said, 'Well, I think we need some harder springs on the front and probably a harder rollbar at the back.'

"I said to him, 'John, bite my ear.' He was wearing an open-faced helmet which he pulled back and asked, 'What did you say?' So I said it again: 'John, bite my ear please,' and he said, 'What the hell are you on about now?' to which I replied, 'I like a bit of passion when I'm being f****d about.' With that, he said, 'Well, the restaurant is open, why don't you go and get some food and tell them I'm paying for it?'" Never slow on the uptake, clearly, John could see that breaking point had been reached; a meal was the only way that team equilibrium could be restored.

Champagne and oil in Reims

The 1966 French Grand Prix was held at Reims. For the works Cooper team, running Cooper-Maserati T81s, it was an epic weekend. Denis Daviss takes up the story. "Cooper fielded three cars in that race. One for Jochen Rindt, one for Chris Amon and one for John Surtees. After every day's practice—and there were three—engineer Alfieri supplied a new specification engine for John Surtees' car; John's engine went into Rindt's car, and Rindt's engine went into Amon's car. Well, as practice was over three days, that made nine engine changes . . .

"The Cooper had these booms that stuck out under the gearbox. The engine had to be lowered in so they used to take quite a while to change—eight hours or so. Fortunately, we had two block and tackle sets, plus the one from Cooper. All three cars were having engine changes at the same time, but we still ended up working most of the night, with just two mechanics on each car.

"A big problem was oil, as each car held approximately four gallons in total. At the end of the race meeting, with these nine engine changes, everything that could hold oil was full. Maserati also wanted all the engines back, plus the three race engines—we had their total stock of engines. So the race engines also had to come out, and what did we do with the oil from them? That was a big problem. The oil tank was right at the front and the oil went back down to the engine through an aluminium pipe, about an inch and a quarter in diameter, in a channel in the bottom of the chassis on each side. It was decided to plug these aluminium pipes with rag and hold it on with tape.

"Unfortunately, just as the first car was being pulled up the ramp, the plugs popped out and covered the transporter floor, the ramps, and the road with oil. Before there was time to clean it up, a French cyclist came riding down the road, slowed to look into the back of the truck, and promptly found his bike sliding away from under him. Not a very happy Frenchman. A few racing giveaways made him smile again, however.

"As John was fastest on the first day, he was awarded 100 bottles of champagne—normal for the French as Reims is in champagne country—which I collected and put under the transporter's full-width bench seat. On arriving at Dover, six of us lined up with our duty-free purchases, and the official asked 'Do you have anything to declare?' I told him we had some champagne that John Surtees had won for fastest lap, but nothing purchased. He then asked how much champagne, and I told him there was enough for one bottle each.'Oh well, that's okay then, off you go,' he replied, thinking I meant enough for the six of us standing there, whereas I had actually meant enough for everyone at the Cooper factory . . ."

Lotus: king of the all-nighter

The undisputed king of the all-nighter within the Formula 1 paddock was Team Lotus; just as most teams were knocking off for the night, its mechanics would be making a start on one of Colin Chapman's legendary job lists, as Arthur "Butty" Birchall recalls. "We used to get people taking the mickey out of us, walking past the door calling 'We're off now, boys!' There was a good rapport with the other teams, but they all knew that Chapman was a pretty hard taskmaster."

Former Lotus Indy and Formula 1 mechanic Hughie Absalom

remembers that the constant round of all-nighters took their toll. "At Lotus, we were fairly close to being zombies. Seven days a week was normal, whether we were going racing or back at the factory, you just seemed to be forever working. When Dick Scammell got more of a say as to how to operate the place, it got a lot better, but before that Chapman would just walk all over us."

Scammell vividly remembers the effect all-nighters had on his ability to concentrate behind the wheel as well. "The first race that I ever went to was at Aintree in 1960. I remember driving up there. It would be terrible today, you wouldn't even be able to talk about it. But everybody worked an all-nighter. I remember setting off with somebody in a van and only being able to drive for about twenty minutes each, we went all the way to Aintree, really falling asleep, which wasn't very clever but that's the way life was."

There were so many Team Lotus all-nighters that it's hard to distinguish them more than forty years on. However, even today, Scammell names the 1967 British Grand Prix weekend as the occasion of probably the team's greatest all-nighter. "When Graham crashed a 49 in the pitlane at Silverstone, we took the wreckage back and had built another car by the following day. That was a huge effort by everybody; a whole mass of people who didn't stop from the time they left Silverstone to the time they got back."

Eamon Fullalove, a fabricator back at the factory, takes up the story. "In practice, Jimmy had pole position and Graham was second. Graham was coming into the pits and hit the pit wall, wrecking the car. That was on the Friday night. We were just going home at about five o'clock when we got a call at the factory to say we had to stay there to rebuild a tub for Graham. The third tub was just done, it had come back from the paint shop. It was painted green, grey inside, but there was nothing in it, only the brake lines. So Graham flew back in his plane and brought a couple of guys with him. Then the truck had to be sent to bring back the wreck. Leo Wybrott and myself started, then Graham got back with a couple of guys, making five of us.

"I asked what we were going to do for food, and Graham said he would go to the shop and get some. He took the two secretaries from Lotus and they all went back to my place, which was not far from the factory, and cooked dinner for us. Then the truck arrived and we had to get the car out and strip it. We worked all night on that. It was a monumental effort, which I don't think could be done nowadays. By Saturday morning we had got the car sitting on the floor. Dick Scammell was running around saying 'Come on guys, you've got to get this in the truck.' We eyeballed it, lined it up, put the strings round it and got the toes right, fired it up—all the functions worked—and loaded it into the truck.

"Meanwhile, Allan McCall, Jimmy's mechanic, had noticed that the engine mounts on his car were cracking; the rivets were coming loose. It needed to be taken for a spot of welding. There was a garage not far from the track, which they towed it to, using Jim Endruweit's station wagon. Allan, in this little garage with a light bulb, by himself, worked all night and fixed that car. The truck was loaded and headed back to Silverstone and I flew back with Graham, still in our dirty overalls.

"When they got back to the track, someone asked, 'What about Allan? He's down in the village with Jimmy's car,' so Jim Endruweit had to go, driving against the flow of people walking in, and the traffic. He finally arrived to find Allan McCall in total panic, thinking they'd forgotten him and were not going to come and get him. They put a rope on and towed Clark's car through the traffic and finally got to the track. Everybody gathered round, working on Graham's car. In the race, Graham had a problem and Jim won. But it earned us the start money and Graham gave the money to all the guys. It was one of those great moments of teamwork."

The 1968 season was a particularly difficult one for Team Lotus. Having had both Jim Clark and Mike Spence die at the wheel of one of Colin Chapman's designs, Graham Hill grabbed the Formula 1 team by the scruff of its neck, restoring spirits by winning both the Spanish and Monaco Grands Prix. However, that year the advent of wings placed a huge strain on drivetrains and suspensions, with the result that new specifications were constantly evolving as the season developed, adding considerably to the mechanics' workload. This situation came to a head during the US Grand Prix meeting at Watkins Glen.

It was common practice in those days for Lotus to field a third car for a local ace in some of the North American races. In this case, it was to be Mario Andretti, though it seems that the news had not filtered through to the mechanics, who were surprised to learn at 10 p.m. on the Friday night that his participation meant that the engines on all three cars had to be changed. With the other teams long since departed, their cars covered and the mechanics tucked up in bed, Lotus mechanics started work, as Graham Hill's mechanic, Bob Sparshott, recounts. "A deal was done with Ford and, of course, he [Andretti] had to have a decent engine, because they demanded he had one. We were short of good engines, and to get Mario a decent one and still give Graham a decent one meant that all three engines had to come out of all three cars, with Jackie's the first car. We did a sort of 'Round Robin' and put a fresh one in Graham's and the best of the bunch—Oliver's engine, which was quite good—into Mario's car. Oliver had to have the worst of the lot, which he didn't like, and whinged and moaned about!

"In the morning, the drivers and the Old Man [Chapman] came back, all fresh and bubbling. We'd gone to have a bacon sandwich and arrived back at the garage as they were all standing around looking at the cars. I'll never forget Jackie [Oliver]; he was looking down at his car, with his hands on his hips, to identify the colours on the springs, because one of the jobs that needed doing on his car was changing the front springs. It was a big job on a Lotus 49, with the inboard front springs. We had prioritized the work and the springs had been left, basically. Jackie didn't say anything to us, not 'Good morning, I see you've been here all night, guys,' nothing at all. He just walked over to the Old Man and said 'Colin, they haven't changed my springs.' Colin went to talk to Bob Dance, I suppose to ask why, took one look at Bob, who was totally wound up, and got the vibes straight away. He turned back to Oliver, put his arm round him and said 'I want you to start practice on those, Jackie.'"

Matters then went from bad to worse, as Sparshott explains. "Jackie wanted to get out quickly. I was still finishing my car because I was down on manpower. Graham arrived and asked if he could help. I asked if he'd mind giving it a polish! You always had to polish the car before it went out, but I literally hadn't got time to do it. So Graham was polishing away and everybody else buggered off out of the tech-shed to begin practice. Before we got out, they were back. Oliver had crashed the car, he'd stood on it and a wheel had broken on about the second lap. It turned out that where we screwed up was that a truck came up in the middle of the night from the airport—as they did all through the night bringing parts for everybody—and a load of wheel halves arrived which got shoved in the corner. Nobody said anything about these wheels, so we didn't know anything about them. Afterwards there was an inquest into why we hadn't put on the new wheels, which came to nought because we didn't know about them, or that they had suspected they were going to break."

The team needed an additional complication like a hole in the head, but the decision had been taken to switch to CV-type drive-shafts for the US race, necessitating further work for the already hard-pressed mechanics. However, help came from an unlikely source, as Sparshott explains. "We were trying to perfect CV joint driveshafts, which was a new thing and, of course, being Lotus, they were never, ever, man enough for the job, they were always just about there. It was one man's job to maintain all of the joints and Bob [Dance] did it for the race on all three cars. That was a whole night's work. They had to be stripped out, measured, regreased and all the seals put back on. They were all glued on, it was a nightmare of a job. I'll never forget, dear old Maurice Phillippe [Lotus designer] helped us, because Trevor Seaman was ill, which left five people for three cars. Maurice said 'I'll stay and give you a hand.' He worked all through the night with us to make the six."

Among Team Lotus mechanics, the annual visits between 1963 and 1969 to the Indianapolis Motor Speedway during the month of May for the legendary Indy 500 race were another source of sleepless nights.

However, as Bob Sparshott recalls, they also provided plenty of opportunities for mechanics from other sections of the team to get involved and even to go to the races. "During the period that we did the Ford Lotus-Cortinas, I was working with Bob Dance, and we were used to fill gaps when the other teams were in trouble. It gave me the marvellous opportunity of going to some of the races in America with the Indy team. We did a couple of races in 1964—Milwaukee and Trenton—because they had some crashes and not enough people to do the work. At that time we were somewhere in America, so were shipped up to help.

"In 1965, Bob, Allan McCall and I were in California with the Cortinas doing Riverside and Laguna, two races on the spin. The team encountered all sorts of trouble at Indy with a written-off car, and the Old Man asked Andrew Ferguson who he could get. He said that some of the lads were in California so he told Andrew to bring us over to Indy and just leave one guy to look after the Cortinas. Poor old Allan McCall got left to look after two cars on his own, it being lower priority than Indy, and Bob and I went off.

"That was a memorable day because we arrived at nine o'clock at night at the airport to be met by Andrew Ferguson. It had been a long day and we were already tired because of working on the Cortinas, so I thought we'd be going straight to the hotel. No fear. We were taken to the garage at Indy, where they were all working away, and Dave Lazenby said 'Put your suitcase over there in the corner, you can check in later when we've finished.' I was thinking 'Geez, when are we going to finish?' We went from the garage at eleven-ish to have something to eat, and only got to the hotel after that.

"That was my first day at Indy and every day after that seemed the same. Bob and I ended up there for the entire period, staying for the race and everything. They had me signalling on the wall for the race. With me being the youngest, they told me they had a great job for me. I thought 'Signalling? Oh that's all right, I've done plenty of signalling.' Then I thought, 'Hang on, I don't like the look of this wall,' it was a very low wall between you and the track, very isolating. Peter Jackson of Specialised Mouldings did the other board, so I never really worked on the Indy team per se though we worked at odd parts of it."

The 1967 event, when the team went with both Type 38s and 42s (originally designed for a BRM H16 Indy engine), was something of an epic in this regard, as Jim Pickles recalls. "We were building the cars and working reasonable hours, but as you got nearer and nearer the time for shipping the cars, we were working longer and longer hours. However, the really heavy stuff didn't start until we got there.

"We found that we were in a shocking condition and were working all hours then. There were problems with both handling and engines: we blew engine after engine after engine—Ford Quad-Cam 4.2-litre, normally-aspirated V8s. I didn't fully realize why but there was something odd somewhere. We ended up borrowing engines from A. J. Foyt and all sorts of other people in the end.

"The cars wouldn't handle either. In fact, the Old Man [Colin Chapman] cleared off one of the benches in the garage and actually laid out a drawing board and redrew the rear end. We had to set that up there, we had no special workshop, and had to rebuild the back end and rewind the springs and all sorts. It was grim. But even though I was knackered, I was still enthusiastic about it. I'd learnt very early on that Chapman was so innovative and, in that respect, to be very much admired, from my point of view.

"At some unearthly hour, three o'clock in the morning or something, the cars were on the deck and we were still working on them. I happened to be working on the front suspension and just nodded off draped over the wheel. I did come to, briefly and looked around; everybody else was draped over wheels or the bench, asleep, too."

Dozing "on the job" was quite commonplace in those circumstances, according to Arthur "Butty" Birchall, particularly at Indy, and he, too, recalls that year as being especially gruelling. "Falling asleep on the floor or on the toilet was nothing unusual because it was bloody hard work. The worst I ever did was 1967 at Indy when we did three 140-hour weeks."

Birchall also remembers 1969 as another difficult year, not just because of the hours but also the logistics of getting cars and mechanics to the USA, and then moving them around without the use of a transporter. "It wasn't much better in 1969, when we tried to get the Type 64 qualified. We did a couple of weeks of 140 hours then. You'd work on the car at Hethel, get it to the airport to be flown out to somewhere like California or Indy for testing, come back, wash and change and pack a bag and then fly out there yourself. You'd then wait at the airport for it to come in and clear customs and then drive it on a U-Haul trailer to where you were going testing. And that was the way it was done. There were no transporters, you'd hire a car and illegally hitch a U-Haul trailer on the back of it and go."

The fact that Colin Chapman and his chief designer, Maurice Phillippe, were constantly innovating and introducing new designs, made for a much heavier workload for Team Lotus mechanics. Another example of this came in 1970, when the Lotus 72 initially proved recalcitrant and the team had to put in nightmare hours to have the car ready for its debut at the 1970 Spanish Grand Prix, as Eddie Dennis recalls. "We reached Jarama and kept on working and had a period where we just didn't sleep Herbie [Blash] was the first one to go, we just lifted him into the truck. He hadn't collapsed but just gone to sleep, passed out really. Later in the meeting, I'd gone somewhere to find a bite to eat, sat down and the next thing I knew was back at the garage. Apparently, I'd been helped back there because I'd passed out."

During that season, almost every one of the Team Lotus Formula 1 crew collapsed from exhaustion at some point. After the poor performance of the Lotus 72 at Jarama, and a subsequent non-Championship race, the decision was taken to modify the cars, missing the Monaco Grand Prix and aiming to complete the work in time for the Belgian Grand Prix at Spa. As another mechanic on the team that year, Dave Sims, explains, it was touch and go whether the car was finished in time. "The rest of the team went on to Spa and left me and Dougie [Garner] behind to finish the latest car with all the mods for Jochen. We had a Thames van and trailer and when we eventually finished at something like four in the morning, set off for Harwich and took the ferry to Zeebrugge. We'd had five days with no sleep and were completely knackered.

"When we reached Spa we had no tickets, no passes, no windscreen sticker, and didn't know where we were going because we approached, somehow or other, from the other side of the circuit to where you would want to go for the paddock. We then got lost inside the circuit; eventually, a marshal led us into the paddock. When we arrived, the Old Man [Chapman] saw us and said 'Get it off the trailer, come on quick, quick, we've got to get it out.'

We'd only started it at Hethel, warmed it up, checked that all the systems were working, it hadn't even turned a wheel. The Old Man said, 'You've got to change the ratios, it's got the wrong ratios' and the last thing I can remember is trying to take the layshaft off the gearbox and that was it, I passed out. Apparently, the Old Man said, 'Put him in the back of the truck, don't let anybody see him.' Eventually I was taken to a hotel."

Evidence of the zombie-like state of Lotus mechanics that year is provided by a story from Dave Sims from that Spa weekend, when they finally made it to their hotel in Stavelot. "There were three of us— me, Joe 90 [Derek Mower] and Eddie [Dennis]—in that room. Derek woke me up in the middle of the night and said, 'Look what Eddie's doing.' Eddie was crouched at the bottom of the bed. We asked him what he was doing and he said, 'I'm changing the rollbar.'"

Later that year, it was Dougie Garner's turn to flake out on the final day of practice for the French Grand Prix at Clermont-Ferrand, although, as it turned out, the end result was a rare period of relaxation. "They took me back to the hotel without me knowing. I woke up in bed the next morning, looked around and everybody else had gone. They had decided to leave me there because I was too knocked out. They all went up for the race and I thought 'Oh God, duty first, must get there.' I got changed into my gear, rushed downstairs and, of course, Clermont-Ferrand was shut. I asked the receptionist to book a taxi to the racetrack but she said there was no chance. I had got all the passes, I could get in easily but getting there was the problem. In the end, I thought 'What the hell do I do?' and it was getting on because I'd slept quite a long time. I was sat in the bar and looked up at the TV and saw they were on the start line. I thought, 'There's absolutely nothing I can do now anyway.' I sat in the bar and watched the race and we won. That was the satisfying part. There I was, drinking a beer and watching the other guys work their socks off. It didn't happen like that very often . . ."

Six out of ten ain't bad
Former Cooper and Honda mechanic Denis Daviss recalls one epic period in 1967 when he experienced almost a week of

all-nighters. "The first Canadian Grand Prix that year was between the German and Italian Grands Prix but it made lots of extra work packing up stuff for us guys. There being no truck drivers in the good old days, I left Byfleet on a Saturday morning in a brand-new Transit van, with used engines from the last race, to pick up fresh ones from Maserati in Modena for Canada, arriving back at Byfleet around nine o clock on the Monday morning. When you consider there was no Mont Blanc tunnel, so we had to drive all the way down to Provence, then cross over round the back of Chamonix and over the Mont Cenis pass, which leads down into Turin, I guess I probably did 1,600–1,800 miles. Arriving at Maserati, I unloaded the engines—I didn't even bother to turn off the engine—popped the other ones in and left. That was another all-nighter, on the road. Back in Byfleet, we had to fit the engines into the two race cars that were leaving on Wednesday evening for Mosport Park in Canada.

"I was one of an advance party of two; the other was a Lotus mechanic by the name of Dougie Bridge. All the teams were sending two cars and the instructions were for one car to be in a box and the other car to travel on top of the box. But when it came to it, the transport aircraft was a Boeing 707 with a door in the side, and these boxes with the cars on top wouldn't fit. Dougie and I removed roll-over bars on what cars we could and, if they were in a box, cut the box down a little bit so that they would be lower and we'd be able to put cars on top. That took us just about all night. The plane being a freighter in a strange airport, there were no services, no food, nothing. It didn't even have any seats in it. During the flight, Dougie and I sat on the boxes and occasionally on the co-pilot's seat when he wasn't, and stole sandwiches from the pilot and his mate.

"When we arrived at Toronto, the following morning, we were parked way over the other side of the airport. We then had to get all this stuff out—manhandle it up the plane and out through this door. Fortunately, only one car got dropped and that was a BRM. It didn't hurt the BRM but it made a big dent in the tarmac . . . Then the organizers dropped another little bombshell: they wanted all the cars on flatbed trucks because they were going to parade them around some shopping malls to pull the crowds in.

So that was another all-nighter, taking all the cars out of the boxes and onto these flatbed trucks and then putting the boxes back together so that they could be used again. Everything was then ready for the race teams to reload the cars for the return flights. The cars went on their open flatbeds to the shopping malls and the boxes went to the racetrack. They were all unloaded and just left there.

"The racetrack was a bit of a surprise, too. Our garage was a long tent, about 15–20 feet—long enough to get a car in, including the nose and back, but that was all. There were no services, no electricity to speak of, nothing at all. It was a question of working on the cars and doing what you could. Practice turned into another mini disaster, when one of our cars had its starter ring disintegrate and it came out through these side-pod extensions, which punched a hole on the engine side and another on the outside. So it was out with the engine and the fuel bag and, using a borrowed van and a trailer from a very helpful Canadian team called Comstock, I hi-tailed it off to their workshop with the car on the back of the trailer.

"With the metal tidied up, I riveted patches over these holes, which again took a little time. On returning to our motel, the Flying Dutchman, I had a look around for some of the workers and found they'd all had a meal, along with a little lubrication, and were not in any real fit state to work, so I took the car back to the racetrack. Finding the tractor used for moving the boxes still had the forks on the back, I re-installed the engine. As there were no lights and the ground wasn't that good in the garages anyway, I found a nice little flat bit of concrete to do that job and, using the headlights of my borrowed van and of some Canadian fans who were camping nearby, I re-installed the engine. It was a bit of a performance because, with headlights, you are always in your own shadow. With the forks, it was a case of gently easing it down on the hydraulics, so that took a little while, too. Without the tractor, it would have been impossible. They were quite heavy lumps, even with people that know what they are doing, it would have taken more than four guys to lower it in and hold it in the right place.

"When the crew arrived in the morning, another two or three

hours saw everything left fitted—new ring-gear, gearbox back on and the electrics connected and the car fuelled and ready to run. In the race, Jochen's car retired with a drowned engine and the patched-up car finished tenth. So that was reasonable.

"With everything packed up, the team went off to fly home, leaving Dougie and myself rear guard to help load the aircraft. Flying with the cars, we arrived at Heathrow at 11.30 p.m. on Monday—another all-nighter, by the way. My team manager, Roy Salvadori, had asked me to give him a call as soon as I got in, no matter what time it was, so I called him and he told me to take a taxi home and asked could I be at the workshop at about nine in the morning? I turned up there at nine in the morning and it appeared that all the mechanics were worn out from all this work they'd done, and were all going to have a long weekend. So, with a spare car on a trailer, I was on my way back to Modena. When you think about it, what with the run to Italy and back, then all the messing around at Heathrow and Toronto, fixing the chassis, and my return to Heathrow, I did a minimum of six all-nighters within a ten-day period . . ."

Bernie's good deed

According to Alan Challis, Grand Prix mechanics have Bernie Ecclestone to thank for the easier time they have of it nowadays, and the fact that they rarely work all-nighters. "His deal of getting practice at the same time of day at virtually every Grand Prix was the best thing that has ever happened to the mechanic in Formula 1, because we used to go to Monte Carlo and practice would be Thursday, Friday, Saturday. You'd have practice one first thing in the morning, the next one would be late at night, and the next day it would be 'crack of sparrow's' in the morning, so you could guarantee you were going to have to work all night."

Another Ecclestone-driven development saved his mechanics considerable work at races, as Roy Topp recalls. "Things changed a bit later on. One of the first organized teams was Brabhams, when they came along with a complete back end, just unscrewed it from the back of the tub and slid the new one on, engine, gear-box, the lot. Unfortunately, you can't do that today but there was nothing in those days to stop you changing what you liked."

It is with understandable envy that people like Tony Robinson, who worked in a time when all-nighters were part and parcel of the job, look at today's Grand Prix mechanics. "It is easier for the racing mechanics of today because they lock up the damn cars twenty-four hours before the race, you can't touch them. In theory, you could go out and get plastered on a Saturday night."

THE BUSINESS OF WINNING

INSIGHTS ON LEADERSHIP BY MARK GALLAGHER, 2014

Good leaders and empowered managers build businesses; poor leaders with bureaucratic regimes destroy them. It may not be the accepted wisdom that successful companies should rely heavily on the ability of one individual to devise, communicate and create the framework for the delivery of his or her vision. In Formula One it is certainly the case that we witness the benefits of entrepreneurial, individual leadership over the malaise created by purely process-driven, committee-style management.

The term "benevolent dictatorship" is often discussed in hushed tones by many people I have worked with, simply because, once you have experienced the compromise and ineptitude displayed by cosy consensus, having someone with whom the buck finally stops, and who will make a decision, is refreshing.

It is my firm belief that every business needs a single, clear leader who has the ability to define the direction that organization should take and then fully empower and support his or her management to unleash their own leadership skills and cascade that throughout the enterprise. The leader of an organization must set the tone and style he or she wishes the organization to follow. Marina Nicholas, CEO of Franco Formula Entertainment, on whose board of directors I sit, runs leadership training courses under the "Navigator" headline, the analogy being that the leader of a company is like the captain of a ship. I relate fully to that—the leader should be determining the destination, setting the course and making sure he or she gets the best performance from the crew for the journey.

My evidence for the inadequacy of committee-led management can be seen in the dramatic inability of four of the world's major car companies to achieve significant sporting or commercial success in Formula One over the last decade. These failures resulted in their ignominious withdrawal, while the entrepreneurial teams flourished and now dominate the industry just as they did during its growth phase of the 1980s and 1990s. Indeed, in the cases of both Ferrari and McLaren, although they are themselves heavyweights in the luxury automotive sector, both were born out of individual entrepreneurship and retain the essence of such.

Ford, Honda, Toyota and BMW each created their own teams, ran them as divisions of their businesses, and yet failed to achieve their objectives within the sport. Harsh words, perhaps, but if the KPIs were the ability to win races and championships, or create profitable businesses, they didn't succeed. If "brand marketing" was the *raison d'être*, that too rings hollow, since in F1 the inability to win leaves a brand associated with the alternative: losing.

BMW and Honda each won a race during the 2000s, but in some ways those singular successes underlined an inability to discover the formula for achieving sustained success. Four companies, spending billions of dollars, were unable to crack the code to achieve and sustain success in the F1 business.

As though to drive this hard-to-swallow message home, in two of these cases the withdrawal from Formula One was achieved by a fire sale of the businesses, only for the new owners to take the very same organizations and become World Champions—not simply win a few races, but actually achieve multiple victories and tie up the title. In the case of Ford, an ignominious foray into the sport in the guise of Jaguar Racing saw an abject lesson in poor leadership, management structure and delivery followed by a takeover by Austrian energy drinks company Red Bull. Within a few short years the approach taken by the producer of a fizzy drink produced a World Championship-winning car, which Henry Ford's successors had been unable to do in spite of the vast resources and capabilities of their global automotive empire.

If that wasn't enough, Honda's sudden withdrawal from the sport in December 2008—ostensibly because of the recession—was followed by British engineer Ross Brawn taking over the business with a group of colleagues and promptly winning the very next year's World Championship in a Mercedes-Benz-powered Brawn. Quite what Honda's management will have made of this, one can only imagine. An estimated $2 billion spend had failed to achieve the glory required by a company in whose DNA racing could be found. From the outset, Soichiro Honda's vision of powered bicycles had grown into a legendary success story, a post-war Japanese business that came to produce some of the best motorcycles and cars and prove their pedigree through racing.

In the white heat of contemporary Formula One, however, Honda was unable to find the winning formula, yet under Ross Brawn's leadership the same team, with technology created while still owned by Honda, took the World Championship by storm and would be snapped up by Mercedes at a relatively knock-down price one year later.

All too often when discussing the reasons for Ford, BMW, Toyota or Honda's difficulties, the answer from senior management would come back that the issue was "Tokyo", "Munich" or "Detroit". The point was clear: the F1 teams were answerable to head office, to an extent that meant there was a layer of management bureaucracy that does not lend itself easily to the fast-paced business of Grand Prix motor racing.

As mentioned elsewhere in this book, David Coulthard often relates that when he joined Red Bull Racing shortly after their takeover of the Jaguar Racing team, he found a culture where "doing reports" in order to justify your existence was more important than developing strategies to ensure success for the business. It was as though the financial reporting for previous quarters or years had become in itself more important than delivering successful outcomes in the future.

Eddie Jordan was a very interesting man to work with, and he possessed leadership qualities that remain overlooked by people within the industry, including former staff, customers, suppliers and drivers. It is not likely, in response to the question, "What was Eddie Jordan best known for?", that many people will state "leadership". And yet I would promote Eddie's reputation in that regard, because our team undeniably achieved both sporting and business success under his leadership. None of it happened by accident, and none of the individual managers can take credit for the success the team achieved.

It's likely that some people will say Eddie was only interested in money, a frustrated rock star better known for his parties than his podiums, certainly unorthodox, sometimes foul-mouthed, with a whacky dress sense, something of a rebel whose approach often appeared contrary to the world of corporate sponsorship.

Eddie could be, to a greater or lesser extent, all of the above, but ultimately he created something out of nothing: a highly profitable business that he successfully sold a large chunk of at the peak of its value, winning races and almost a World Championship in 1999—more, in other words, than four major car companies ever achieved. He was a businessman, and he provided a leadership that could be inspiring, sometimes frustrating, but ultimately set the tone for a team that achieved a great deal.

It was an off-hand comment by myself to none other than Dietrich Mateschitz, founder and owner of Red Bull, that prompted me to be brought up short and reflect on Eddie's achievements. I made a comment, the kind we all make from time to time, half-serious, half-light-hearted, about "EJ's manner of doing business", to which Mateschitz responded by saying that Jordan had done more than most and was someone he admired. An accolade indeed, given that was only a few weeks after Mateschitz himself had taken over Jaguar Racing from Ford Motor Company and set Red Bull Racing on a path to F1 glory.

I first came across Eddie Jordan in the paddock of a Formula Atlantic race in Kirkistown, Northern Ireland, in 1978. I asked him for his autograph; he told me to "F★★★ off." I was sixteen, a teenage fan; he was thirty and driving for Marlboro Team Ireland. It wasn't exactly the response I was expecting.

The next time was in the Lisboa Hotel in Macau in 1987 when I was reporting on the Macau F3 Grand Prix and supporting my friend Martin Donnelly, who would win the main event and in doing so launch himself towards an international career leading to F1. One of my contracts meant I was providing editorial services and media support to Marlboro, through its Marlboro World Championship Team sponsorship programme. EJ realized that my race reports were influential in so far as the Marlboro management would use this information in their quest to pick and choose which teams in the lower formulae in which to place their drivers.

"Gallagher, I need to see you," was EJ's greeting on seeing me.

He was vying with rival F1 team West Surrey Racing to land a young Northern Irish driver by the name of Eddie Irvine, fully backed by Marlboro, and the deal he offered me was simple. If I put in a glowing report about Eddie Jordan Racing and helped Marlboro's agency based in Chiswick, London, to select his team as the best option for Irvine, there'd be £5,000 in it for me. A commission, an inducement, a fee or perhaps simply a bribe: to a twenty-five-year-old freelance journalist with a large overdraft and rent to pay, it was attractive. I am glad to say I didn't take it, since I didn't know EJ other than my previously brief encounter at Kirkistown, and I did know that my contract with Marlboro was important to me.

This unpromising start to our relationship did not hint at things to come because, by 1990, EJ was on the cusp of entering Formula One. Championship wins in Formula 3 and Formula 3000 had demonstrated the capabilities of his team, with Eddie doing the deals and team manager Trevor Foster ensuring that the money was spent wisely with immaculately maintained, well-engineered cars and talented drivers providing a winning combination. In parallel to his race team, EJ had developed with his lawyer Fred Rodgers a driver management business, Eddie Jordan Management, which had nurtured talent and brought to F1 drivers including Johnny Herbert, Martin Donnelly and Jean Alesi.

EJ always had vision and, yes, that included making money. Some might say it was his only motivation, but that's simply wrong. There are many, easier ways of making money than going motor racing; in fact it is a harsh and unforgiving industry, which has buried many a dreamer. Just as profit can be a dirty word to some, EJ's commitment to making money, whether for his team or himself, was often criticized. Yet it was difficult to understand this, since as a self-made man he knew that in the expensive world of international motor racing you have to work seriously hard to make deals work and to make money. Money is the lifeblood of an industry where everything needed to compete to win comes at a high price: cars, engines, tyres, gearboxes, transporters, top engineers and mechanics. Eddie, though, had a passion for what he did, for motor racing, and on top of that he had a passion for making money and therefore doing business. And, ultimately, he knew that the best place to make serious money would be in Formula One, if he worked hard, built a team that could deliver on the track and win customers off it.

After a failed attempt to buy Team Lotus, EJ opted to take his own team into F1 in 1991, and it was at the Italian Grand Prix in Monza, 1990, that I found myself typing out a one-page press statement announcing the arrival of Jordan Grand Prix. Compared to today's sophisticated PR wire services, distribution of the release was rather straightforward; I stood in the middle of the F1 paddock and handed releases to passers-by. Our first press conference, to launch the Jordan–Ford 911 as tested by ex-F1 star John Watson, was attended by around twenty of the media, and infamously caused the late, but celebrated, French journalist Jabby Crombac to wonder why EJ was bothering. Failure seemed inevitable.

Twenty years later, Eddie Jordan has a tattoo under the watch on his arm. It reads simply FTB. The acronym means "F*** The Begrudgers" and gives an important insight into the complex motivations that drove him to create a highly successful, profitable Formula One team that did indeed make him serious money. If making money and going motor racing were key motivators, so too were the fear of failure and a deep desire to prove the critics wrong. Irish people are not alone in begrudging their peers their success, but EJ knew from years of experience that to achieve success often means attracting adulation and jealousy in equal measure. Not only that, but your competitors often don't simply want to beat you; they want to see you fail. FTB was a motivation, and Jabby Crombac's article would never be forgotten by Eddie. "Why I am bothering? I'll show you."

As a leader, Eddie could be many things: difficult, and sometimes impossible because he had very fixed ideas about what he wanted. He could be prone to fierce outbursts, and be highly litigious, but he could also be funny, charming, inspirational and stylish. Frankly, he was different, and his leadership style reflected this as he set out to differentiate Jordan Grand Prix from the morass of competitors. He succeeded.

Importantly, he liked to remind us that it was his name that hung over the door to the business and that, since the buck stopped with him, he would also have the ultimate say in key decisions.

His tiny Irish-registered team based at Silverstone entered F1 with a spectacular green livery, top-quality brand sponsors in the form of 7UP, Fujifilm and Marlboro, and a quick car designed by a tight-knit team under technical director Gary Anderson. They brought a well-conceived, competitive product to market, and in its first season Jordan would finish fifth in the Formula One World Championship for Constructors, out of seventeen entrants. It would score multiple points finishes, embarrass the factory Ford-based Benetton team, have one driver sent to prison as the result of a traffic incident in London, and replace him with a relatively unknown German driver by the name of Michael Schumacher whose backers would pay for the privilege of seeing him make his F1 debut for Jordan. You cannot say that was anything other than a memorable debut season.

Add to that the infectious enthusiasm that the Jordan team brought to F1, with EJ's sometimes manic sense of fun and his ability to take the qualities of being Irish and apply them to the creation of an F1 team brand, and the stage was set for a decade and a half of highs and lows that would see Jordan among the most recognizable brands in the sport.

Although it's going too far to state that there was a carefully crafted strategy from the outset, there was certainly an approach that EJ took from the start. He was different, so the team would be different; the Jordan brand would be differentiated.

The Irish traditions of being welcoming, open, friendly and up for a bit of "craic" or gossip became values the team espoused. Friday morning breakfasts for the media became *de rigueur* and ground-breaking; for a travel-weary British press corps, a proper fry-up in the Jordan hospitality unit set the weekend off, and inevitably led to some banter with EJ, a running commentary on current events, and the odd bet or two. A love of music and of storytelling would lead to languorous and sometimes riotous evenings with media, sponsors, suppliers and officials. Win, lose or draw, Jordan promised noisy parties, a sprinkling of celebrities, and a glass or three.

This was a powerful combination. We couldn't sell sponsorship on the basis of winning races and World Championships; that would take time. It was clear that the experienced opposition sold sponsorship on the premise of achieving success on the track, and perhaps some corporate hospitality or PR to support it. But that was an all-or-nothing strategy and, since there can only be one race winner on the Sunday, the majority of teams and thus their sponsors can easily feel a deep anticlimax. To finish second is to be the first of the losers.

At Jordan the sales pitch was different: the team will try hard and work to win, as it had done in Formula 3 and Formula 3000. But it would guarantee other things: a strong media profile, and an engagement that would offer rewards beyond merely the rolling of the dice to see whether we could win or not. I used to say to sponsors that the difference with Jordan was that you could call us on a wet Tuesday in February and ask to borrow our F1 car or arrange a driver appearance and we'd say "yes".

Did this come from Eddie? I would say yes, because his determination to win deals, make them happen and make money meant we worked hard at that and we knew that going the extra mile to win the customer over was supported from the top down. Sometimes, in later years, he'd kick back hard if he thought we were overservicing customers; he had a keen sense of the value that the team was delivering.

If his charisma, sense of fun and ability to entertain and frustrate in equal measure were qualities that come to mind, so too were a relentless work rate and energy that often left the rest of us wondering where it came from. If he seemed to disappear to his yacht or house in Spain for half the summer, it was no less true that he was working the phone and the fax, and networking with celebrities and businesspeople like crazy. He had never done a deal, he would say, sitting "on my arse in the office waiting for the phone to ring".

Departing Kidlington airport near Oxford early one morning, EJ's HS125 private jet took us to Amsterdam's Schiphol Airport for a sponsorship contract signing with DiverseyLever, part of the global Unilever group, and a deal I had worked on with agent Graeme Glew. Eddie, immaculately suited, was charm personified when we met the management, signed the contract, enjoyed light refreshments and celebrated the start of another million-dollar deal. From there we headed to Germany and a meeting with the owner of Liqui Moly, producer of auto-motive lubricants, fluids and cleaning products, a difficult meet-ing where the owner complained that a small sponsorship of the Jordan F1 would be a "chickenshit" deal and he worried that he'd be treated accordingly—not an easy meeting, but we perse-vered. On arriving back at the airport, Eddie announced we had a third, previously undisclosed meeting, which required us to fly to a small regional airfield, touching down for less than an hour. I didn't know who he was meeting, and never did. He could be extremely discreet when needed, since sometimes deals required the utmost secrecy. The meeting was over quickly and we were gone. Ten hours and three countries later, we were on our way back to Kidlington.

Or not. Partway home EJ realized that Coventry City football club were playing away at Newcastle and so, with the pilots having been informed and spoken to air traffic control and Newcastle Airport, we diverted north and landed shortly after 7 p.m. We were collected by a chauffeur-driven car and deposited outside the ground, were met by a Coventry City runner with tickets for the directors' box, grabbed a sandwich and sat down with one minute to go before the start. By 10 p.m. we were out of the ground, back to Newcastle Airport and bound for Kidlington to arrive by 11 p.m. and get home by midnight. Just another day with EJ, and an insight into his relentless work ethic.

With the passage of time we all suffer from looking back with rose-tinted glasses, but there's never been any question in my mind that the success Jordan achieved was down to Eddie's unique blend of leadership qualities that excited, motivated, cajoled and sometimes browbeat staff into looking for the next deal, the next result, the next PR triumph for a team that achieved more column inches than many who had won more on the track. If you needed to see EJ's enthusiasm for the job, his excitement, witness his dance to the podium when Jordan won its first Grand Prix in Spa-Francorchamps, Belgium, in August 1998. It became the first team in the history of the sport to achieve its maiden victory with a one-two, and on a day when we had our fan club, our factory staff and the majority of our major sponsors present: a unique turnout to see the team's moment of glory.

Three months later Warburg Pincus concluded a deal to purchase half the equity in Jordan Grand Prix for around £50 million, making Eddie a very rich man. He and CFO Richard O'Driscoll had read the market perfectly, sold at the peak, and flipped the company onwards. It annoyed a lot of people but, FTB, it was good business for Eddie, and who can begrudge him that? In business terms, for an entrepreneur, it was a super-successful outcome after decades of hard work.

With money came many things: new friends, new projects and, perhaps, the beginning of a decline in the Jordan F1 business. Although 1999 was to be the zenith of the team's fortunes on and off the track, with two Grand Prix wins, third in the World Championship and a host of sponsorships, by 2000 it was clear there was a malaise setting in.

I have a theory about why that decline began, but in part I think the very fact of the Warburg Pincus deal having happened altered the motivations and rationale behind Eddie running a Formula One team. He was the leader of the business, but his motivations began to change. Simply put, the private equity people didn't understand the business and began to drive it in directions that were alien to the management, EJ included. By the end of 2005 the business was sold on to a Russian businessman, and Eddie was gone from F1: a loss to the sport, but with fifteen years of enormous success behind him.

As leaders go, Eddie Jordan showed what can be achieved through a combination of charisma, graft, focus on the bottom line, and inspiring staff, customers and suppliers to want to work with him to achieve the goal.

Two observations about his style.

One was his list of deals and tasks, neatly written, edited, scored out or underlined, in his Filofax. Every day, when he was in the office, I'd get the call to join him, often with Ian Phillips or Richard O'Driscoll, to run through the list of deals, issues, opportunities and challenges—always with a view to managing how we were doing and where the next incremental deal could be clinched. It was very disciplined and relentless and kept us all focused on what mattered: securing customers and growing the business.

The other was his daily phone call to his mother Eileen back in Dublin. Around eight o'clock in the morning he would often be on the phone to his mother for a quick chat, staying in touch. In the midst of the high-tempo work rate demanded by Formula One, with constant travel, lots of pressures and deadlines, he always remembered to call his mother, and I felt that showed something about him as a leader. Ultimately he had very grounded human feelings, and he wasn't afraid for us as managers to see that side of him; a little bit of humanity goes a long way.

In Formula One the businesses led by an individual with vision and entrepreneurial flair, supported by a highly professional management team who can deliver, inevitably win out over those with top-heavy, committee-style structures where the vision and power are pushed out and downwards. If the failure of Ford or Toyota to build winning teams was evidence of the latter, the consistent success achieved by McLaren and Williams illustrates the former.

In both instances their leaders, Ron Dennis and Sir Frank Williams, forged their skills in the fires of disappointment brought on by their early forays into the complex, expensive and unforgiving world of running an international motor racing team. It would be too easy to describe Ron Dennis's experience with the Rondel F2 team or Frank Williams near bankruptcy caused by the collapse of Frank Williams Racing as mere failures. As we have seen many times in business, it is these trials that often shape entrepreneurs and develop their skills, lessons from the school of hard knocks.

Ron Dennis today presides over arguably the most successful business in Formula One. Success can be measured in many ways in this business: success on the track, as we have seen, and profitability off it, and in these respects McLaren has proven itself time and again. However, there have been successful, profitable F1 teams in the past, so perhaps what makes McLaren the standout organization has been its ability to attract, sustain and develop long-term customer relationships, expand its offering as a business to include electronics, advanced technologies and business processes that can be applied in other industries, and now a fast-developing automotive business that is seeing McLaren take its bitter on-track rivalry with Ferrari into the high-end sports car marketplace.

By 2015 McLaren Automotive aims to be producing 4,000 vehicles a year from its state-of-the-art production centre near Woking in Surrey, UK, while its Formula One team continues to thrive. Its new partnership with GlaxoSmithKline brings F1 decision making and business processes to bear in one of the world's largest companies, and its Applied Technologies business takes the knowledge gleaned from fifty years of race-bred technology and applies it in areas as diverse as designing world championship-winning bicycles and data analytics to aid London's Heathrow Airport.

All this stems from a Formula One team that became a business, and ultimately from the vision of Ron Dennis supported by a strong board and skilled management team.

I first met Ron Dennis in 1984 when I applied for a job as marketing executive at McLaren International and was interviewed by him in his pristine office in the team's headquarters situated, at that time, in Station Road, Woking. He was businesslike and polite, asked fairly straightforward questions of a twenty-two-year-old Irish economics graduate with one year's experience in the advertising department at *Autosport* magazine, and ultimately decided I wasn't the one for the job. I was disappointed not to get it, but impressed even to have been interviewed and had a glimpse behind the scenes of what was already a highly successful F1 team.

The next time I met him in person was for a one-to-one interview in the McLaren motor home in 1989. Ron was having breakfast, a bowl of something very healthy, and my hopes for the meeting were damaged as I fell up the step into the inner sanctum of the motor home and banged my head on the ceiling. Ron found that very amusing, and his staff told me after that, "Everyone does it, and Ron quite likes it."

Pride dented, I spoke to him at length about his team's continued high fortunes after a year in which they had won fifteen of the sixteen Grands Prix thanks to the combined capability of the team, the McLaren MP4/4 car, its Honda engine, and two drivers by the name of Ayrton Senna and Alain Prost. Something of an A team, that, and Ron had pulled it all together. Interestingly, he told me that he felt they should have won all sixteen races in 1988 and, of course, he was right. Only a hasty mistake by Senna had caused them to lose in Italy.

Ron is known for being extremely determined in the manner he runs his business and obsessive about attention to detail. Sometimes his critics make fun of these attributes, but, as with those who said Eddie Jordan was only interested in money, this shows a lack of understanding about what's really going on. Eddie's interest in money meant he was determined to run a highly profitable business, while Ron's attention to detail is not a flaw but part of his inherent desire to ensure that everything McLaren does is very precisely planned and executed as well as possible.

If the Formula One World Championship was awarded for presentation alone, I suspect McLaren would have won it every year. Everything about its presentation—the cars, the garages, the trucks, the team's motor homes, the livery, the team clothing, its headquarters—screams attention to detail. The immaculate presentation suggests a very clinical approach to the engineering disciplines that lie at the heart of success in Formula One: a strong message.

McLaren has been the benchmark team in terms of presentation, perhaps matched only by Ferrari's upmarket Italian design. Both have undoubtedly benefited from the input of Philip Morris's Marlboro marketing managers, the brand backing McLaren between 1973 and 1994 and supporting Ferrari to the present day.

If Marlboro's brand guidelines demanded clean lines—the crisp white and red chevron with easily read black-on-white lettering, and best-in-class designs for everything from press kits to drivers' race suits and hospitality units—then they can only have contributed to Ron Dennis's belief in the importance of presentation. Prior to taking over the running of the McLaren team in 1982, Ron's Project 4 Motorsport organization had already attracted Marlboro backing for its championship-winning BMW M1 sports car racing programme, and established a reputation for building cars that were reliable, high-performing and impressively turned out, the key ingredients for any engineering business keen to attract customers and build for the future.

Attracting commercial partners was therefore an established skill when Ron Dennis secured what would become a pivotal business relationship with businessman Mansour Ojjeh, a Saudi-born French national whose company Techniques d'Avant Garde was known in the aviation sector as well as one-time owner of the elegant watch brand brought about by TAG's acquisition of the Swiss brand Heuer in 1989.

In taking over the running of McLaren at the behest of Marlboro in 1981, Ron Dennis and his new partner Ojjeh convinced none other than Porsche to design a V6 turbo engine with which McLaren would be powered for the following three years. Funded by TAG, this unit would power McLaren to success in the World Championship for Drivers in 1984 with Niki Lauda and 1985 with Alain Prost.

Marlboro had identified in Dennis a man who had the attention to detail and commitment to continuous improvement necessary to win in Formula One. These attributes were supported by another ingredient common to all success in this sector: a desire to innovate. The MP4/1 featured the first all-carbon-composite chassis, a development that in one fell swoop changed the direction of F1 car design, combining great strength with lightness. In order to produce this vehicle McLaren entered into yet another partnership, this time with the US aerospace company Hercules, and under the technical direction of John Barnard the car set new standards in terms of safety and performance, paving the way for the development of the F1 cars we see today.

Another aspect of McLaren's development was that the initially successful business partnership between Ron Dennis and John Barnard ultimately came to an end in 1989 when the pair split, Barnard leaving to join Ferrari. Whether intentionally or not, this helped Dennis to move away from a technical structure over-reliant on the skills of one man, a facet within F1 teams that has been commonplace.

It may be one thing to have someone leading the charge at the head of the company, but ultimately the need for structure, process and sustained performance means that the spread of management skills must remain evenly balanced. A contemporary example of the dangers here lies in Red Bull Racing, a team that many people feel has been dominant because of the technical leadership of one man—Adrian Newey—without whom the team would retreat from its position as industry leader.

Newey, too, was once a McLaren technical director, joining in 1998 after a successful, championship-winning, seven-year spell at the Williams team. He achieved much at McLaren, including presiding over the championship-winning cars piloted by Mika Häkkinen and David Coulthard, and when he departed many felt that Ron Dennis had made an error in not doing all in his power to retain him. As with the departure of Barnard, however, McLaren proved that it had strength in depth, subsequently winning the World Drivers' Title with Lewis Hamilton in 2008.

In 2004 McLaren moved into a new headquarters facility near Woking, and if ever the world needed to see an example of Ron Dennis's single-minded approach, attention to detail, belief in innovation, commitment to growth and desire to move constantly to the next level the "Paragon" facility design by Britain's Sir Norman Foster is it. Originally spread over a 150-acre site, the futuristic design houses all the facilities required for the design, manufacture, development and operation of championship-challenging F1 cars, and spacious accommodation for all the other disciplines. The McLaren Technology Centre was born, and stands as testimony to Ron Dennis's vision for the business.

As ever at McLaren there was a keen emphasis on presentation with a view to impressing the customers, but on a level that had never previously been seen in F1. The lines of previous championship-winning cars, innovative designs and heritage models from the days of McLaren's Can-Am and Le Mans sports car racing programmes help visitors to become immersed in a long tradition of excellence. This is underlined by the ability to see today's cars being developed and manufactured behind floor-to-ceiling glass walls. In 2013 the team celebrated fifty years since founder Bruce McLaren incorporated the company. McLaren's heritage is an extremely important part of its business story.

The fact that visitors can see precisely what the workers are doing, the culture of their work environment, the cleanliness and the immaculate presentation underlines the deep conviction that Ron Dennis has in his business and his pride in his staff. While many companies would hide their operations away, McLaren gets them out in the open, free to view, as a demonstration of self-confidence.

When visiting for a luncheon with director of communications Matt Bishop I was struck by many things: yet again the cleanliness and presentation, the fact that lunch was served by professional staff from yet another of McLaren's diverse businesses—Absolute Taste—and the provision of ample meeting space. Not for McLaren the discourtesy of "Let's see if we can find a free room" such as you get in many businesses. However, on that day, the item that struck me most was the butter; chilled, served on a dish, and shaped in the form of the McLaren logo. Even in the presence of this transient product, soon to melt or be consumed, the attention to detail and pride in the brand were present.

McLaren's approach is not to everyone's taste in Formula One. I myself have often teased them for being somewhat "bland" or "corporate" in a sport where noise, excitement and the feverish support of fans around the world appear to lend themselves more to the passion of the racing red Ferraris. Think also of the "extreme sports" lifestyle marketing of Red Bull or even the brash, showy, rock'n'roll approach that Jordan took. This misses the point; McLaren's customers are multinational corporations and, for a customer-focused business, that comes first.

If the positioning of the McLaren brand has been a tad boring to some, and underlined at times by media criticism of Ron Dennis's tendency to verbosity—coined "Ron-speak" in the industry—this has rather missed the point about McLaren as a business. For McLaren, more than any team in F1, has created an identity around attracting, sustaining and developing customers, and in this respect Ron Dennis's attention to detail has worked right down to the bottom line.

Helped in no small part by the demands of Marlboro as its sponsor, McLaren under Ron Dennis has, since the 1980s, maintained an impressive array of customers. Having been involved in sponsorship throughout my career, I take the view that the majority of Formula One teams have failed to maximize their income potential from sponsorship simply because at no stage did they put in the effort required to research potential customers, woo them through presenting relevant, compelling, business-led propositions, and then deliver on all their obligations to those sponsors.

I recall many in the media, and other teams, decrying McLaren for creating a McLaren Marketing entity under Ekrem Sami, one of the lynchpins of the McLaren business. But in making the effort to carve out "marketing" this showed that Ron Dennis understood a business imperative that passed others by, namely that if you are serious about securing customers you need to get serious about resourcing the sales, marketing, client services, hospitality and communications departments necessary to deliver customers to the organization and deliver results for the customer.

Having put that structure in place McLaren went on to show its understanding of another fundamental: that the decisions to buy global commercial sponsors are not made by the media or fans hungry for a brash, racy image, but by senior executives at board level of multinational businesses where a corporate approach is more likely to be understood. When senior executives aged in their late forties, fifties or sixties are making multimillion-dollar commitments to associate their closely guarded and valuable brands with a company that races cars, it is understandable that they will relate to a highly polished, corporate identity that wants to deliver real value rather than a team selling "car racing" and "publicity".

This may be disappointing for those who would still like Formula One to be risqué, but times have changed and the approach to securing and maintaining customers has had to evolve. There are many in our sport who yearn for the "old days"—most often the 1960s and 1970s when the race tracks of the world were graced by a cosmopolitan mix of drivers with strong personalities and a wide range of ages, drivers with the quiet brilliance of Scotland's Jim Clark, the taciturn nature of Australian "Black Jack" Brabham or the quintessential English qualities of Graham Hill.

It is understandable, for many of us in Formula One grew up

watching those drivers race, and it was an apparently more glamorous affair than the one we view today. That, however, is mainly myth, since it was also a time when sponsorship was in its infancy, the sport was at best semi-professional, and running a Formula One team required patronage because no business model worth the name existed. It was also a sport that sometimes seemed to kill as many drivers as it made famous.

The T-shirts that emerged a few years back with the legend "The 1960s—when sex was safe and racing was dangerous" raised a few laughs and emphasized our rose-tinted view of the past. There was a similar reaction when the latest in a line of biographies of British star James Hunt was published in 2009, including a photograph of him wearing race overalls on which he had sewn a badge reading "Sex—Breakfast of Champions".

But if F1 back then was sexy, dangerous and glamorous, it also wasn't very corporate; it wasn't a business. As that changed, so too did the approach to wooing customers, and the progression of sponsorship through the decades has followed a fixed path. If in the 1970s and 1980s it can be said that a great many sponsorships were decided by individual executives of major companies for whom corporate governance, accountability and responsibility were unknown tenets, there is no doubt that this has changed dramatically over the last two decades.

As the sponsorship industry has grown and matured, with global companies and their brands now advised by an entire industry of sponsorship consultants, marketeers and communications specialists, so too the managements of those companies have had to adhere to stricter corporate discipline. Making a decision to sponsor a Formula One team "because the CEO likes it" is much less common that it used to be. There are some who maintain those days are over; they are not, and never will be, because there will always be self-made entrepreneurs who enjoy a particular sport presiding over their own businesses and deciding how to spend their company money. Witness Dietrich Mateschitz at Red Bull, for one. But it is no longer the mainstay of sponsorship, and so teams have had to evolve a more sophisticated approach to wooing and retaining customers than merely appealing to the fans within the boardroom.

The ravages of the global recession following the financial crash of 2008 will unquestionably have made the quest for new business more difficult, and corporate decision makers will inevitably have to be more careful than ever to follow good corporate process and governance before signing off on major sponsorships. It will be those teams that can demonstrate and reflect their understanding of this that will stand a chance of benefiting from the upswing when it comes, and in that regard McLaren is likely to lead the field.

In 2004, while head of commercial affairs at Jaguar Racing, I was advised to go and meet a senior executive of the Royal Bank of Scotland on the basis that there was evidence of a potential Formula One sponsorship. The background to this was that Fred Goodwin, RBS's CEO, was known to be a great fan of Formula One and was allegedly using the good offices of close friend Sir Jackie Stewart to secure headline sponsorship of a major team.

The meeting duly took place, and I found myself in RBS's main London office seated opposite Howard Moody, the director of corporate communications, who explained to me that RBS was not considering F1 sponsorship for a variety of reasons, including the cost, the fact that Bernie Ecclestone's Formula One Management (FOM, the commercial rights holder of the Formula One World Championship) TV figures did not appear realistic, and that half of the bank's customers were women, who did not like F1.

I then got to the point and said that we were informed that Jackie Stewart was working on a sponsorship opportunity with Fred Goodwin, and Howard responded by saying that they were indeed good friends but that did not mean RBS would be sponsoring Formula One. He was emphatic on that point, and unsurprisingly the meeting was not a long one.

Fast-forward a few months and in January 2005 RBS announced title sponsorship of the Williams F1 team, with Jackie Stewart acting as brand ambassador, and none other than Howard Moody quoted in the press statement as advocating Formula One as a logical step for their brand.

There can be two ways of looking at this. One is to say that I had been quietly sold a dummy; the other is to imagine that Fred

Goodwin's now infamously known dictatorial style won through and the company was instructed to sponsor Williams. This was at a time when RBS's fortunes were riding high, two years before it would begin a dramatic slide into achieving the largest loss in UK corporate history, subsequent near collapse, and rescue by the British taxpayer.

Whatever the case, the rise and fall of RBS as both global banking giant and major customer of a Formula One team provide the ultimate argument that all future sponsorships of this size, and certainly from the financial services sector, will come under greater corporate scrutiny than before. This scrutiny of executive decision making cuts across every area of business, of course, but in the world of Formula One—where potentially vast sums of money can change hands because of the scale of the sport—this is going to be increasingly the case when it comes to justifying sponsorship.

McLaren's approach to creating a corporate F1 team identity addresses some of these fundamental questions. It has set out to create an environment that its customers can relate to and in which they feel comfortable doing business. This has been one of the master strokes of Ron Dennis's leadership, developed and executed with the help of lieutenants such as Sami.

Consider the colour palette used in McLaren's corporate identity and the livery of its F1 cars. The silvers, blacks and greys with a flash of red on its logo promote a corporate image, reflecting the grey-suited executives who represent its customer base.

First to move away from the motor homes, which were converted buses used as hospitality units in the paddocks of Formula One races in Europe, McLaren created a portable building—the Communications Centre—in 1999, with a suite of offices that reflected technology, innovation, attention to detail and success.

That was replaced in 2006 by the enormous Brand Centre, a three-storey building that takes a fleet of trucks to carry it—and all for the five days of a Grand Prix, and only in Europe. The very name "Brand Centre" shows what McLaren is all about: working with and building brands—its own brand, or the customer's brand. It's about adding value, promoting excellence, and ultimately winning the race off the track to achieve the most important goal of all: profitability.

Before finishing with McLaren, with its uber-corporate approach to Formula One and relentless desire to sustain and develop its customer relationships, there are two final points worth making.

The first is that McLaren enjoys some of the longest-standing unbroken customer relationships of any team in Formula One. Hugo Boss has been with the team since 1983, Exxon Mobil since 1982, and they have enjoyed a relationship directly or indirectly with TAG Heuer for the same thirty-year period. As a metric for client-centricity, customer relationship management and customer satisfaction, there can be no simpler test than the longevity of relationships. Through decades, with the natural ebb and flow of business and the inevitable cycles that brings including changes of management and strategy reviews, these companies have repeatedly chosen to remain customers of McLaren. The team has not won every race, won every championship or dominated the sport, but it has been relentless in its pursuit of sporting success on the track and commercial success for itself and its customers off it.

The second is that this most corporate of teams knows success, celebrates success, and in a unique way. Winning in Formula One usually means that hot, sweaty, dishevelled team personnel, drivers and managers are photographed in less than photogenic circumstances in the very moment of achieving their goal. As part of its partnership with Vodafone, McLaren introduced victory shirts, not in the corporate silvers and greys, but in the mobile phone giant's "rocket red", which every member of the team changes into when they win. While the rest of the pit lane wearily sets about packing their equipment away, hot and bothered, the McLaren crew are visibly successful, trumpeting their victory, posing for the cameras in their fresh tops, faces beaming.

As if to illustrate, or even illuminate, the attention to detail, this visible sign of celebrating success, showing the world that they are winners, extends to the McLaren headquarters in Surrey. Outside the $700 million facility, on a busy roundabout adjacent to Farnborough Airport, the McLaren company sign stands proudly.

When they win, the sign is lit with the same orange hue as that of the T-shirts worn by the team personnel at the track. Even the factory celebrates the success of its products.

Dietrich Mateschitz is seldom written about in the international business press, and indeed I have often wondered how, in this age of Silicon Valley celebrities such as Gates, Jobs and Zuckerberg, when billionaire businesspeople are so often feted and wooed by media celebrity, the founder and owner of Red Bull has managed to remain relatively low-key.

One reason is undoubtedly that, although a giant of the soft drinks industry, Red Bull is also a privately owned company and thus does not have its results put under the media microscope each quarter.

When Red Bull first appeared in Formula One as a personal sponsor of Austrian driver Gerhard Berger, few imagined that the sweet, fizzy drink with a taste some described as being akin to cough medicine would go on to achieve dominance in one sector of the global beverages market and embarrass giants of the car industry by achieving complete success as a team owner in Formula One.

From my perspective there are two distinct strands to the Red Bull story: the development of the brand into the success story it is today, and the manner in which this upstart Austrian drinks company was able to take over the largely failed Jaguar Racing Formula One team from Ford Motor Company and turn it into a winner. I was there the day Mateschitz addressed the workforce at Jaguar Racing in late 2004, hot on the heels of Red Bull's takeover, and he told a relieved but tense staff about his vision for the team and his ambitions for the future. If any of us doubted him, we only had to look at what he had achieved with Red Bull in the two decades since it was founded to know that this was a man who knew how to get things done.

Having worked in marketing for Unilever, Mateschitz later joined German cosmetics company Blendax, where he had an international marketing role for products including toothpaste. It was while on a business trip to Asia that he came across an invigorating tonic drink known as Krating Daeng, which a Thai company was selling widely in the region. Popular for its ability to boost energy thanks in part to its caffeine and taurine constituents, it featured a pair of fighting gaur, or Indian bison, as a logo.

I was fortunate in 2004 to meet with Peter Huls and Roland Concin, two of Mateschitz's right-hand men, and during a day-long introduction to the world of Red Bull they explained how he had taken this regional product and turned it into a global brand phenomenon. Huls was head of engineering and technology, Concin responsible for operations, and they explained how Mateschitz had taken the original drink and had it modified to appeal to a Western palate before setting about producing it. They also explained how product differentiation was key from the start, including the choice of slim, tall can with its cool blue and silver colours.

The Red Bull production lines at the Rauch factory in Rankweil illustrated the scale of Red Bull's success less than two decades after Mateschitz had the vision to create an "energy drink" and in so doing invent an entirely new sector of the soft drinks industry. The automated production lines were busy when we walked around, each producing 90,000 cans of Red Bull an hour, or twenty-five a second. And there were four lines.

Later I met Jurgen and Roman Rauch, whose family business is primarily known for producing fruit drinks, and who recalled the moment when this man named Dietrich Mateschitz came to them with the idea of producing his "new" product in their factory. His initial order quantities were so small they really didn't want to know, but when the original batch quickly sold out and he came back for more they began to realize Red Bull might be worth the effort.

From the outset Mateschitz's vision for Red Bull was to eschew traditional advertising techniques and instead to promote the product by creating events and getting cans of Red Bull into the hands of consumers. Event-based marketing was a core philosophy from the outset and, in embracing extreme sports, motor sport and aviation, Red Bull was very much reflecting the personal energy and enthusiasm Mateschitz himself had for skiing, Formula One and flying. As a lesson in leadership it illustrates that, when you combine your passions with your business, success can follow.

Many people in Europe will remember the early days of Red

Bull's rollout when Minis began appearing in our streets with a large Red Bull on the rear deck of the converted car. Guerrilla marketing was part of the Red Bull game plan, targeting everything from major events to traffic jams where promotional staff on scooters would hand out the product to frustrated motorists.

I asked Mateschitz about this when we met in 2004, and he told me that the whole point of their marketing is to motivate people to try the product. Their own research showed that some consumers simply did not like the taste at all—as many as 50 per cent—but that left half the world's population who might come to regard a can of Red Bull as a convenient way of getting the same caffeine rush as from a cup of coffee, and the more regular users will drink perhaps four to six cans a week.

From the start Red Bull was associated with high energy, youthfulness and vitality; it was edgy. Mateschitz was very clear about the brand values. Extreme sports showed cool young people performing outlandish feats on skis and snow boards, from base jumping to free climbing. Drink Red Bull and this "is" you.

Aviation sport was by no means a mainstream activity to reach consumers, but by using base jumping and free-fall parachuting, and then extending it into the creation of the Red Bull Air Race, which visited major cities around the globe, Mateschitz was again combining high energy and underlining the slogan adopted in their advertising: "Red Bull gives you wings."

Other events soon followed, from the motocross-based stadium events called Red Bull X Fighters to the very popular Red Bull Soap Box derby and amusing Red Bull Flugtag, which involves members of the public trying to build homemade, human-powered aircraft. There have been more than two dozen Red Bull-created event formats, each underlining the brand's values and putting the product into the hands of consumers worldwide. Each activity was carefully crafted so that it could travel well from market to market and be scalable.

Probably the most well-known event was Red Bull Stratos, when on 14 October 2012 Austrian free-fall parachutist and wing flyer Felix Baumgartner jumped from a special capsule floating at 128,100 feet above New Mexico. He broke the sound barrier during his free fall, reaching a terminal velocity of 833.9 mph, setting new world records for the highest free fall and highest human balloon flight. More importantly it was widely broadcast, with eighty television stations in fifty countries taking the live feed and rolling news channels replaying the jump endlessly. There were fifty-two million views online. It took Red Bull to new heights in terms of its message of "energy", being on top of the world and, of course, giving you "wings". It also contributed directly to Red Bull's record year, selling 5.2 billion cans worldwide.

Mateschitz's prowess for creating a brand marketing phenomenon is well established, therefore, but perhaps his greatest achievement in showing how his vision and leadership can drive success whatever the industry comes in the form of Red Bull Racing.

Jaguar Racing, owned by Ford Motor Company and formed as the result of the US firm's acquisition of Stewart Grand Prix from three times World Champion Jackie Stewart in 2000, was markedly unsuccessful and something of an embarrassment to the brand it represented. Between 2000 and the end of 2004 the team scored only two podium finishes, in the hands of Eddie Irvine, but failed to win a race, score a pole position or finish higher than seventh in the Constructors' Championship.

Ford, frustrated by the team's lack of success and unwilling to underwrite the budget for 2005, put the team up for sale with the threat that if a buyer could not be found then the team would close with the loss of several hundred jobs.

I had just joined the team to handle commercial affairs and found myself dispatched to China to work on a rather complex deal whereby Ford would sell the team to a Chinese entity. In so doing we would create a Ford Team China team, which would take the F1 liability off Detroit's desk and at the same time build a joint venture relationship with investors in China. Somewhere in Shanghai, to this day, lies a quarter-scale Jaguar F1 car painted in Ford Team China colours, much to the confusion of anyone who might find it.

Fortunately back in Europe a more obvious deal materialized when Red Bull, which had been sponsoring the team via its driver Christian Klien, came to the table with the idea of buying Jaguar Racing outright. Tony Purnell, Jaguar Racing's diligent team principal and CEO, was thankfully able to conclude an agreement with Mateschitz and the deal was done; Red Bull Racing was the result.

Those hundreds of jobs had been saved too, but what, everyone wondered, would a Red Bull Formula One team be like? If Ford and Jaguar failed to produce a winning car, what chance would a producer of energy drinks have to compete against the likes of Ferrari, McLaren and Williams? The only thing that seemed certain was that the parties would be good, for even in Milton Keynes the staff knew that Red Bull had a legendary reputation when it came to staging events.

When Mateschitz came to the factory a few days later he addressed the staff and explained very clearly what his vision for the business would be. He explained how, in the 1960s, he had become a big fan of Formula One. He liked the technology, the purity of the racing, the challenge that was involved. He said he liked the lifestyle around Formula One, but that it had all become a little bit serious, and so Red Bull would be putting the fun back into the sport and doing things differently from the rest. He also said that he wanted to win, to find the right people and spend the money in the right areas. He was very clear that Red Bull did things properly, and that Formula One would be no different.

Christian Horner and Adrian Newey, the team principal and technical director of the Red Bull Racing that came to dominate Formula One in 2010–13, weren't even in the audience that day. They had yet to be employed. This moment was ground zero for the creation of Red Bull's F1 project.

With his vision outlined, Mateschitz began to make the changes that he felt were absolutely necessary to turn Jaguar Racing from a loss-making, unsuccessful embarrassment into a team capable of winning. One of the early changes was to replace the senior management, notably Tony Purnell and David Pitchforth, with a fairly brutal "clear your desks" approach. New management was installed, with Horner brought in by Mateschitz's motor sport adviser Helmut Marko. A successful and ambitious team boss in lower formulae, Horner was regarded as a reliable lieutenant who could be trusted to deliver on Mateschitz's vision.

David Coulthard was recruited as one of the drivers for the 2005 season, and "DC" brought with him some vital qualities. He knew how to win and had worked with two of the most successful teams in the sport—Williams and McLaren. Understanding the culture of a winning team was important, and Mateschitz wanted to get the right people into key roles as soon as possible.

By the end of the team's first year changes had already been made to the way things were done, but some of this "pain" was balanced by the recruitment of Newey from McLaren. As a multiple championship-winning car designer, and someone whom DC had worked with at both Williams and McLaren, he was a key appointment. He was then empowered to put in place the people, structure and technology necessary to haul Red Bull Racing up the league table.

Newey's first full design for Red Bull came in 2007, since the 2006 car was already "done" when he joined, but in the meantime DC had scored the team's first podium finish in Monaco 2006. Regular points-scoring results and more podiums began to come the team's way, but 2009 became the breakthrough year with an inaugural victory in the Chinese Grand Prix. By season's end the team and lead driver Sebastian Vettel were second in both the Constructors' and the Drivers' World Championships, but little hinted of what was to follow: complete dominance of Formula One in the four seasons that followed, with successive titles from 2010 to 2013 inclusive.

Behind all of this success is Mateschitz's vision, outlined on that day in November 2004 when he stood in front of the relieved staff in their Milton Keynes factory. He had a clear vision of what he wanted to achieve, put in place the right people for the job, empowered them, provided the resources necessary to invest in the areas that mattered, and used Red Bull's global marketing power to build the team into a giant of the sport both on and off the track. In so doing he created not only a winning racing team, but a winning business, which is now itself sponsored by major global brands including Infiniti cars, Total oil, Pepe Jeans, Geox apparel and even Rauch—the Austrian fruit juice company that gave Mateschitz the production capability he so desperately needed all those years ago.

Working with Eddie Jordan and Dietrich Mateschitz, competing against Ron Dennis's McLaren, teaches a lot about leadership. In each case they eat, sleep and breathe their businesses and are relentless in their quest for success. They are consistent and clear in their vision for the business, require their staff to buy into that and drive businesses where everyone is responsible for their role and accountable for their actions.

EJ was focused on the bottom line, in his quest for the next deal. Ron Dennis wanted nothing left to chance; everything is about attention to detail. And in Dietrich Mateschitz we had someone who combined his business with his passions, found the right people to deliver for him in each area, and in so doing showed how to be the best in the world, from drinks to Formula One.

Lessons in leadership

Leadership requires a broad range of skills, and although not every leader can possess every quality it is worth reflecting on those attributes that have made successful business leaders in Formula One.

There are some common features. Although it is a much over-used word these days, having a genuine "passion" for the industry helps. Having that innate interest in the subject makes it much easier to make the commitment necessary to become the very best. It is clear that the really successful leaders work extremely hard at what they do, put in the hours and find that the returns are proportionate.

Eddie Jordan, Ron Dennis and Dietrich Mateschitz achieved success as the result of a lifetime's commitment to what they do. There was never any half-measure; each in his own way was relentless in his quest to achieve success. In each case it was a question of working harder than the competition in every area, driving towards having the best-engineered product, developing the most powerful brand and landing the most lucrative deals.

They set the tone for their businesses, and expected their employees to buy into the vision they set out. Whether it was Ron Dennis's attention to detail, EJ's focus on sales or Mateschitz's belief in the importance of the brand, each drove his business with great conviction and focus.

Great leadership requires that the leader is fully empowered. In each of the examples I have given the leaders of these top

Formula One businesses were self-made entrepreneurs. As Eddie Jordan said, his name was "over the door" of the business, and this translated into his leadership style; the business meant everything to him personally, and ultimately he knew the buck stopped with him. Leadership without authority is impossible, and it is for this reason that Formula One's most successful teams have traditionally been entrepreneurial in nature with strong, centralized leadership.

To summarize:

Centralized leadership works. Formula One teams have seen the advantages of strong centralized leadership supported by managers empowered to run the business functions. A "benevolent autocracy" is seen to be preferable to consensus-led management.

Empowered leaders can transform a business. As we saw with Jaguar–Red Bull and Honda–Brawn, empowered leadership can deploy resources faster and more efficiently, and create a culture of innovation and continuous improvement.

Be yourself, but look after the fundamentals. A leader can be unorthodox, like Eddie Jordan, but a strong work ethic and relentless sales drive were combined with a strong human touch that inspired personnel and drove success on and off the track.

Be different. Both Eddie Jordan and Dietrich Mateschitz saw the importance of differentiating their brands, standing out from the competition, and deploying the "fun factor" as an important weapon in nurturing staff loyalty and attracting customers.

Have greater attention to detail than your rivals. McLaren has always been a team with a pin-sharp focus on engineering excellence combined with world-class presentation. Aiming to be the very best in every single area has been a core message.

Understand who your customers are and focus accordingly. Designing the business to attract customers may seem obvious. McLaren's success in retaining clients for decades reflects their ability to match their proposition continuously to customer demands.

Don't be afraid of failure. Many leaders make mistakes, or suffer failures, on their way to the top. Ron Dennis, Frank Williams and Eddie Jordan all endured tough times early on in their careers; it didn't put them off, because they had the determination to keep trying.

Put the right people and resources in place. The McLaren story showed how Ron Dennis assembled the best customers, technical suppliers and investors to enable his business to create great products and place innovation at the heart of his business.

Communicate your vision to your staff. As Dietrich Mateschitz showed in his address to the staff of Red Bull Racing upon its creation in November 2004, it's important to be clear about what you are doing, why you are doing it and how you plan to go about it.

Believe in your brand. As leader you are "brand ambassador in chief", and it's important to have complete belief in your brand, its values and its proposition. Jordan, McLaren and Red Bull were very clear about their core values, attributes and goals.

GLOSSARY

A: from Accident Data Recorder
to Auxiliary Driving Features

Accident Data Recorder (ADR):
A module that collects accident data. Its installation has been required in every Formula One car since 1997. The ADR logs speeds and deceleration rates that occur in an accident. This data is analyzed to obtain findings about possible causes of the accident in order to further increase safety in Formula One.

Acoustic Signals:
Ten minutes before the start, an acoustic signal indicates that everyone except drivers, officials and technical staff have to leave the starting grid. Three minutes before the start there is another acoustic signal. At this time, the tyres must be properly fitted. Those who have not fitted their tyres within three minutes before the race starts receive a ten-second penalty. One minute before the start, the engines are started. By the time the fifteen-second signal is given, all team personnel must have left the grid and taken all equipment with them. If a driver then requires further assistance, he must raise his arm and he will be pushed into the pits.

Adjustable Rear Wing:
According to the moveable bodywork regulations introduced in 2011, drivers of suitably equipped cars can adjust the rear wing from the cockpit, altering its angle of incidence through a set

range. Also known as DRS (Drag Reduction System), it can be used at any time in practice and qualifying (unless race direction is suspending its use due to poor weather conditions or yellow flags in the activation zone), but only in the designated DRS zones. During the race it can only be activated when a driver is less than one second behind another car at pre-determined points on the track. The system is then deactivated once the driver brakes. In combination with ERS, it is designed to boost overtaking.

Aerodynamics:
The study of the interaction of air with solid bodies moving through it. The basic rule when designing cars for Formula One is simply to create as much downforce and as little air resistance as possible.

Air Box:
The air inlet behind the driver's head. The airbox channels the air necessary for the combustion process to the engine.

Allianz:
As one of the world's largest Financial Services Providers, Allianz has been committed to Formula One since 2000. In 2007, Allianz adopted a three-tiered approach to its Formula One engagement—combining its status as "Official Global Partner of Formula One" with trackside branding at key races, featuring the message "Drive Safely" and a classical team sponsorship (since 2011 with the MERCEDES AMG PETRONAS Formula One Team). Allianz, providing coverage for 50 million motor insurance customers around the world, effectively communicates its expertise in risk management and road safety via this sponsorship platform. Further proof of the insurer's commitment to road safety comes in the shape of the additional sponsorship for the official Formula One Safety Car and Medical Car, as well as a personal sponsorship with ex-Formula One driver and Allianz Safety Ambassador Christian Danner.

Allianz Center for Technology (AZT):
The AZT, based in Munich today, was founded as a full subsidiary of Allianz in Berlin in 1932. It has built an eighty-year reputation as a successful accident and damage research institute dedicated to practical knowledge and thorough analysis. Core areas of its expertise are risk management, consultancy services and damage reduction. The AZT conducts weekly crash tests to improve the safety of its 50 million motor insurance customers and other motorists. Eighty to ninety crash tests per year contribute to the development of technologies that help to reduce the incidence of traffic accidents and to minimize the resulting damage.

Apex:
The point at which the ideal racing line touches the inner radius of a corner.

Aquaplaning:
When there is more water between the tyres and the road than can be displaced by the tyre tread, the car "floats" and consequently cannot be controlled by the driver. Formula One races can be stopped if there is a danger of aquaplaning. Under very wet conditions, the Official Formula One Safety Car is generally used to keep the field at a lower speed.

Autoclave:
A pressure vessel, in which vacuum packed composite components are cured at 100–200° C for ten to twenty minutes up to twenty-four hours—depending on the piece and its purpose. This procedure lends the composite components their high strength while maintaining low weight.

Auxiliary Driving Features:
Traction control, automatic transmission or launch control are examples of auxiliary driving features. An expert team commissioned by the FIA may check at any time during the race weekend whether a car's electronics contain banned auxiliary driving features. In the 2004 season, launch control and automatic

transmission were banned and since 2008 traction control is also no longer permitted.

B: from Balaclava to Briefing

Balaclava:
Fireproof face mask made of Nomex® brand fibre, a flame retardant synthetic fibre. It is worn under the helmet.

Black Flag:
The black flag—together with the respective car number—is shown to drivers who should stop at their pit or near the pit lane entrance. If for any reason a driver does not respond, the flag should not be shown for more than four successive laps. When the stewards decide to show the black flag to a driver, they also immediately inform his team. A black flag with an orange circle informs drivers of technical problems with their car that could endanger themselves or others. They must go in for a pit stop immediately. They can then rejoin the race following repairs. The black/white flag together with the car number warns a driver about unsportsmanlike behaviour.

Blistering:
Formation of blisters on the tyres, caused by excessive use. The negative consequence is reduction in grip.

Blue Flag:
The blue flag is used when a faster vehicle approaches a lapped car from behind. The slower car has to make way immediately. The blue flag is also used at the pit lane exit to signal that a car is approaching on the track.

Boots:
Formula One shoes are ankle boots made of soft, cushioned leather. They have thin rubber soles with good grip to prevent drivers' feet from slipping off the pedals.

Brake Balance:
To gain a better balance when braking, the driver can adjust the brake force distribution between the front and the rear axle even during the race via a button on the steering wheel or via a lever at the cockpit wall.

Brake Discs:
Formula One brake discs are made of carbon. The discs may not be thicker than 28 mm and their diameter may not exceed 278 mm. Carbon brake discs and pads require an operating temperature between 350 and 550°C and reach up to 1,000°C during a braking process. Disc brakes were introduced in 1955.

Brakes:
The regulations call for two separate, independent hydraulic braking circuits operated from a single pedal. One circuit has to operate the brakes on the front axle, and the other the brakes on the rear axle, where a brake-by-wire system is in use from the 2014 season. This system electronically converts the driver's input on the brake pedal into braking force at the rear axle. Only one brake calliper and a maximum of six pistons are permitted per wheel. Brake callipers must be made of an aluminium alloy. Anti-lock braking systems (ABS) are not allowed. Full braking will bring a Formula One car from 200 to 0 km/h within 65 metres, all within 1.9 seconds. Anti-lock systems are prohibited, as are cooling systems using fluids. Force distribution may not change during the braking process.

Briefing:
At the meeting with the drivers and representatives from their teams convened by the race director before every Grand Prix, the discussions focus on current issues such as special features of the respective track or changes to the rules or weekend format. At the team briefings, the team managers, engineers and drivers set out the strategies for each day of the Grand Prix weekend. The subsequent review of the race day by this group, which forms the basis for future strategies and technical enhancements, is called the debriefing.

C: from Carbon to Cylinder

Carbon Fibre
A construction material for Formula One cars. The monocoque, for example, is made of epoxy resin reinforced with carbon fibre. These materials, when laminated together, give great rigidity and strength, but are very lightweight.

Carbon-fibre-reinforced Plastic (CFRP):
CFRP covers composite materials such as carbon and Kevlar® which, when combined with epoxy resins, provide high rigidity and strength and an extremely low weight. Many parts are produced from these materials, e.g. the monocoque.

Chassis:
The central part of a Formula One car, with the main component being the monocoque. All the other components are connected to the strong, lightweight monocoque. The chassis walls must be at least 3.5 mm thick with 2.5 mm reserved for a casing with DuPont™ Kevlar® brand fibre. The geometry of the chassis suspension must not be modified while driving.

Checkered Flag:
The checkered flag (black/white) indicates the end of a practice session, qualifying session or race.

Chicanes:
Tight corners that race organizers use to break up long, straight stretches of a circuit for safety reasons. Chicanes force drivers to reduce their speed and serve as overtaking opportunities, too.

Cockpit:
This is the driver's workplace. The cockpit must be designed so that the driver can get out easily within five seconds. The width of the cockpit must be 50 cm at the steering wheel and 30 cm at the pedals. The opening must have a minimum length of 85 cm. For safety reasons, no fuel, oil or water lines may pass through the cockpit. The cockpit temperature may reach an average temperature of 50°C.

Computational Fluid Dynamics (CFD):
CFD makes the airflows surrounding the vehicle visible on the computer, and at the same time shows the effects of individual vehicle parts on each other and on the aerodynamics. The engineers can simulate these effects without even having to build the parts first. That saves time and money.

Computer Aided Design (CAD):
Intelligent computer programs provide efficiency and speed and make the designers' work much easier. Drawing boards have long been a thing of the past in modern racing factories.

Concorde Agreement:
This agreement specifies the rights and obligations of the teams and the FIA. It also calls for unanimity for important decisions. The sixth Concorde Agreement so far—after 1981, 1987, 1992, 1997 and 1998—was signed on 1 August 2009 and remained in effect until 31 December 2012. As of now, there is no current Concorde Agreement.

Crash Barrier:
Safety measure at track locations where there is no space for run-off zones.

Crash Tests:
The FIA specifies and defines two main types of crash tests: static and dynamic. The crash tests were introduced in 1985. Since the cars start with more fuel due to the refuelling ban, the FIA has issued stricter rules for crash tests. As of 2012, only cars that have passed the mandatory crash tests can be used in test drives. There are tests for front, side and rear constructions. They are carried out under the supervision of the FIA, usually at the Cranfield Impact Centre in Bedfordshire, England. The front impact crash test is done at a speed of 15 metres per second, the lateral at 10 m per second and the rear at an impact speed of 11 m per second. The deceleration measured on the chest of the dummy may not be in excess of 60 g within three milliseconds. A fourth dynamic impact test relates to the steering column which must collapse

under a simulated head impact. The safety cell must remain undamaged after all the dynamic tests have been performed. The quick release for the steering wheel must also remain fully functional. In addition to the dynamic crash tests, the front, side and rear structures must withstand collateral pressure during static crash tests. The roll-over bar is tested in three directions: laterally with 5 tons, longitudinally with 6 tons and vertically with 9 tons. Deformation may not exceed 50 mm.

Curfew:
Since 2011 a curfew has been imposed on team members who work directly on the cars. During two six-hour time periods prior to the start of practice on Friday and Saturday they may not be present in the trackside facilities. The two time periods start eleven hours before the start of the first practice session on Friday and nine hours before the third practice session on Saturday, respectively. Each team is entitled to six exceptions during the course of the season.

Cylinder:
Component in the engine where the power is generated. The upward and downward movement of the piston and the combustion of the fuel air mixture take place in the cylinder.

D: from Differential to Drivers

Differential:
A differential is connected between the drive wheels to compensate the speed differences between the outer and inner wheels when cornering.

Diffuser:
Air outlet at the rear of the car's underbody that has a strong influence on the aerodynamic properties. Rising to the rear, the tail ensures a controlled airstream on the underbody which generates low pressure under the car and supplies the downforce critical to fast cornering. The double diffusers introduced in 2009 have been prohibited since 2011. As of 2012 it is no

longer permitted to blow fumes under the underbody to increase downforce.

Dimensions:
Whereas the length of a Formula One car is up to the designers, the width is limited to 1.80 m. The maximum width of the front wing must not exceed 1.65 m. The rear wing may not exceed 0.75 m in width and it may have no more than two wing elements. Car height is limited to 95 cm, measured from the lowest point.

Downforce:
Pressure that propels the Formula One car downward. It is generated by low pressure conditions under the body of the car, as well as by the angle of attack of the front and rear wings, and enhances the grip. Especially on slower circuits, this effect permits higher cornering speeds.

Drag Reduction System (DRS):
The Drag Reduction System (DRS) is a method to aid overtaking by altering the angle of the rear wing flap to reduce drag. Drivers are able to activate the system in designated DRS zones around a track. In practice and qualifying they can do so at will, during a race only if they are within one second of the car in front at the DRS detection point. DRS is deactivated again the first time the driver uses the brakes after activation. DRS may not be used during the opening two laps of a race. In wet conditions and with yellow flags race direction may choose to suspend the use of DRS.

Drive:
Two-wheel drive is the limit. Automatic and continuously variable transmissions are prohibited. As of 2014, ERS, the predecessor of which (KERS) has been permitted since 2009, has to be integrated into the power unit.

Driver's Seat:
The entire seat is one single unit and specially tailored to the respective driver. Drivers can be extracted from the cars together

with their so-called rescue seats. In the normal seating position, the soles of the driver's feet must not protrude from the centre of the front axle. Since 1971 the cockpit must be designed in such a way that the driver can be rescued within five seconds. Since 1999, regulations have stipulated that the seat may no longer be installed as a fixed part of the car. The risk of damaging the driver's spine when removing him from the car is thus eradicated.

Drivers:
Each team can use four drivers per season. The drivers may be substituted on a race weekend up to the start of qualifying. Any later substitutions due to *force majeure* are at the discretion of the stewards. In the first and second free practice sessions the race teams may use two additional drivers, who must also be in possession of a super license, but not more than two cars.

E: from Electronic Control Unit (ECU) to Energy Store

Electronic Control Unit (ECU):
The unit that controls all the electronic processes in a Formula One car. The ECU has been standard since 2008 and is designed by a manufacturer specified by the FIA. The current ECU supplier is McLaren Electronic Systems.

End-plate:
Vertical border area on the front and rear wings that helps to streamline a car's aerodynamics.

Energy Recovery System (ERS):
For 2014, the notion of hybrid energy recovery has shed a letter (KERS has become ERS) but become significantly more sophisticated. Energy can still be recovered and deployed to the rear axle via a Motor Generator Unit (MGU), however this is now termed MGU-K (for "Kinetic") and is permitted twice the maximum power of the 2013 motor (120 kW or 161 hp, instead of 60 kW or 80.5 hp). It may recover five times more energy per lap (2 MJ) and deploy ten times as much (4 MJ) compared to its 2013 equivalent, equating to over thirty seconds per lap at full power.

The rest of the energy is recovered by the MGU-H (for "Heat"); an electrical machine connected to the turbocharger. Where the V8 offered one possible "energy journey" to improve efficiency via KERS, there are up to seven different efficiency enhancing energy journeys in the ERS system.

Energy Store:
The Energy Store does exactly what it says on the tin; storing the energy harvested from the two Motor Generator Units (MGUs) for deployment back into those same systems. It is capped in terms of maximum and minimum weight: the maximum (25 kg) setting engineers an aggressive target, while the minimum (20 kg) means weight reduction will not be chased at all costs.

F: from Factory Shutdown to Fuel

Factory Shutdown:
All competitors must observe a factory shutdown period of fourteen consecutive days in August, during which time their wind tunnels and Computational Fluid Dynamics (CFD) facilities must not be used for Formula One activities.

Fading:
Technical term for the gradual loss of the brake effect after relatively long, heavy use. Occurs less with the modern carbon brakes than with conventional steel disc brakes.

F-duct:
This aerodynamic modification was introduced in 2010. A channel (duct) conducts air to the rear wing where it causes the flow to separate. This reduces downforce and aerodynamic drag, enabling the vehicle to achieve a higher end speed. The F-duct has been prohibited since 2011.

Fédération Internationale de l'Automobile (FIA):
The FIA is the international automobile umbrella organization and draws up the technical and sporting regulations for Formula

One. It is based in Paris. Since 23 October 2009, Jean Todt has been the new FIA president. The FIA was founded in 1904.

Fire Extinguisher:
Every Formula One car must have a fire extinguisher that spreads foam around the chassis and engine area. It must be operable both by the driver and from outside the car.

Flags:
The cars are fitted with diodes that transmit the flag signals from the marshals to the drivers in the cockpits.

Formation Lap:
Thirty minutes before the start of a Formula One race, the pit lane is opened and the drivers may drive one or several formation laps—in this instance the pit lane needs to be used. Vehicles that fail to finish the formation lap and are unable to reach the starting grid on their own power are not allowed to participate in the race. Fifteen minutes before the start, the pit lane is closed. If a car is still in the pits, it has to start from the pit lane.

Formula One:
The term "Formula One" was not introduced until after the Second World War. It was intended to identify top class motor racing. The first Formula One World Championship took place in 1950 under the direction of the FIA. The first race in the World Championship was the British Grand Prix on 13 May 1950.

Formula One Commission:
This commission consists of representatives from the teams, race organizers, engine manufacturers, sponsors, tyre manufacturers and the FIA. The commission decides whether changes to the regulations suggested by the FIA's technical committee should be implemented.

Formula One Teams Association (FOTA):
The FOTA was the association of all Formula One teams and was chaired by McLaren team principal Martin Whitmarsh. The

FOTA represented the racing teams in negotiations with the FIA and, for example, made suggestions to regulation changes. It was founded on 29 July 2008 and dissolved prior to the 2014 season.

Free Practice:
During practice sessions on Friday and Saturday before a Grand Prix, the lap times are recorded, but they have no influence on the starting order or the result. The teams use them as an opportunity to set up their cars for the respective track and work on the tyre wear.

Front Wing:
Creates downward pressure on the front area of the Formula One car and is thus an important part of the aerodynamics. Details of the front wing sometimes change for every new race—according to how much downward pressure is required for the respective circuits. Apart from that, the drivers make adjustments to the front wing, mainly modifying the angle of the second flap.

Fuel:
Only super unleaded petrol may be used in Formula One. It corresponds to a large extent to the fuel available at a conventional filling station with a minimum of 87 octane. However, the fuels contain additives that ensure faster and better combustion; in some cases, they are also lighter than commercially available petrol; 5.75 per cent of the petrol must originate from biological sources. Each team can choose its supplier independently, but it must submit two five-litre samples of the petrol used to the FIA before the season for test purposes. From 2014, only 100 kilograms of fuel are allowed in the race. Before that, teams were free to carry as much fuel as they thought fit.

G: from Gloves to Gurney Flap

Gloves:
Like the racing overalls, these are made of Nomex® fibre, a fire-resistant material. The close fitting gloves with suede leather palms provide the necessary sensitivity for steering.

Graining:
Due to excessive use, tyres show signs of corrosion and the rubber compound begins to disintegrate. This is referred to as graining. The negative consequence is reduction in grip.

Grand Prix Drivers Association (GPDA):
Association representing the interests of Formula One drivers.

Gravel Trap:
Run-off zone at a racing circuit that quickly slows down cars that have gone off the track. It is filled with small gravel stones of between 5–16 millimetres diameter and is about twenty-five centimetres deep.

Green Flag:
The green flag indicates the track is clear. The green flag is also waved after a spell of yellow flags.

Grip:
Describes how much the car adheres to the ground. High grip means high cornering speeds. Main factors of grip are the aerodynamics, the downforce created by the vehicle and the tyres' properties. Without grip, a vehicle will begin to slide or skid.

Ground Clearance:
The distance between the underbody and the surface of the track.

Ground Effect:
The contact force generated by an aerodynamically shaped underbody. In the late seventies and early eighties, sills were attached to the sides of the cars to create a vacuum underneath the vehicle that held it down on the track. The enormous resulting grip allowed for extremely high cornering speeds. The pure ground effect cars developed in the seventies were banned by the FIA for safety reasons in late 1982.

Gurney Flap:
L-shaped counterflap on the trailing edge of a car's wing, which was invented by the American race driver Dan Gurney.

H: from Hairpin to Helmet

Hairpin:
Very narrow turn. The most famous hairpin is the former Loews hairpin in Monaco, which is now known as the Grand Hotel hairpin.

Head and Neck Support (HANS):
Since the 2003 season the drivers have been given additional head and neck protection. The Head and Neck Support system consists of a carbon shoulder corset that is connected to the safety belts and the driver's helmet. In case of an accident, HANS is intended to prevent a stretching of the vertebrae. Additionally, it prevents the driver's head from hitting the steering wheel. The HANS was invented by Jim Downing and Robert Hubbard.

Head Support:
The removable padding on the inside of the cockpit. The cockpit is fitted with removable padding around the driver's head, designed to absorb any impact. The two side pads must be at least 95 mm thick and the rear pad between 75 and 90 mm.

Helmet:
The helmet is made of carbon, polyethylene and Kevlar® and weighs approximately 1,250 g. Like the cars, it is designed in a wind tunnel to reduce drag as much as possible. Helmets are subjected to extreme deformation and fragmentation tests. As of 2011, the most vulnerable part of the helmet, the visor, is reinforced with Zylon strip to increase its impact performance and to provide even more effective head protection. Only helmets tested and authorized by the FIA may be used.

I: from Internal Combustion Engine to International Sporting Code

Internal Combustion Engine (ICE):
The Internal Combustion Engine (ICE) is the traditional, fuel-powered heart of the Power Unit; previously known simply as the engine. For 2014 this took the form of a 1.6-litre, turbo-charged V6 configuration, with direct fuel injection up to 500 bar of pressure. Where the V8 engines could rev to 18,000 rpm, the ICE is limited to 15,000 rpm from 2014. This reduction in crank-shaft rotational speed coupled with the reduction in engine capacity and number of cylinders, reduces the friction and thus increases the total efficiency of the Power Unit. This down-speed-ing, down-sizing approach is the key technological change at the heart of the ICE structure.

International Court of Appeal:
The FIA's Court of Appeal is composed of professional judges, and its twenty-three members are appointed for a three-year term. In order for the court to make a legally binding decision, the presence of at least three judges is required, none of which may be of the same nationality as the parties involved. A Formula One team that is unwilling to accept a decision by the racing commis-sioners can appeal to the FIA's International Court of Appeal. In this case, a declaration of intent must be submitted within an hour of the decision. The FIA, too, can send a decision by the commissioners to the Court of Appeal. Since December 2009, the teams have been allowed to nominate one of the three judges.

International Sporting Code:
The FIA code that contains all the regulations governing interna-tional racing.

J: like Jump Start

Jump Start:
A jump or false start is committed by drivers whose cars start moving before all the lights on the starting grid have gone out.

This is determined by sensors on the starting straight. A jump start normally results in penalties imposed by the race stewards.

K: from Kerbs to Kevlar®

Kerbs:
Raised kerbstones lining corners or chicanes on racing tracks. The kerbs provide additional safety as the drivers must reduce their speed when driving over them.

KERS (Kinetic Energy Recovery System):
KERS was used for the first time in the 2009 season. For the following year the teams agreed to suspend its use for cost reasons. In 2011 the system returned to Formula One, but was suspended again for the 2014 season with ERS taking its place. KERS recovered kinetic energy (which is normally wasted) under braking and made it available to the driver for about 6.6 seconds as a 60kW boost when he pushed a button. This power boost could either be used once or in quantities during a lap. To prevent heavier drivers from being disadvantaged, the minimum weight of the car including the driver had been raised.

Kevlar®:
Highly durable artificial fibre used in the covering of the headrest. Combined to form a composite with epoxy resin, it has high strength, but is very lightweight.

L: from Logistics to Lollipop

Logistics:
The tour of Formula One around the globe demands sophisticated logistics. For every race, around 120 crates of different sizes have to be packed with the help of a twenty-page checklist. The two race cars are always part of the cargo—plus spare parts and tools, wheels and the pit lane equipment. The luggage also includes several engines. PCs and notebooks, secure data lines and radios are all part of the basic equipment of every team. For European races the equipment is transported to the

venues by trucks, for races on other continents by chartered cargo planes.

Lollipop:
The signal pole with a sign saying "Gear" on one side and "Brake" on the other. During a pit stop, the chief mechanic posted in front of the car uses the sign to show the driver when he should apply the brake and when he should shift gear and drive off.

M: from Manufacturers to Motor Sport Safety Development Fund

Manufacturers:
Any manufacturers wanting to enter Formula One must prove to the FIA that they have designed and built the chassis of their racing cars. They are also obliged to compete in all the races in a particular season and to prove that they possess the necessary technical and financial means.

Marshals:
Officials posted along the side of the track. They wave the flag signals and secure any possible accident sites; they also rescue any cars that have broken down.

Medical Car:
The Official Formula One Medical Car is staffed by a driver and the Formula One rescue coordinator of the FIA. Like the Safety Car, it is on standby at the exit of the pit lane during every practice session and race. Since 2009 it has been driven by the former Formula Three Champion Alan van der Merwe. Dr Ian Roberts, the official Formula One physician, is also on board.

Medical Centre
Every Formula One race and test circuit must have a state-of-the-art emergency service facility staffed by experienced physicians. A rescue helicopter must always be on standby, ready for lift-off.

MGU-H:

The Motor Generator Unit-Heat (MGU-H) is a new electrical machine that is directly coupled to the turbocharger shaft. Waste exhaust energy that is in excess of that required to drive the compressor can be recovered by the turbine, harvested by the MGU-H, converted into electrical energy and stored in the Energy Store. Where the MGU-K is limited to recovering 2 MJ of energy per lap, there is no limit placed on the MGU-H. This recovered energy can be used to power the MGU-K when accelerating, or can be used to power the MGU-H in order to accclerate the turbocharger, thus helping to eliminate "turbo lag". This new technology increases the efficiency of the Power Unit and most significantly provides a method to ensure good driveability from a boosted, downsized engine.

MGU-K:

The Motor Generator Unit-Kinetic (MGU-K) has double the power capability of the previously used KERS motors and operates in an identical way. Some of the kinetic energy that would normally be dissipated by the rear brakes under braking is converted into electrical energy and stored in the Energy Store. Then, when the car accelerates, energy stored in the Energy Store is delivered to the MGU-K which provides an additional boost up to a maximum power of 120 kW (approximately 160 hp) to the rear axle for over thirty seconds per lap.

Monocoque:

French for single shell. A safety cell made of carbon-fibre composite that forms a protective shell around the driver. In some parts the monocoque even has sixty layers of carbon fibre. The "drivers' life insurance" is surrounded by deformable structures that absorb energy in an accident. The molding and binding process takes place within an autoclave at high levels of pressure and heat.

N: from NACA Duct to Nose

NACA Duct:
The NACA duct is a common form of low-drag intake design, originally developed by the National Advisory Committee for Aeronautics (NACA). It is a triangular air inlet on the surface of the car body.

Nomex®:
Artificial fibre that undergoes thermal testing in the laboratory. It is subjected to an open flame with a temperature of between 300–400°C that acts on the material from a distance of 3 cm— only if it fails to ignite within ten seconds can it be used for racing overalls. The drivers' and pit crews' underwear, socks and gloves are also made of Nomex®.

Nose:
Front part of a Formula One car, subjected to various crash tests for safety reasons. The nose also functions as a protruding crash structure protecting the monocoque. For the 2014 season, the nose height was reduced to 185 instead of 550 millimetres due to safety reasons.

O: from On-board Camera to Oversteering

On-board Camera:
A mini TV camera on board the racing car, which can be attached near the airbox, the rear mirror or the front or rear wing. It provides live footage throughout the race weekend.

Overall:
Protective suit with elastic cuffs on wrists and ankles made of two to four layers of Nomex® for drivers and pit crews. A completed multi layered overall undergoes fifteen washes as well as a further fifteen dry-cleaning processes before it is finally tested. It is subjected to a temperature of 600–800°C. The critical level of 41°C may not be exceeded inside the overall for at least eleven seconds.

Oversteering:
When oversteering, a car's rear wheels lose grip and break away. In order to get through the corner, the driver must decrease his steering angle or, in the case of extreme oversteering, even steer in the opposite direction—called opposite lock.

P: from Paddle to Push Rod

Paddle:
Manual gear shifter behind the steering wheel.

Parc Fermé:
Restricted area of the pit lane in which the FIA's technical stewards inspect the cars after each race to make sure they conform to technical regulations. Team members are not admitted to this area. As soon as a vehicle leaves the pit lane in qualifying, the Parc Fermé rule comes into effect as well but it is not linked with a certain area. Up to the race the teams are only allowed to make minor changes to the car. For more extensive work such as repairs of accident damage FIA approval must be obtained. If modifications are made to the set-up (in case of expected weather changes, for instance), the respective driver must start from the pit lane. Three-and-a-half hours after the end of qualifying the vehicles are wrapped in special tarpaulins in the team garages and the wraps are sealed by the FIA. Overhead cameras are used for night-time monitoring to verify that the race teams comply with the ban on working on the cars. Five hours before the start of a Grand Prix race the wraps may be removed, but the Parc Fermé rules remain in effect.

Penalties:
The stewards can (amongst other things) impose time penalties, disqualification or a ban for subsequent races in case of violations of the rules. The race director can recommend time penalties for a false start, causing a collision, forcing another car off the track, not responding to a blue flag, and deliberately hindering another driver. The final decision for a stop-go and drive-through penalty, as well as places added to a driver's grid

position, is made by three race stewards, the official Formula One race referee jury. Since 2010 they have been assisted by a race steward from the national automobile association and experienced former Formula One drivers or experienced drivers from other categories. The stewards may use the video footage and radio communications of the race teams to make their decisions. If drivers commit sport-related or technical violations during qualifying, the racing commissioners can cancel all their qualifying times. Additionally there are penalties for technical defects.

Pit Lane:
This is where changes to the car take place. During practice sessions the speed limit in the pit lane is 80 km/h, as it is during qualifying and races. On street circuits, where pit lanes are especially narrow, the speed limit is reduced to 60 km/h for practice, qualifying and race alike. The pit order acts in accordance with the teams' position in the Constructors' Championship of the previous season.

Pit Stop:
During a regular pit stop in a race, a team of twenty-seven mechanics changes the tyres on the car and possibly performs further mechanical or aerodynamics settings. Between 1994 and 2009, cars were also refuelled during pit stops. Refuelling has been banned since the 2010 season. It takes a well-trained crew less than three seconds to change all four tyres on a modern Formula One car.

Points:
Since the 2010 season, the first ten drivers in each race have been awarded points for the championship ranking. The winner of the Grand Prix is awarded twenty-five points, while the drivers that follow receive eighteen, fifteen, twelve, ten, eight, six, four, two and one respectively. The same points system is used for the Constructors' Championship. Both cars of each team can collect points in one race. For the last race of the 2014 season, double points were awarded for both the Drivers' and

the Constructors' Championship, for the first time—although this scoring system was subsequently dropped for the 2015 campaign.

Pole Position:
First place in the starting order for the race, which is given to the fastest driver in qualifying. In 2014, Formula One introduces the Pole Position Trophy for the driver who earns the most Pole Positions during the season.

Power Unit:
In regulatory terms, the Power Unit comprises six different systems: the Internal Combustion Engine, Motor Generator Unit-Kinetic (MGU-K), Motor Generator Unit-Heat (MGU-H), Energy Store (ES), Turbocharger and the Control Electronics. The change in terminology reflects the fact that this new powertrain is far more than simply an Internal Combustion Engine. Where the previous V8 format utilized a KERS hybrid system which was effectively "bolted on" to a pre-existing engine configuration, the new power unit has been designed from the outset with Hybrid systems integral to its operation. The 1.6-litre V6 engine delivers approximately 600 bhp and gains additional 160 bhp through its ER system for about thirty-three seconds per lap.

Pull Rod:
A suspension layout where the suspension rockers are pulled. Long forgotten, it returned to Formula One in 2009, at first on the rear suspension. 2012 saw the return of vehicles with a pull rod layout on the front suspension for the first time in eleven years.

Push Rod:
This layout, where the suspension unit is operated with a strut, is used in the front and rear of most Formula One cars.

Q: like Qualifying

Qualifying:
The starting order for the race is determined during qualifying.
The driver with the fastest lap time qualifies for the best starting
place: pole position. Qualifying is conducted in shoot-out format
in three individual sessions. In the first session the slowest seven
drivers are eliminated, and seven more in the second one. The ten
fastest drivers fight for the pole position in the third session. A
driver who on his fastest lap in Q1 fails to post a time that is
within 107 per cent of the fastest driver in qualifying is not
allowed to contest the race. The top ten drivers of qualifying have
to start the race on the tyres they have used for their fastest lap in
the final qualifying session.

R: from Race Director to Run-off Zone

Race Director:
The FIA race director supervises the safety measures on the race
weekend and makes improvements when necessary. Additionally,
he decides whether the Safety Car should be deployed or whether
the race should be stopped. If a driver does not behave in a
sportsmanlike manner or if he endangers a competitor, the race
director can recommend a penalty. The current FIA race director
is Charlie Whiting from the UK.

Race Distance:
The smallest number of laps needed to exceed a distance of 305
km (exception: Monaco, 260 km). The maximum duration of the
race is two hours. The entire duration per Grand Prix, including
possible interruptions, has been limited to four hours since 2012.

Race Stop:
If weather conditions are so poor as to endanger safe driving (e.g.
heavy rain, snow, fog) or if a vehicle is blocking the track, a red
flag signals that the race has been stopped. If a race is stopped
during the first two laps, it is started again. If this is not possible,
no points are awarded. If a race is stopped after the first two laps,

it is restarted. In this case, half of the total points are awarded. The total number of points is awarded if 75 per cent or more of the race are completed.

Racing Line:
Also known as the ideal line, the racing line is the imaginary line on which the circuit can be driven in the fastest possible time. Due to the rubber build-up, this is also usually where the grip is best.

Rear Light:
Decreases the risk of pile-ups. When using wet weather tyres, the rear light must always be switched on. The red rear light must be positioned between 30 and 35 cm above the floor of the car. The rear light also lights up when a car is slowing down ahead of a corner while recovering energy on the rear axle.

Rear Wing:
Also known as rear wing assembly. Creates downward pressure, mainly upon the rear axle. The rear wing is adapted to the conditions of the tracks (the steeper it is, the more downforce is created). The settings and angles of the surfaces can be additionally modified. These modifications are part of the set-up. To facilitate overtaking manoeuvres, the regulations allow the drivers to adjust the rear wing from the cockpit since 2011.

Red Flag:
The red flag is shown simultaneously at the starting line and around the circuit when a practice or a qualifying period or race is stopped.

Refilling/Refuelling:
Nitrogen and compressed air are the only gases that may be replenished during the race. Since 2010 refuelling during races has been banned. This has a big effect on race strategy as drivers have to pay more attention to tyre and brake conservation. As from 2014 the maximum amount of fuel for the race is 100 kilograms, drivers have to save fuel, too, to make it to the finish line.

Regulations:

The FIA draws up the sporting and technical regulations for Formula One. The technical regulations primarily aim at two important things: speed should be controlled in the interest of safety, while simultaneously retaining the ongoing technical development so critical to the nature of Formula One. In addition, safety is to be guaranteed in the event of an accident. To achieve these aims, the following factors have been limited: engine capacity, fuel composition, tyre size, tyre contact surface, minimum weight and width of the cars. The sporting regulations primarily control the "procedure of a Grand Prix weekend, such as qualifying, deployment of the Safety Car and the podium ceremony".

Roll-out:

The first test drive of a new racing car, usually at a private test.

Roll-over Bar:

If a car rolls over in an accident, the roll-over bar, a curved structure above the driver's head made of metal or composite materials, is intended to provide the driver with better protection.

Rubber Build-up:

This is a knock-on effect of the slow erosion of tyre surfaces. When tyres are driven on asphalt, the surface rubs off and leaves behind a layer of rubber on the road, which accumulates over the course of the racing weekend and progressively enhances grip. This erosion is influenced both by the vehicle set-up and the abrasive properties of the asphalt.

Run-off Zone:

Run-off zones are empty spaces directly beside the actual racetrack. They are supposed to passively or actively decelerate cars that are out of control and prevent a collision with track walls or barriers. Only recently, an increasing number of asphalted spaces have been introduced at various circuits as drivers have a better chance of regaining control of their vehicle. Previously, gravel pits were more common. Although gravel has a decelerating effect, the chances of controlling the car are fairly low and the danger of getting stuck is rather high.

S: from Safety Belt to Suspension

Safety Belt:
The safety belt used by the driver in the cockpit is also known as a six-point harness and can be opened with a single hand movement.

Safety Car:
The Official Formula One Safety Car is deployed when the race director wants to reduce speed for safety reasons—for instance, after an accident or because the track is waterlogged after heavy rainfall—"whenever there is an immediate hazard but the conditions do not require the race to be interrupted". It moves onto the track regardless of what the current race leader's position is and all cars line up behind it. During a Safety Car phase there is an absolute ban on overtaking and the drivers have to reduce their regular lap time. The Official Formula One Safety Car was introduced in 1992. Since 2012 it has been a Mercedes-Benz SLS AMG with 571 horsepower, weighing 1,620 kg.

Safety Car Driver:
Since 2000, the Official Formula One Safety Car has been driven by Bernd Mayländer, a German former touring-car driver. He has raced in Formula Ford, the Porsche Carrera Cup, the FIA GT Championship and the Deutsche Tourenwagen Meisterschaft (DTM). His co-driver is Peter Tibbetts.

Scrutineering:
Scrutineering, the technical approval of the cars, takes place on the day before the first free practice session of a Grand Prix. The scrutineers of the FIA check whether the vehicles comply with regulations.

Set-up:
General vehicle tuning for all the adjustable mechanical and aerodynamic parts (wheel suspension, wings, etc.). Specifically, the

term describes the various possibilities for adapting a Formula One car to the conditions of a particular circuit. Included are, among other things, modification to the tyres, suspension, wings and engine and transmission settings.

Shake-down:
The final test drive of a newly prepared car before the team departs to a Grand Prix.

Sidepods:
Side cladding of the cockpit which is integrated in the monocoque. The sidepods contain crash structures that absorb the forces arising from an accident or impact. The radiators are also located in the sidepods.

Skid Block:
A plate made of plastic or wood fitted to the underbody of a racing car. It is intended to prevent a strong suction effect, limiting excessively high speeds, especially in the corners, for safety reasons. It also acts as protection for the underbody.

Slicks:
Slicks are tyres without tread patterns. In 2009 slicks were reintroduced to Formula One in order to facilitate overtaking. Slicks provide around 20 per cent more grip compared to the grooved tyres used from 1998 until 2008.

Slipstream:
Low pressure area behind a Formula One car created by air currents. Driving in the slipstream can provide a boost to a car's speed, making it the ideal position for a pursuing vehicle to start an overtaking manoeuvre.

Spare Car:
Replacement cars are no longer permitted in Formula One. However, if a driver irreparably damages his car the team can prepare another car. If the driver changed his car between qualifying and race, he must start from the pits.

Speed Limiter:
The cruise control feature used in Formula One pit lanes. It is activated by pressing a button on the steering wheel. Speed is then reduced down to the pit lane limit.

Speed Limits:
At most of the tracks a speed limit of 80 km/h will be imposed in the pit lane during all sessions. However, this limit may be amended by the stewards following a recommendation from the FIA Formula One safety delegate, for example in Monaco, Melbourne or Singapore, where the pit lane is particularly narrow and a speed limit of 60 km/h is imposed. Except in the race, any driver who exceeds the limit will be fined €200 for each km/h above the limit (this may be increased in the case of a second offence in the same event). In the race, speeding leads to a drive-through penalty.

Stabilizers:
Rotary or torsion bars that connect right and left wheel suspensions flexibly to each other. The so called "rollbars" help to reduce the rolling movement of the chassis along the longitudinal axis and so provide more precise handling during load shifts.

Starting Grid:
Formula One uses a standing start. The deciding factor for the starting position of a driver is his time from qualifying. The driver with the fastest lap time starts from pole position. The cars line up at offset eight-metre intervals. Each row of the starting grid has two race cars, one slightly in front, with a distance of eight metres to the next row.

Starting Lights:
As soon as the last car is positioned on the starting grid, the five starting lights go on successively at one-second intervals. The race starts when all the lights go off at once.

Starting Number:
The maximum number of teams that may race in Formula One is thirteen, with two cars each. All cars have to be fitted with the

starting number of the driver. For the 2014 season drivers were able to choose their starting number as Formula One moved from pre-assigned starting numbers to personal starting numbers the drivers are supposed to keep for their whole career in Formula One Only the reigning World Champion can choose to run number one. However, he may go with a different number, too. In this case, number one will not be given to another driver. The highest possible starting number is ninety-nine.

Steering Wheel:
A Formula One car's steering wheel is the control centre for the driver. He steers, clutches, changes gear and is able to influence electronic functions by means of numerous buttons. A small screen displays current car statistics. The design and the arrangement are adjusted to suit the individual driver.

Strategy:
Formula One teams can use their own race strategy. The number of scheduled pit stops is optional. However, the drivers have to cover at least one stint (section up to the next pit stop) on tyres with both prescribed compounds. Typically, there are two to three pit stops. Depending on the race situation (for instance a Safety Car period), the teams may also change their strategy.

Super Licence:
Formula One driving licence issued by the FIA. In the interest of safety, it is only granted on the basis of good results in the junior series or, in exceptional cases, if other proof of ability can be supplied. It may also be granted under provisional terms. To get a Super Licence, a driver must demonstrate that he is capable of handling a modern Formula One car over a full race distance of about 300 kilometres in an appropriate amount of time.

Suspension:
Several years ago, the wheel suspension was the Achilles' heel of a Formula One car, but the use of composite materials has since made it extremely robust. Basically, double arms are used at the front and rear, and each team gives them a different aerodynamic shape.

T: from Tank to Tyres

Tank:
The fuel tank is a fibre-reinforced rubber hull that must yield flexibly when deformed. It must fulfil the FIA's rigid criteria. To avoid damage, the tank is also located within the monocoque and is thus encased in the survival cell, the car's best-protected area. For the 2014 season, a car only can use 100 kilograms of fuel in the race.

Team Order:
The clause prohibiting team orders was removed from the Sporting Regulations in 2011.

Technical Committee:
These FIA experts lay down the Formula One regulations. Every team's technical director is a member of the Technical Committee. The Committee makes recommendations to the FIA Formula One Commission. The decisions made by the Commission are in turn forwarded to the FIA's World Motorsport Council and must then be approved by the FIA's general assembly.

Technical Delegate:
The FIA technical delegate leads the team of technical inspectors (scrutineers). They check whether the cars meet the regulations. If the technical delegate does not think a car conforms to the rules, he submits a report to the racing commissioners, who are authorized to impose penalties.

Telemetry:
A system allowing a large quantity of data, e.g. concerning chassis and engine, to be recorded in the car and transmitted to the pits. There, the data is analyzed so as to determine any faults (a loss of brake fluid or a slow puncture, for example) at an early stage and to be able to improve the car's set-up.

Test Runs:
During test runs Formula One teams try out new developments and various set-ups on the car. However, the teams are substantially

limited since 2009. Teams may not exceed 15,000 test km during a calendar year. From one week prior to the first race until 31 December, teams may only test during four two-day tests and during the Young Driver Test at the end of the season. In addition, each team may use four days for aerodynamics tests. Two promotional events, with a maximum distance of 100 kilometres each, are allowed without counting towards the total kilometre tally.

Time Penalty:
This is a penalty during the race for drivers who have violated regulations. Once his team has been informed by the racing commissioners, the driver must drive through the pit lane within the next three laps. He may not stop there to change tyres. Entering and leaving the pit lane costs the penalized driver valuable time. If the penalty is imposed during the last five laps, the driver no longer has to sit it—instead, he will have twenty-five seconds added to his final time. In more severe cases, race direction may impose a stop-and-go penalty, which requires the driver to park in front of his pit garage for ten seconds before rejoining the race. Again, no work on the car is allowed.

Torque:
Generated in the engine by the combustion pressure acting on the crankshaft via the pistons and the connecting rods. The maximum torque is a benchmark for the power and usability of the engine and the acceleration capacity of a racing car.

Traction:
This term describes the ability of a race car to apply its engine's power to the track.

Traction Control:
An electronic system, also called anti-slip control. It uses sensors to detect whether the wheels are spinning and then automatically reduces the engine power. This guarantees ideal acceleration, especially at the start, when leaving a corner and on wet tracks. Traction control has been prohibited since 2008.

Transmission:
A Formula One car may have a maximum of eight forward gears.
A reverse gear is prescribed. The gear wheels in the transmission
must be made of steel.

Turbocharger:
The turbocharger is an energy recovery device that uses waste
exhaust energy to drive a single stage exhaust turbine that in turn
drives a single stage compressor via a shaft, thereby increasing
the pressure of the inlet charge (the air admitted to the engine for
combustion). The increased pressure of the inlet charge offsets
the reductions in engine capacity and RPM when compared to
the V8, thus enabling high power delivery from a down-speeded,
down-sized engine. The turbocharger is the key system for
increasing the efficiency of the Internal Combustion Engine.

Turbo Engines:
The first turbo engine was used in Formula One in 1977. In
qualifying, these engines boasted up to 1,400 bhp. They were
banned from Formula One in 1989 and re-introduced for the
2014 Formula One season. The current engine format boasts
1.6-litre V6 engines which deliver approximately 600 bhp at
15,000 rpm. The new Formula One power units are closely
linked to the ERS which provides additional 160 bhp for about
thirty-three seconds a lap.

Tyres:
The teams have the choice between two compounds that differ in
terms of hardness and three tyre specifications (Dry, Intermediate
and Wet) on a Grand Prix weekend. Both tyre compounds have to
be used during a race. The specification is selected depending on
the weather and track conditions. Per weekend, a total of thirteen
dry, four intermediate and three wet tyre sets may be used. The
front tyre must be between 305 and 355 mm, the rear tyre between
365 and 380 mm wide. The diameter of the rim must be between
328 and 332 mm, the diameter of the wheels must not exceed 660
mm (with Dry tyres) and 670 mm (with Wets). Since 2011 the
tyres have been supplied by the Italian manufacturer Pirelli.

Tyre Stack:
Tyre stacks have been mandatory at racetracks since 1981. The tyre barrier consists of two to six rows of conventional car tyres that are bolted together and connected by rubber bands. This achieves an optimal absorption of the impact energy.

Tyre Warmer:
The tyres require an operational temperature of around 100°C to achieve optimal effectiveness. In order to rapidly reach this temperature, special electric blankets pre-heat the tyres up to 60–80°C. Cold tyres do not develop enough grip. If they are too hot, they wear out quickly.

U: from Understeering to Underwear

Understeering:
When front wheels lose grip but the back ones do not.

Undertray:
The aerodynamically shaped lower surface of a racing car creates an airflow, which in turn generates a vacuum under the car that provides better grip. However, continuous air ducts are banned in Formula One and are prevented by the skid block, which splits the airflow.

Underwear:
Under the racing overall, drivers wear a T-shirt, boxers, socks and a balaclava. All the underwear is made of fire resistant material made of Nomex® brand fibre.

V: like Valves

Valves:
The task of the engine-controlled valves is to open or close the inlet and outlet ducts at the right moment and so to allow the gases into or out of the combustion chamber. Each valve consists of a stem and a disc.

W: from Weight to World Champion's Title

Weight:
A Formula One car including the driver in full racing gear, with oil and brake fluid, must weigh at least 691 kg whilst on the track. The vehicles' construction weight is actually less. The teams can achieve a better weight distribution using additional weights, thus improving the handling. The technical commission of the FIA may, at any time, send cars to the electronic scales located at the entrance to the pit lane.

Wet Weather Tyres:
In wet weather, cars use special tyres that are better able to displace water from the track and optimize grip.

Wheels:
Formula One uses relatively small wheel sizes of thirteen inches. Most road cars are fitted with wheels ranging between sixteen and twenty inches. Instead, the wheel rims are much lighter due to the use of magnesium. The width of the rear wheels must be between 365 and 380 mm, and between 305 and 355 mm at the front. The maximum wheel diameter is 660 mm for dry-weather tyres and 670 mm for wet-weather tyres.

Wheel Tethers:
Each wheel is connected to the chassis by means of high-performance tethers (PBO, Zylon). They are intended to prevent the wheels from flying off in the case of an accident. Each tether has to withstand a load of seven tons.

White Flag:
The white flag indicates that a slow vehicle is on the circuit.

Wind Tunnel:
The holy shrine of every Formula One team and indispensable for the development of a race car. Aerodynamic studies are carried out in the wind tunnel. Using various flow speeds, the engineers can simulate various car speeds and can test the

effects of new vehicle parts or the aerodynamic behaviour of the entire car in various racing situations. Since 1 January 2009 teams may only test with models that do not exceed a scale of 60 per cent for cost reduction reasons. Additionally, tests in the wind tunnel must not exceed wind speeds of 50 m per second. For 2014, the wind tunnel rules have been tightened again in terms of, for example, how much time teams can spend evaluating in the wind tunnel.

Winglet:
Additional wing located on the car body just in front of the rear wheel. Prohibited since 2009.

Wings:
Fixed surfaces that are intended to increase downforce. The wings serve to press the car downwards more firmly. The secret of wing adjustment lies in finding the best compromise between high speed on straights (low downforce) and optimal performance in corners (high downforce). The front wings are 1,650 mm, the rear wings 750-mm wide. To facilitate overtaking, adjustable rear wings (DRS) have been permitted since 2011.

Wishbones:
The components connecting the wheel suspension and the chassis. Wishbones are mounted at right angles to the vehicle's longitudinal axis. These pivoting rods, which have also acquired aerodynamic significance, must be made of extremely strong materials.

World Champion:
With seven titles to his credit, Michael Schumacher is the most successful racer in Formula One history.

World Champion's Title:
In Formula One, two World Championship titles are awarded— one for the drivers and one for the manufacturers. The Drivers' Title has existed since 1950, and the Constructors' Title was

introduced in 1958. For the drivers, the points won in all the races are added up. If several drivers have the same points total, the title is determined by the final positions they achieved: the number of first places, followed by the number of second places, etc. In the Constructors' division, the points that both of the team's drivers earn each race are added up.

X: like X-wing

X-wing:
Additional wings developed by the Tyrrell team and first used in 1997. The X-wings created high levels of downforce. For safety reasons, the FIA banned them before the Spanish Grand Prix in 1998.

Y: like Yellow Flag

Yellow Flag:
The yellow flag waved once indicates danger. When this is shown, drivers should reduce speed, refrain from passing and be prepared to deviate from their ideal racing line. If the yellow flag is waved twice, drivers have to be prepared for a full stop. The yellow flag with red stripes informs drivers that oil or water is on the track.

Z: from Zip to Zylon

Zip:
While the zip on a driver's overall is hidden behind the layers of Nomex® fibre, it has to be able to withstand the same temperature that the suit will take in the event of a fire. It must not melt or transfer heat close to the driver's skin.

Zylon:
The safety cells' flanks are protected by a six-millimetre layer of carbon and zylon. The drivers' crash helmets feature a Zylon strip across the top of the visor since 2011 in order to enhance protection from flying objects. Zylon is also used for bullet-proof

vests and is intended to prevent objects such as splinters from entering the cockpit. Padding has absorbed impact energy on the inside since 2002. Occupants of production vehicles are protected by airbags and side-impact protection in side crashes.